Previous names and/or well established names	Recent and/or accepted names
Torulaspora delbrueckii	*Saccharomyces delbrueckii; fermentati; florenzanii; rosei; saitoanus; vafer*
Saccharomyces kloeckerianus	*Torulaspora globosa*
pretoriensis	*pretoriensis*
transvaalensis	*Pachytichospora transvaalensis*

Previous names and/or well established names	Recent and/or accepted names
*Selenotila intestinalis**	*Metschnikowia lunata*
peltata	*Selenozyma peltata*
*Torulopsis bacillaris**	*Torulopsis stellata**
*bovina**	*Saccharomyces telluris*
candida; famata*; C. famata*	*Debaryomyces hansenii*
*colliculosa**	*Torulaspora delbrueckii*
*dattila**	*Kluyveromyces thermotolerans*
*domercqii**	*Wickerhamiella domercqii*
glabrata;	*Candida glabrata*
*holmii**	*Saccharomyces exiguus*
*pintolopesii**	*telluris*
*Trichosporon aculeatum**	*Aciculoconidium aculeatum*
*capitatum**	*Geotrichum capitatum*
*fermentans**	*fermentans*
*inkin**	*Sarcinosporon inkin**
*penicillatum**	*Geotrichum penicillatum*
Saccharomyces cerevisiae	*Saccharomyces aceti*
	bayanus
	beticus
	capensis
	chevalieri
	cordubensis
	coreanus
	diastaticus
	globosus
	heterogenicus
	hienipiensis
	inusitatus
	italicus
	norbensis
	oleaceus
	oleaginosus
	prostoserdovii
	uvarum
Saccharomyces amurcae; cidri	*Zygosaccharomyces cidri*
bailii; acidifaciens;	*bailii*
elegans	
bisporus	*bisporus*
eupagycus; florentinus	*florentinus*
microellipsoides	*microellipsoides*
montanus	*fermentati*
mrakii	*mrakii*
rouxii	*rouxii*
Torulaspora delbrueckii	

production without modern scientific knowledge and yeasts were essential if unknown agents. One can understand how the native crops, barley and grapes, became so important to these highly intelligent and industrious peoples whose records illustrate beautifully how their arts and crafts were interwoven with law, religion and agriculture. Not all ancient peoples developed fermented foods and beverages as did the Egyptians. Nevertheless in countries where the maize plant was indigenous, other fermented products were produced. Some of these maize-based foods are still made by primitive Indian tribes in remote areas of Central and South America. A form of fermented dough (pozol) is made, in which yeasts and bacteria are essential constituents (Ulloa & Kurtzmann 1975). Among the yeasts present is a strain very similar to modern baking yeast (*Saccharomyces cerevisiae*) and there are others which could become pathogenic to man (e.g. *Candida parapsilosis*) or spoil his food and drink (e.g. *C. tropicalis*). Ulloa & Kurtzmann (1975) also showed that the process encourages beneficial yeasts and bacteria whilst suppressing undesirable organisms. Again, this process must have evolved by trial and error without the benefit of modern scientific knowledge.

Today more than ever, yeasts and similar organisms are inseparable from the activities of mankind, but they are agents for good or evil, for the wrong organism in the wrong place can mean the difference between profit and loss or living and dying.

2. Definition

Yeasts and yeast-like organisms are microfungi which live either as saprophytes or parasites. A simple, precise definition of the term 'yeasts' is impossible because it is now known that many unicellular yeast forms have a complicated life-cycle (e.g. genus *Rhodosporidium*, see Ch. 11, this volume). These life-cycle organisms are placed within either the ascomycetous or basidiomycetous fungi (Table 1 and Ch. 11, this volume) hence the terms 'yeast-like organism' or 'yeast-like fungi' are commonly used for many yeasts, but in this chapter the word 'yeast' will be used for both unicellular and more complicated forms. In the latter cases there will be some qualifying information given.

3. Description

Yeast cell shapes range from oval (Fig. 1*a*, *b*, *f*), circular (Fig. 2*a*, *b*, *c*), cylindrical (Figs 1*d*; 2*a*), lemon-shaped (Fig. 1*e*), ogive-shaped (Fig. 3*d*)

An Introduction to Yeasts and Yeast-like Organisms

R. R. DAVENPORT

University of Bristol, Long Ashton Research Station, Long Ashton, Bristol, UK

Contents

1. History

THE EARLIEST BEGINNINGS of the 'yeast story' are unrecorded. It is presumed that micro-organisms must have had some effect on man's food even when he lived by gathering berries and fruits. Later when agricultural practices became established, records often referred to the growing of crops which were processed into beverages and foods, and it is now known that yeasts were an essential part of many of these commodities. The splendours of Ancient Egypt included an abundance of papyri, pyramidal and tomb texts, and stone and wooden sculptures as well as mechanical toys and models depicting everyday life (Mackenzie 1925*a*, *b*). Among the latter were exquisite model bakeries and brew-houses. An enthralling scene is one of the tomb panels of Nakhti (*ca.* 4000 years ago). Here is the vintage scene, the gathering of grapes from a cultivated arched vine, the pressing of grapes, juice collection and storage in amphorae. A pyramidal text of 2700 BC lists the menu for the celestial morning meal for a Pharoah. Among the 1000 quantities of each fresh and cooked foods listed, was a 1000 each of beer and bread (Mackenzie 1925*b*). The great skills and craftsmanship of Ancient Egypt stretch back even further. Some 6000 years ago an Old Kingdom aristocrat's menu included 16 different kinds of bread and cakes, six kinds of wine and four types of beer (Mackenzie 1925*a*). Truly these ancient peoples had a refined system of food and beverage

TABLE 1

Characteristics which place yeasts within the fungi *

Lack photosynthetic pigment, e.g. chlorophyll[†]
Lack locomotion[‡]
Possess a rigid cell wall[§]
Nucleus present — difficult to demonstrate
Single cell phase as part of vegetative cycle
Distinct aerial asexual spores (conidia) are not produced[∥]

*Modified data from Phaff *et al.* (1966).
[†]Exception ? *Prototheca* = colourless alga-like organism.
[‡]Some ascospores have whip-like appendages, e.g. *Nematospora.*
[§]Protoplasts can be produced — regeneration of cell wall.
[∥]Ballistospores considered as asexual state by some authorities.

to triangular (Fig. 2e). Vegetative reproduction is by budding (Figs 1a, b, c, d; 2a, b, c, d, e; 3d, e), fission (Figs 2f; 3f) or bud-fission (Fig. 1e) which is characteristic of bipolar budding yeasts (e.g. genus *Hanseniaspora*). Other vegetative forms include arthrospores (Fig. 3c), ballistospores (Fig. 3a), chlamydospores (Figs 3b; 7e), endospores (Fig. 5c), pseudomycelium (Figs 2d; 5a, b; 8a) and mycelium. The ascus, the sexual structure of the ascomycetous yeasts, is a sac-like structure which usually has 1–4 ascospores: rarely there are 8–16 (e.g. genus *Lipomyces*) and one species, *Kluyveromyces polysporus* is unique with several hundred ascospores. During a current investigation, a multisporous isolate (up to 24 ascospores formed) of *Sacch. cerevisiae* (Fig. 8b) was obtained (Davenport, unpublished). Usually *Sacch. cerevisiae* strains contain 1–4 ascospores and on rare occasions up to 9 (Santa Maria 1959). Plurispored asci have also been reported for some *Pichia* species (Santa Maria 1966). The shapes of ascospores include: spherical, elliptical, oval (Fig. 4e), kidney, hat (Fig. 4a), saturn, helmet, walnut, needle (Fig. 4d), doughnut (Fig. 4b, c). Moreover, ascospore walls may be smooth (Fig. 4b, c), rough, warty (Fig. 4f), thick, double or starchy and the contents may contain an oil droplet. The basidomycetous yeasts (shooting spore types) usually have a complex life-cycle where the common state is the unicellular 'budding yeast' but special environmental conditions, and often an opposite mating type, are required to produce various stages within the cycle. The structures formed are dikaryotic mycelium, teliospores, promycelium and sporidia and sometimes basidiospores; these are very similar to those seen with other basidomycetous fungi. Some of these features are given in Ch. 11. Some yeasts do not show any sign of sexual structures; this may be due to the loss of 'sporulation' during laboratory cultivation, or because an opposite mating type is required or special environmental conditions are essential. Such yeasts are classified within the imperfect fungi whilst the remainder are assigned to either the ascomycetous or basidiomycetous genera. It is important to know that many yeasts have sexual and asexual states and this means that yeasts are often named for each state since both have usually been found and described independently. This situation can cause problems with classical taxonomy which is, in any case an artificial means of classifying yeasts. Table 2 gives examples of the relationships between

TABLE 2

Some examples of relationships between imperfect and perfect genera within the yeasts and yeast-like organisms

Perfect — basidiomycetous	Imperfect	Perfect — ascomycetous
	Brettanomyces	*Dekkera*
Filobasidium (=Leucosporidium) 1 *	*Candida*	*Debaryomyces*
Syringospora 2		*Hansenula*
		Hyphopichia 3
		Kluyveromyces
		Pichia
		Saccharomycopsis 4
		Clavispora 5
Filobasidiella 7	*Cryptococcus*	*Sporopachydermia 6*
Sirobasidium 8		
	Geotrichum	*Dipoascus 9*
	Kloeckera	*Hanseniaspora*
Rhodosporidium 10	*Rhodotorula*	
	Sporidiobolus	
Aessosporon 11	*Sporobolomyces*	
	Torulopsis	*Debaryomyces*
		Hansenula
		Saccharomyces
		Wickerhamiella 14
Aciculoconidium 12	*Trichosporon*	
Sarcinosporon 13		

*References: where no number is given see Lodder (1970).
 1 Fell (1970); Olive (1968); Rodrigues de Miranda (1972).
 2 van der Walt (1970*a*).
 3 von Arx & van der Walt (1976).
 4 van der Walt & Scott (1971); von Arx (1972, 1974); Yarrow (1972).
 5 Rodrigues de Miranda (1979).
 6 Rodrigues de Miranda (1978).
 7 Kwon-Chung (1977, 1978).
 8 Moore (1979); Flegel (1976).
 9 Weijman (1979).
10 Fell (1970).
11 van der Walt (1970*b*); Fell (1970). *B. dendrophila* also has a perfect state in *Aessosporon*.
12 King & Jong (1976).
13 King & Jong (1975).
14 van der Walt & Liebenberg (1973).

imperfect and perfect genera. One can observe that some imperfect genera only have their perfect states in one genus [e.g. genus *Kloeckera* (no spores) = genus *Hanseniaspora* (with spores)]. In contrast, other non-sporing genera have sporing states within several genera [e.g. genus *Candida* (no spores) has sporing states within six genera]. This situation also emphasizes the heterogeneity of the yeasts.

Yeast macromorphology is also diverse; growth on solid media can be coloured various shades of black, brown, red, pink, orange, yellow, cream or white with textures including smooth, mucoid, rough or tough. Yeast pigments include melanin (*Aureobasidium pullulans*), carotenoid (e.g. genus *Rhodotorula*) and pulcherrimin (restricted to *Metschnikowia pulcherrima* and some *Kluyveromyces* species). The colour and texture of yeast colonies and streaks varies with the kind of yeast and the prevailing environmental conditions such as medium, nutrient status and growth temperature. Table 3 gives the temperature requirements and effect of heat on yeasts. It can be seen that some yeasts are restricted in their requirements but these are trends, hence dissimilar observations may be encountered due to environmental conditions (e.g. different chemical, physical and biological parameters) as well as strain differences. Likewise for the same reasons only trends can be given for the effect of heat on yeasts (see Table 3).

TABLE 3

Temperature requirements and the effect of heat on yeasts and yeast-like organisms

Temperature requirements*	Range (°C)	Examples
General	0-37	*Sacch. cerevisiae*
Restricted		
Obligate psychrophiles	0–15 up to 25	*C. curiosa*
Facultative psychrophiles (or psychroduric)	0–15 up to 25	*Rh. mucilaginosa*
Psychrophobic	20-28 minimum 42-45 maximum	*T. pintolopesii*
Thermoduric	grows well at 45	*K. fragilis*

Effect of heat on yeasts[†]	Lethal temperature (°C)	
Trends		
Vegetative cells	55-60	
Ascospores	65-70	
Exceptions		
Cold-tolerant species	< 55	
Strains exposed to industrial environments (i.e. heat processing)	> 70	

*Compiled from data of Phaff *et al.* (1966); do Carmo-Sousa (1969); van Uden & Vidal-Leiria (1970); Stokes (1971).
[†]Data from Ingram (1955); Stokes (1971); Put (pers. comm.).

The optimum pH range for yeast growth is between pH 3·5 - 6·5 (Ingram 1955) but there are exceptions, e.g. *Schizosaccharomyces* spp. are restricted to around pH 5·45 (Battley & Bartlett 1966) and *Torulopsis pintolopesii* (now *Sacch. telluris*) to pH 0·9 - 1·1 when grown at 37°C (van Uden & Vidal-Leiria 1970). In general most yeasts and yeast-like organisms can grow within the range pH 3 - 8 (Ingram 1955) or between pH 3 - 11 (Davenport, unpublished). This is important because the growth response is the result of other accompanying environmental factors as well as pH. These include availability of nutrients and water, temperature and the presence of either naturally occurring or added yeast inhibitors. In addition the presence of other micro-organisms may have a profound effect. Therefore pH effects should not usually be seen in isolation but rather as interaction components.

Factors affecting nutrition and growth of yeasts may be summarized as follows (based on data from Barnett & Pankhurst 1974; Ingram 1955; Phaff *et al.* 1966; Phaff & Starmer, this volume):

Carbon sources: no photosynthetic pigments; fermentation and assimilation of some compounds (about 13); assimilation only (wide range > 80 compounds).

Nitrogen sources — assimilation: ammonium ions are used (exception, *Cyniclomyces guttulata*); amino acid patterns (usually vary between strains); inorganic nitrate and nitrite by some species (adequate vitamins must be present); proteolytic activity (confined to a few strains of some species).

Vitamins — confined to the B group: absolute requirements (i.e. one or more necessary); no requirements (i.e. all needs synthesized); linked with nitrogen utilization.

Fats: some yeasts produce fats (e.g. *Lipomyces starkeyi*); few yeasts show lipolysis (e.g. *Saccharomycopsis* — formerly *C. lipolytica*, the strongest lipolytic species known).

Not all yeasts behave in the same way in their nutrition and growth. This means that there is some restriction in the types of yeasts found in various habitats, and different growth responses, from the data given above, can be used both in identification and yeast control procedures.

Oxygen stimulates all yeasts and promotes replication. Some genera (e.g. *Cryptococcus*) only have such an oxidative metabolism, but many yeasts, after starting to grow with an aerobic phase, proceed to the anaerobic phase (fermentation). This happens when free oxygen is depleted and carbon dioxide accumulates (e.g. genus *Saccharomyces*). When fermenting yeast cultures are aerated, fermentative metabolism is decreased and part of the glucose is respired to carbon dioxide and water (Pasteur effect). Many yeasts, in particular *Brettanomyces* species, show the opposite of this, known as the negative Pasteur effect (Custers 1940; Wikén *et al.* 1961; Scheffers 1966, 1967, 1979). Scheffers (1966) found the Pasteur effect had different mechanisms so he proposed the term 'Custers effect' for the inhibition of alcoholic fermentation under anaerobic conditions.

The importance of water for yeasts is two-fold: first, in the amount required for growth, and second, in their growth characteristics in various environments with different water contents (Table 4).

TABLE 4

Micro-organisms and water*

Approximate range of a_w *values for selected micro-organisms*		
Bacteria	0·91 – 0·75	e.g. *Escherichia coli* 0·96 : *Staphylococcus aureus* 0·86
Yeasts	0·88 – 0·60	e.g. *Tr. pullulans* 0·91 *Zygosacch. rouxii* 0·62 (osmophile)
Filamentous fungi	0·80 – 0·65	e.g. *Botrytis cinera* 0·93 : *Aspergillus glaucus* 0·70 (xerophile)
Growth characteristics		
Bacteria and yeasts	Can grow on damp surfaces	
	Need a moist surface or a nutrient liquid	
Filamentous fungi	Resist drier conditions than bacteria and yeasts	

Some species of yeasts produce extracellular polysaccharide materials in environments of low water content, e.g. *Rh. glutinis* in dry soil and on certain leaves.

*Compiled from Jay (1970); do Carmo-Sousa (1969).

4. Distribution and Dissemination of Yeasts in Environments

Davenport (1975) showed that most habitats had a yeast microflora comprising two components. The first was a resident collection of yeasts which possessed suitable characteristics both to survive and reproduce; therefore one could isolate these forms consistently. The second group consisted of those yeasts which lacked these characteristics: this transient microflora was solely dependent on dissemination for survival. This was principally by animal vectors, although air and rain with high particle counts did give more yeasts, since the particles acted as rafts for yeasts and other organisms. There is a wide distribution of many different forms of yeasts in many divergent environments, for example those shown in Figs 5, 6 and 7. Most yeast distribution investigative studies have been limited to results of examining natural habitats or spoiled products. The established techniques used were those which have changed little since the early beer yeast microbiology of the late nineteenth and early twentieth century. Hence there could well be a different total yeast ecological picture if one used both environmental and organism characteristics for ecological

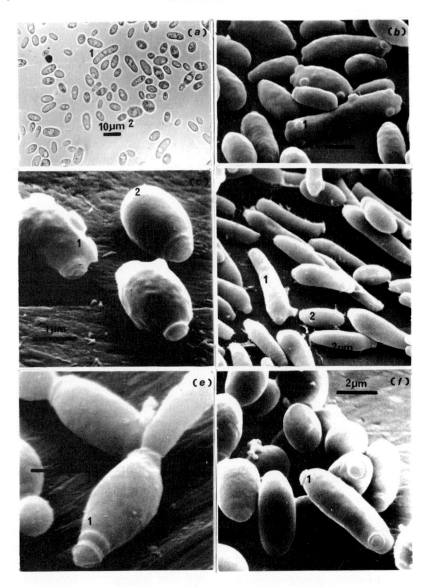

Fig. 1 *(a) Sacch. cerevisiae* (formerly *Sacch. uvarum*) from a cider fermentation. Note the mixture of cells (1, elongated; 2, oval) which divide by multilateral budding. *(b) Sacch. cerevisiae* (same strain as *a*) showing bud scars. *(c) Zygosacch. bailii*, a common osmophilic (xerotolerant) yeast. 1, mother cell with bud scars (typical of multilateral budding yeasts); 2, detached daughter cell from one of the bud scar sites. *(d) C. rugosa* type species (NCYC 130). 1, multipolar budding, at or near the 'shoulder' — not along the long axis; 2, detached daughter cell. *(e) Kl. apiculata* (imperfect state of *H'spora valbyensis*) from grapes. Cells show bipolar budding (i.e. at ends of the cells only) or bud-fission hence annular ring scars (1). *(f) Br. anomalus* type species (NCYC 449). A multipolar budding yeast producing many cell shapes including lemon (see Fig. 3*d*), oval, cylindrical and irregular.

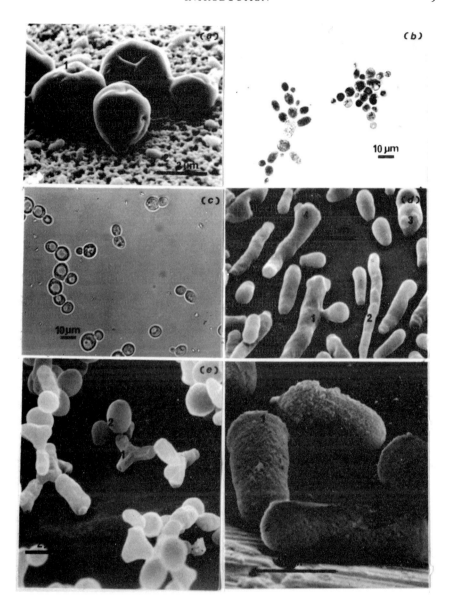

Fig. 2 *(a) Sacch. cerevisiae* (1) and acetic acid bacteria (2) from a cider fermentation. *(b) Sacch. cerevisiae* (brewer's strain — NCYC 1036) showing multipolar budding. *(c) Sacch. cerevisiae* (baker's strain — NCYC 487). Note the slight difference in morphology compared with Fig. 2b. *(d) H. anomala* type species (NCYC 18). Mixture of cells — multipolar budding yeast. 1, cylindrical; 2, cylindrical cell forming pseudomycelium; 3, oval cell with bud scars; 4, cylindrical cell with bud scars. *(e) Trig. varibilis* type species (CBS 1040). 1, triangular cells; 2, oval cells. Note multipolar budding on oval cells and apices of the triangular cells. *(f) Schiz. pombe* type species (CBS 351). Reproduction by binary fission. 1, fission just completed; 2, a single cell.

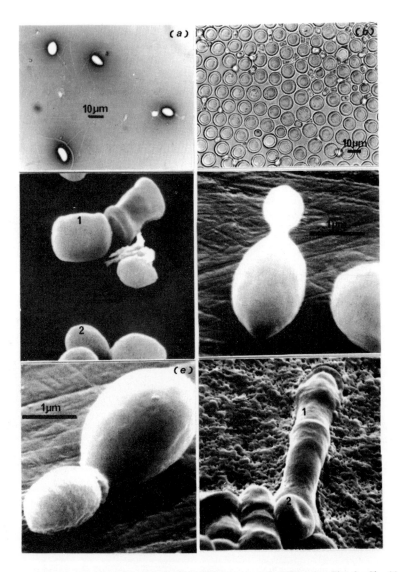

Fig. 3 *(a) Sp. roseus,* from an apple leaf. Ballistospores only. (See also Fig. 2, Ch. 11.)
(b) C. pulcherrima (imperfect state) *(M. pulcherrima,* perfect state — see Fig. 3*a,* Ch. 11)
from grapes. Note, these cells have a very high lipid content, i.e. pulcherrimin cells
(a form of chlamydospore formation (1)). *(c) Hyphopichia burtonii,* type species
(NCYC 439). Yeast-like organism exhibiting many features of certain filamentous
ascomycetous fungi. 1, arthrospore formation; 2, yeast budding cells. *(d) Br. anomalus,*
type species (NCYC 449). Ogive shaped cell, i.e. pointed at one end. (See also Fig. 1*f.)*
(e) Sp. roseus, from an apple leaf (same yeast as *(a);* See also Fig. 2, Ch. 11) Note this is
different from *Br. anomalus (d). (f) Schiz. octosporus,* type species (NCYC 131).
1, elongated cell with cross-wall — binary fission would take place here as shown in
Fig. 2*f*; 2, an ascospore released from an ascus.

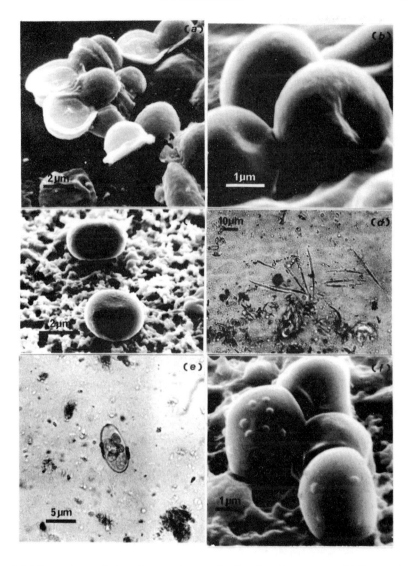

Fig. 4 *(a) P. membranaefaciens,* from a wine sample. Hat-shaped ascospores. *(b) Zygosacch. bailii,* from a wine sample. Smooth, round ascospores. *(c) Schiz. pombe,* type species (CBS 351). Smooth, round ascospores. Very similar to *Schiz. octosporus* (see Fig. 3*f*). *(d) Nem. coryli,* from an infected coffee bean (Davenport, unpublished data). Needle-shaped ascospores. *(e) Cyniclomyces guttulata,* from rabbit faeces. Large cylindrical ascospores. *(f) S'codes ludwigii,* from cider. Large, round ascospores — no ledge discernible but 2 papillate and 2 smooth ascospores.

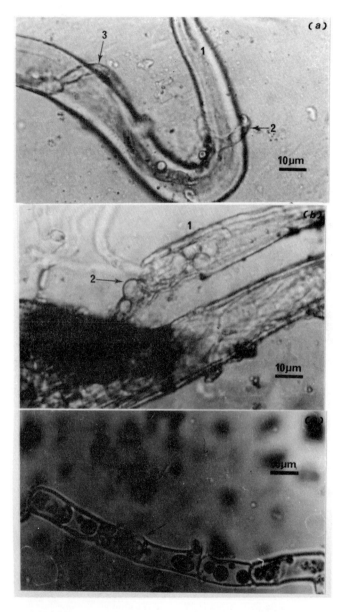

Fig. 5 *(a) Aureobasidium pullulans*, on the upper surface of an apple leaf (from Davenport 1968). 1, leaf hair; 2 & 3, filamentous cells. *(b) Aureobasidium pullulans*, growing out of a leaf hair (apple) on apple juice extract agar (from Davenport 1967). *(c) Aureobasidium pullulans*, from grape leaves, growing on potato dextrose agar. Note vegetative endospores growing within the pseudo-hyphae (from Davenport 1973, 1975).

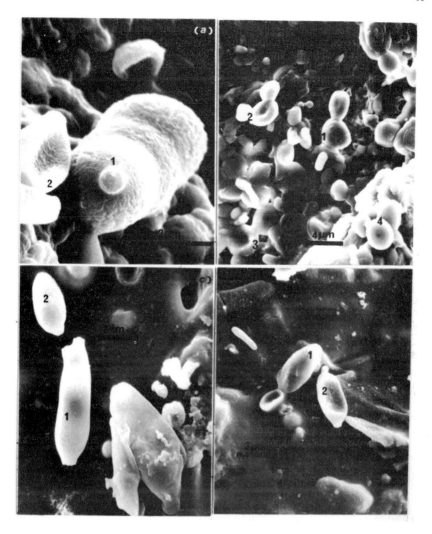

Fig. 6 *(a)* Budding yeast-like organism on a mummified apple surface. 1, mother cell with a bud scar; 2, detached daughter cell. *(b)* Collection of micro-organisms on the surface of Stilton cheese. 1 & 2, budding yeasts with mature daughter cells; 3, conidia of *Penicillium roqueforti* (blueing organism); 4, yeast mother cell with a new bud. *(c)* Cherry tree exudate with a yeast (probably a *Candida* species). 1, mother cell with terminal and shoulder bud scars; 2, a detached daughter cell. *(d)* Cherry tree exudate (same sample as *(c)*). 1, multi-polar budding yeast (mother and daughter cells); 2, bipolar budding yeast (*Kloeckera/Hanseniaspora* species).

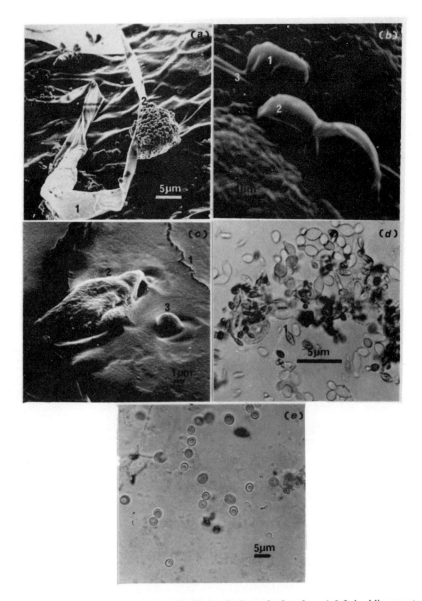

Fig. 7 *(a)* Yeast microcolony on a grape leaf hair. *(b)* Grape leaf surface. 1 & 2, budding yeasts and hyphal strand. *(c)* Grape surface with an apiculate yeast (*Kloeckera/Hanseniaspora* species). 1, bud scar; 2, mother cell. *(d)* Grape must yeasts. 1, small apiculate yeast (? *Kl. apiculata/H'spora valbyensis*). 2, large apiculate yeast (? *Kl. corticus/Han'spora*). *(e)* Yeasts from the gut contents of a hare; note the chlamydospore-like cells of a *Candida* species.

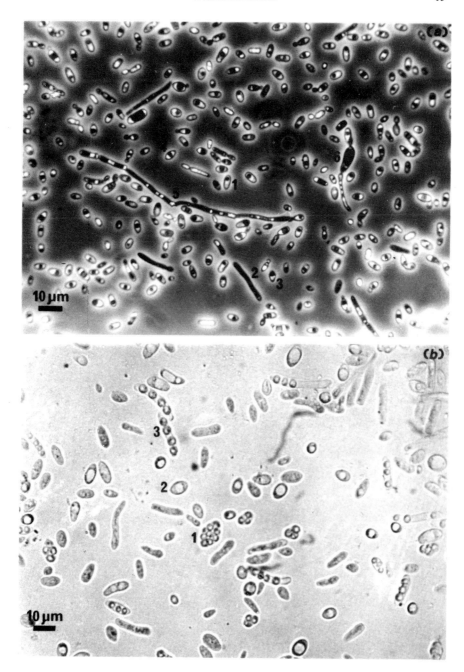

Fig. 8 *(a) Br. bruxellensis* isolated from a lambic beer sample. Note the various cell shapes with typical multipolar budding. 1, oval cell; 2, cylindrical cell; 3, ogive-shaped cell (pointed at one end); 4, apiculate or lemon-shaped (note bud fission); 5, pseudomycelium; 6, irregular. *(b) Sacch. cerevisiae.* Multispored strain from a lambic beer. 1, large ascus (9 ascospores); 2, oval shape; 3, cylindrical ascus (4 ascospores).

surveys. This approach has a further advantage in that it can be integrated with taxonomy (Davenport 1974), assessing resident and transient species as well as determining the yeast's role or significance in an environment (Davenport 1973, 1975, 1976) (Table 5).

TABLE 5

Environments and the distribution of yeasts and yeast-like organisms*

---

1. General: Genera which contain species likely to be found in a wide varied selection of habitats:
   Aureobasidium, Candida, Cryptococcus, Debaryomyces, Hansenula, Kluyveromyces, Leucospiridium, Pichia, Rhodotorula, Saccharomyces, Sporobolomyces, Torulopsis, Trichosporon.
2. Restricted:A. Genera confined to a single or very few habitats:
   Ambrosiozyma (insects); *Arthroascus* (soil/tree exudate); *Brettanomyces, Dekkera* (beverages); *Bullera* (plants); *Filobasidium* (grass/beverages); *Hanseniaspora, Kloeckera* (fruit products); *Lipomyces* (soil & human skin); *Metschnikowia* (plants & some aquatic habitats); *Nadsonia* (tree exudates); *Nematospora* (plant pathogens); *Pachysolen* (tanning fluids); *Pithrosporum* (animal ears & hairs); *Schizosaccharomyces* (high sugar environments); *Sporidiobolus* (plants); *Sterigmatomyces* (aquatic & plant sources); *Schwanniomyces.*
   B. Genera, having a single species, confined to a single or very few habitats:
   Cit. matritensis (high sugar environments); *Cyniclomyces guttulata* (rabbits & hares); *Guilliermondella selenospora* (tanning fluid); *Hormoascus platypodis* (bark/beetles); *O. margaritiferum* (tree exudate); *Pa. tannophilus* (tanning fluids); *S'codes ludwigii* (tree exudate & beverages); *Schizobl. starkey-henricii* (soil); *Sympodiomyces parvus* (marine yeast); *Trig. varibilis* (wine); *Wickerhamia fluorescens* (squirrel dung); *Wingea robertsii* (insects).
   C. Examples of species, within common genera, found in a single habitat:
   C. aquatica (freshwater lake); *C. cacaoi* (fermenting cacao); *Cr. melibiosum* (bark beetle); *T. anatomiae* (formalin tanks — human cadaver storage); *H. petersonii* (human cadavers); *P. chambardii* (tanning fluids); *Sacch. norbensis* (olive oil production).

From Table 5 some very important sub-groups can be determined. For example, the pathogenic yeasts and yeast-like organisms can be grouped thus: (1) those few which are definite pathological types causing superficial and/or systemic lesions, e.g. *Cryptococcus (Filobasidiella) neoformans* as an animal pathogen (Gentles & La Touche 1969; Mendoza & Patino 1974) and *Nematospora coryli* as a plant pathogen (do Carmo-Sousa 1970); (2) a wide selection of species which are opportunistic forms since they only give pathological symptoms in special circumstances, e.g. *T. candida (Debaryomyces hansenii)*, an occasional pathogen of animals (Gentles & La Touche 1969), or *Geotrichum candidum*, which sometimes causes a disease of tomato fruits (Butler 1960); (3) by tradition rather than science, certain medical microfungi which have definite yeast-like and filamentous phases, depending on environmental conditions, are usually excluded from the yeasts (Lodder 1970). Such 'yeast-fungi' include *Blastomyces dermatitidis, Histoplasma farciminosum, Histoplasma capsulatum* and *Sporotrichum schenckii.* These are not the only filamentous microfungi which can form yeast-like (i.e. budding) cells with environmental changes, and others include *Mucor* species (Romano 1966) and various other fungi reported in a review by Richmond (1975). Thus, the boundaries between the 'yeasts' and other microfungi are diffuse.

The other sub-groups concern spoilage. Table 6 summarizes the products and sites while Table 7 gives the principal industrial spoilage yeasts from foods. Some of these yeasts, *Sacch. cerevisiae* (Figs 1*a*, *b*; 2*a*, *b*, *c*) *Zygosaccharomyces bailii* (Fig. 4*b*), *C. parapsilosis, C. krusei, C. valida, Hansenula anomala* (Fig. 2*d*) and *Deb. hansenii,* are also common contaminants of alcoholic and non-alcoholic beverages where other unwanted pests include *Brettanomyces* spp., *Saccharomyces* spp., *C. zeylanoides, Pichia farinosa, P. fermentans, Saccharomycodes ludwigii* (Fig. 4*f*), *Schizosaccharomyces pombe* (Fig. 4*c*), *Torulaspora delbruekii, Zygosacch. fermentati, Zygosacch. florentinus* and *Zygosacch. microellipsoides.* Of course not all yeasts are detrimental agents; it is now commonplace for many industries to use yeasts successfully for production of beverages and foods. Some yeasts, other than industrial strains of *Sacch. cerevisiae* and *C. utilis,* have been used recently for industrial needs (examples in

TABLE 6
Spoilage yeasts and yeast-like organisms*

Products/sites	Genera	References
Beverages		1
Foods	Various	2
Processing machinery		3
Petroleum products		
Coolants/cooling systems	*Candida, Torulopsis, Rhodotorula*	4
Fuels/fuel systems		5
Lubricants		6
Metal surfaces		7
Pharmaceutical preparations	Various	8

*References
1 Beech & Davenport (1971); Kodama (1970); Rainbow (1970); Walker & Ayres (1970); Sand & van Grinsven (1976); Rose (1977).
2 Walker & Ayres (1970).
3 Hood & Rankine (1977); Sand *et al.* (1976).
4 Hill (1975); Miller & King (1975).
5 Hill (1975); Park 1975.
6 Hill (1975).
7 Miller & King (1975).
8 Beveridge (1975).

TABLE 7

Principal industrial spoilage yeasts and yeast-like organisms from foods*

	Brined products & fermented foods	Meat (red), fish & poultry	Dairy products	Frozen foods	Acid foods	Cereal grains	High sugar products	High salt products	Photomicrograph Fig. numbers
Aureobasidium pullulans		+					+		5a, b, c;1b
Tr. pullulans		+							1c, d
Rh. glutinis		+	+						
†K. lactis/T. sphaerica			+						
†K. fragilis/C. pseudotropicalis			+						
†Deb. hansenii/T. famata	+	+					+		
Sacch. cerevisiae				+					1a, b; 2a, b, c
Zygosacch. bailii	+				+		+		
Zygosacch. rouxii	+				+		+	+	1c
†Lod. elongisporus/C. parapsilosis		+							
†P. kudriavzevii/C. krusei	+	+				+			
†P. membranaefaciens/C. valida	+								4a
C. rugosa		+							1d
C. tropicalis						+			
†H. anomala/C. pelliculosa						+	+		2d
†Hyphopichia (E.) burtonii						+			3c

*Data from Walker & Ayres (1970).
†Sporing or perfect states.

TABLE 8

Recent industrial uses of yeasts and yeast-like organisms

Uses/substrate or treatment	Organisms	References*
Feed yeasts		
Confectionery effluent	C. utilis	1
Citric acid waste	C. scottii	2
	C. vartiovaarai	
Whey also for alcohol and methane production	K. fragilis	3
Cassava also for alcohol	S. fibuliger,	4
	C. utilis	
Waste water		
Agricultural	C. ingens/P. humboltii	5
Agricultural and industrial	Many species cited	6
Phenol degradation	Tr. cutaneum,	7
	C. tropicalis	
Alcohol and feed yeast	C. tropicalis,	8
	C. utilis	

*References:
1 Forage (1978).
2 Braun *et al.* (1979).
3 Reesen & Strube (1978).
4 de Menezes (1978).
5 Henry (1975, 1976).
6 Wiken (1972).
7 Neujahr (1978).
8 Rychtera *et al.* (1977).

TABLE 9

Recent examples of activities exhibited by yeasts and yeast-like organisms

Activity		Organism	Reference*
Degradation of	cellulose	*Aureobasidium* & *Trichosporon* spp.	1
	pectin	*Sacch. uvarum, Sacch. cerevisiae*	2
	xylan	*Trichosporon* spp.	3
	herbicides	*L. starkeyii*	4
Assimilation of	hydrocarbons	Various	5
	methanol	*C. boidinii, P. pastoris, P. methanolica*	6
	Paraquat	Various	7
Miscellaneous effects	killer yeasts	Various	8
	spray residues/ fermentation rates	*Saccharomyces* spp.	9
	production of riboflavine & flavine adenine dinucleotide	*Eremothecium ashbyii*	10
	formation of *l*-tetradecene from tetradecene	*C. parapsilosis*	11

*References:
 1 Dennis (1972); Stevens & Payne (1977).
 2 von Dechau & Emeis (1975).
 3 Stevens & Payne (1977).
 4 Anderson & Drew (1972).
 5 Bos & de Bruyn (1973).
 6 Kichiro *et al.* (1974).
 7 Davenport (1975).
 8 Philliskirk & Young (1975); Stumm *et al.* (1977).
 9 Adams (1954); Alexandri *et al.* (1974).
10 Osman & Ragab (1978).
11 Souw *et al.* (1978).

Table 8). No doubt as yeast knowledge advances other species will be exploited for the benefit of mankind. Thus, Table 9 gives some recent examples of activities exhibited by yeasts — some of these are potentially important commercially.

5. Identification

Sneath & Sokal (1973) stated "traditional taxonomy attempts fulfil too many functions and as a consequence fulfil none of them well". They showed that theoretically and practically the attempts were not adequate to classify, to name, to indicate degree of resemblance and to show relationship by descent, all at the same time. To some degree the same things could be said about yeast taxonomy. Certainly among many non-yeast specialists two common beliefs are held. In the first instance, the yeast name is supposed to be the key to all aspects (i.e. its importance, its true identity, its relationship with other yeasts and sometimes even its control!). Secondly, the yeast name is regarded as the key to guide one through the literature so that relevant information can be abstracted for use. For some yeast work the International Code of Botanical Nomenclature is cited (Lodder 1970). However, there are some instances where rules, correct and proper for taxonomy, have caused both dismay and confusion with investigators other than taxonomists. For example, the yeast names *Sacch. cerevisiae* and *Sacch. carlsbergensis* have long been recognized as two brewing species, where the former includes top fermenting or ale-producing strains and the latter are bottom fermenting or lager-producing strains; this state existed until van der Walt (1970c) re-set the limits of certain *Saccharomyces*. Only the oldest legitimate epithet (*uvarum*) was retained, in accordance with the articles of the International Code of Botanical Nomenclature (Lanjouw *et al.* 1966), so *Sacch. uvarum* became the correct name for all strains previously known as *Sacch. carlsbergensis* and *Sacch. uvarum*. Furthermore, recent investigations on the genus *Saccharomyces* have altered the naming yet again, so that all strains of *Sacch. uvarum* are now strains of *Sacch. cerevisiae* (Barnett *et al.* 1979) and this species name now includes many other former *Saccharomyces* species.

Classification is concerned with the orderly arrangement of units so that like units are collected together, the end result being a series of groups, each of which contains units that resemble one another. Thus classification is a positioning process which can be operated by either non-numerical (Guilliermond 1912, 1920; Lodder & Kreger-van Rij 1952; Lodder 1970) or numerical (Barnett 1971; Campbell 1971, 1972, 1973; Grower & Barnett 1971; Barnett & Pankhurst 1974; Barnett *et al.* 1979) systems.

Identification is concerned essentially with the examination and determination of yeast features which are then compared with descriptions of known organisms. There are many identification systems; none is ideal and methods used vary from simple observations and tests (Davenport 1973, 1974) to sophisticated biochemical or physiological techniques, e.g. serology (Tsuchiya *et al.* 1965, 1974; Campbell 1971), $G + C$ ratios (Nakase & Komagata 1971), cell wall analyses (Gorin & Spencer 1970) and coenzyme-Q system analysis (Yamada *et al.* 1976). Sorting and comparison can be carried out by a variety of means including cataloguing at culture collections, identification keys and tables (Beech *et al.* 1968; Davenport 1973, 1974 and unpublished data), punch cards, Stripdex unit (Davenport 1974) and computer (Barnett 1971; Grower & Barnett 1971).

6. Characteristics to be Determined and Identification Procedures

The delimitation of yeast genera is usually based on a selection of morphological features with one or very few physiological and/or biochemical characteristics. The latter sets of characteristics are used to differentiate species. There is not a standardized set of tests to which all yeasts are subjected. A choice of tests is made to suit the needs of either classical taxonomy, as in Lodder (1970), or the identification of individual organisms such as industrial and medical strains. Thus the yeast characteristics determined, within the broad sphere of yeast taxonomy, include morphology (macroscopic and microscopic features); cultural behaviour on solid and liquid media; sexual characteristics; physiological and biochemical studies (e.g. assimilation and fermentation tests); ecological observations with supplementary tests (e.g. medical yeasts) and industrial characterization of strains (e.g. flocculation tests). Some modifications of routine methods are given later (Ch. 14, this volume).

It is important to know that not all tests are of equal importance; for example, the nitrate assimilation test is essential to distinguish between *Saccharomyces* species (nitrate negative) and fermenting *Hansenula* species (nitrate positive), but the same test is of little significance with other genera such as *Candida, Torulopsis* and *Rhodotorula*. In addition, some supplementary test responses are outstanding features of a few selected yeast species (e.g. *Saccharomycopsis lipolytica*, lipolysis; *Brettanomyces/ Dekkera* species, strong acid formation).

7. Integrated Micro-ecological and Taxonomic Studies

Davenport (1975, 1976, 1977) showed how environmental data and selected taxonomic procedures could be used simultaneously to give a more meaningful account of the presence of yeasts in their habitats. This approach also: (1) yielded information to assist isolation and identification studies; (2) provided studies of interactions; (3) aided development and application of simplified identification systems; (4) emphasized the importance of different morphological forms and their sites in relation to the environment (Figs 5, 6, 7). Other yeasts are shown in Fig. 7*b, c, d, e*. Thus by using some of these plates, one can present an integrated scheme with direct observation and accompanying cultural techniques (Table 10).

TABLE 10

Integration of direct observation, tentative identification and confirmation studies for yeasts

Observation (Fig. Nos.)	Tentative yeast genera	Confirmation studies
(6a) Multipolar budding, large cell	*Aureobasidium* or *Candida* (i.e. bud-fission and fission genera can be excluded)	Rough colonies and film formation
(6b) Mixed population	*Candida, Torulopsis* and other multipolar budding yeasts	Use many methods, e.g. antifungal agents, aerobic and anaerobic incubation, selection of media
(2; 6d; 7c, d) Bipolar budding (bud-fission)	*Kloeckera/Hanseniaspora*	Use media with up to 100 mg/l cycloheximide — suppresses most other likely yeasts.
(7e) High lipid containing cells	*Metschnikowia/Candida*	Large chlamydospores on cornmeal agar. Biotin and ferric ammonium citrate in various media give intense pulcherrimin pigment, *M. pulcherrima*; no pigment, *M. reukfaufii* (Beech & Davenport 1971)

TABLE 11

Simplified identification* of yeasts and yeast-like organisms which use Paraquat†as a sole nutrient source

	Colony type	DBB test‡	Mycelium type	Arthrospores	Ascospores	Fermentation	Nitrate	Starch formation	Melibiose	Inositol
Tr. cutaneum	R	B	H:P	+	−	−	−	−	V	+
Geotrichum fermentans	R	A	H:P	+	−	W	−	−	−	−
Hyphopichia burtonii	R	A	H:P	+	+	W	−	−	−	−
Cr. flavus	S/M	B	−	−	−	−	−	−	+	+
Cr. laurentii	S/M	B	−	−	−	−	−	+	+	+
Cr. albidus	S/M	B	−	−	−	−	+	+	V	+
Cr. gastricus var. A	S	B	−	−	−	−	−	−	−	+
T. ernobii var. A	S	A	−	−	−	−	+	−	−	−
C. melinii	R	A	P	−	−	−	+	−	−	−
C. muscorum	R	B	P	−	−	−	−	−	−	−

*R, rough; S, smooth; M, mucoid; B, basidiomycetous yeast; A, ascomycetous yeast; W, weak; V, variable; H, true hyphae (i.e. separate threads); P, pseudomycelium (chains of enlongated cells only).

†Paraquat (1,1¹-dimethyl-4,4¹ bipyridylium dichloride) — a herbicide, 100 parts/10^6 in washed agar — colony growth after 7 days at 25°C.

‡Colour differentiation test between ascomycetous and basidiomycetous yeasts (data from Davenport 1975).

Table 11 gives a simple identification scheme for yeasts, isolated from soil environments, which were able to use the herbicide Paraquat as a sole nutrient source (Davenport 1975). Thus a simple key for species identification can be constructed by using a combination of ecological and yeast characteristic observations. This method has a further advantage in that only a few tests are required and the time involved is minimal. These principles of the integration of ecological and taxonomic data for simplified identification have been successfully used for many types of yeast studies during the last 15 years.

8. Summary

This chapter illustrates the complexity of studies on yeasts and yeast-like organisms. To begin with, it is still widely accepted that these form a very heterogenous collection of organisms within the plant kingdom (*Thallophyta*). However, the distinction between algae and certain yeast-like forms is faint, since colourless mutants of *Chlorella* can readily be obtained and these are indistinguishable from yeasts of the genus *Protheca* (Lacaz, pers. comm.). Furthermore, one can observe the increase in numbers of habitats where yeasts have been found though, in the majority of instances, the ecological significance of the species isolated is not yet fully determined. Thus, until a more standardized classification system is used there are bound to be difficulties in the naming of yeast isolates. Most people concerned with determining the taxonomic position of a wide variety of yeasts find the standard manual (Lodder 1970) unsuitable since the identification system is only workable for very small groups of yeasts. Therefore simplified yeast identification systems can be constructed (Barnett & Pankhurst 1974; Barnett *et al.* 1979; Beech *et al.* 1968; Davenport 1968, 1970, 1973, 1975, 1976 and unpublished data).

Finally, there have been few yeast ecological studies where an extensive investigation of all possible factors influencing yeast distribution has been related to either the environment or to the organisms. Therefore it is essential that yeast ecological and taxonomic studies should continue in order to obtain vital knowledge that will eventually further the use of beneficial yeasts and the control of harmful ones.

Truly, yeasts and yeast-like organisms are amongst the essential and potentially important microbes for mankind.

9. References

ADAMS, A. M. 1954 Some effects of captan spray residues on sweet cherry fermentations. *Report of the Horticultural Products Laboratory, Vineland, Ontario, Canada.*

ALEXANDRI, A. A., VASILESCU, D. & CHEMAL, M. 1974 Citeva aspecte privind influenta unor fungicide asupra fermentatiei alcoolice. *Analele Institutului de Cercetari Pentru Protecţia Plantelor* **10**, 461–472.

ANDERSON, J. R. & DREW, E. A. 1972 Growth characteristics of a species of *Lipomyces* and its degradation of paraquat. *Journal of General Microbiology* **70**, 43–58.

BARNETT, J. A. 1971 Selections of tests for identifying yeasts. *Nature, London* **232**, 221–223.

BARNETT, J. A. & PANKHURST, R. J. 1974 *A New Key to the Yeasts*. Amsterdam & London: Elsevier.

BARNETT, J. A., PAYNE, R. W. & YARROW, D. 1979 *A Guide to Identifying and Classifying Yeasts*. Cambridge: Cambridge University Press.

BATTLEY, E. H. & BARTLETT, E. J. 1966 A convenient pH gradient method for the determination of maximum and minimum pH for microbial growth. *Antonie van Leeuwenhoek* **32**, 245–255.

BEECH, F. W., DAVENPORT, R. R., GOSWELL, R. W. & BURNETT, J. K. 1968 Two simplified schemes for identifying yeast cultures. In *Identification Methods for Microbiologists, Part B* ed. Gibbs, B. M. & Shapton, D. A. pp. 151–175. Society for Applied Bacteriology Technical Series No. 2. London: Academic Press.

BEECH, F. W. & DAVENPORT, R. R. 1971 Isolation, purification and maintenance of yeasts. In *Methods in Microbiology* Vol. 4, ed. Booth, C. pp. 153–182. London: Academic Press.

BEVERIDGE, E. G. 1975 Microbial spoilage of pharmaceutical products. In *Microbial Aspects of the Deterioration of Materials* ed. Lovelock, D. W. Society for Applied Bacteriology Technical Series No. 9. London: Academic Press.

BOS, P. & DE BRUYN, J. C. 1973 The significance of hydrocarbon assimilation in yeast identification. *Antonie van Leeuwenhoek* **39**, 99–107.

BRAUN, R., MEYRATH, J., STUPAREK, W. & ZERLAUTH, G. 1979 Feed yeast production from citric acid waste. *Process Biochemistry* **14**, 16–20.

BUTLER, E. E. 1960 Pathogenicity and taxonomy of *Geotrichum candidum*. *Phytopathology* **50**, 665–672.

CAMPBELL, I. 1971 Antigenic properties of yeasts of various genera. *Journal of Applied Bacteriology* **34**, 227–242.

CAMPBELL, I. 1972 Simplified identification of yeasts by a serological technique. *Journal of the Institute of Brewing* **78**, 225–229.

CAMPBELL, I. 1973 Computer identification of yeasts of the genus *Saccharomyces*. *Journal of General Microbiology* **77**, 127–135.

CUSTERS, M. Th.J. 1940 *Onderzoekingen over het Gistgesiacht* Brettanomyces. Thesis, Technological University, Delft.

DAVENPORT, R. R. 1967 The microflora of cider apple fruit buds. *Report Long Ashton Research Station for 1966* pp. 246–248.

DAVENPORT, R. R. 1968 *The origin of cider yeasts*. Thesis, Institute of Biology, London.

DAVENPORT, R. R. 1970 *Epiphytic yeasts associated with the developing grape vine*. M.Sc. Thesis, University of Bristol.

DAVENPORT, R. R. 1973 Vineyard yeasts — an environmental study. In *Sampling — Microbiological Monitoring of Environments* ed. Board, R. G. & Lovelock, D. W. pp. 143–174. Society for Applied Bacteriology Technical Series No. 7. London: Academic Press.

DAVENPORT, R. R. 1974 A simple method, using Stripdex equipment, for the assessment of yeast taxonomic data and identification keys. *Journal of Applied Bacteriology* **37**, 269–271.

DAVENPORT, R. R. 1975 *The distribution of yeasts and yeast-like organisms in an English vineyard*. Ph.D. Thesis, University of Bristol.

DAVENPORT, R. R. 1976 Distribution of yeasts and yeast-like organisms from aerial surfaces of developing apples and grapes. In *Microbiology of Aerial Plant Surfaces* ed. Dickinson, C. H. & Preece, T. F. pp. 325–351. London: Academic Press.

DAVENPORT, R. R. 1977 Integrated taxonomic and ecological studies of yeasts and yeast-like organisms. In *Proceedings of the 5th International Specialized Symposium on Yeasts, Keszthely, Hungary* ed. Novak, E. K., Deak, T., Török, T. & Zsolt, J.

DE MENEZES, T. J. B. 1978 Saccharification of cassava for ethyl alcohol production. *Process Biochemistry* **13**, 24–26.

DENNIS, C. 1972 Breakdown of cellulose by yeast species. *Journal of General Microbiology* **71**, 409–411.

DO CARMO-SOUSA, L. D. 1969 Distribution of yeasts in nature. In *The Yeasts* Vol. 1, ed. Rose, A. H. & Harrison, J. S. pp. 79–105. London: Academic Press.

DO CARMO-SOUSA, L. D. 1970 *Nematospora* Peglion. In *The Yeasts — A Taxonomic Study* 2nd edn, ed. Lodder, J. pp. 440–447. Amsterdam & London: North-Holland.

FELL, J. W. 1970 Yeasts with heterobasidiomycetous life cycles. In *Recent Trends in Yeast Research* ed. Ahearn, D. G. pp. 49–66. Atlanta: Georgia State University.

FLEGEL, T. W. 1976 Conjugation and growth of *Sirobasidium magnum* in laboratory culture. *Canadian Journal of Botany* **54**, 411–418.

FORAGE, A. J. 1978 Recovery of yeast from confectionery effluent. *Process Biochemistry* **13**, 8–12.

GENTLES, J. A. & LA TOUCHE, C. J. 1969 Yeasts as human and animal pathogens. In *The Yeasts* Vol. 1, ed. Rose, A. H. & Harrison, J. S. pp. 107–182. London & New York: Academic Press.

GORIN, P. A. J. & SPENCER, J. F. T. 1970 Proton magnetic resonance spectroscopy, an aid in identification and chemotaxonomy of yeasts. *Advances in Applied Microbiology* **13**, 25–89.

GROWER, J. C. & BARNETT, J. A. 1971 Selecting tests in diagnostic keys with unknown responses. *Nature, London* **232**, 491–493.

GUILLIERMOND, A. 1912 *Les Levures*. Paris: Encyclopédie Scientifique.

GUILLIERMOND, A. 1920 *The Yeasts* translated by Tanner, F. W. pp. 131–146. New York: John Wiley.

HENRY, D. P. 1975 *Candida ingens* as a potential fodder protein. *Australian Veterinary Journal* **51**, 317–319.

HENRY, D. P. 1976 A new role for anaerobic fermentation in single cell protein. *Search* **7**, 161–163.

HILL, E. C. 1975 Biodeterioration of petroleum products. In *Microbial Aspects of the Deterioration of Materials* ed. Lovelock, D. W. & Gilbert, R. J. pp. 127–136. Society for Applied Bacteriology Technical Series No. 9. London: Academic Press.

HOOD, A. V. & RANKINE, B. C. 1977 Microbiological contamination in wine bottling. *Reprint No. 152 of The Australian Wine Research Institute, Grapegrower and Winemaker, Annual Technical Issue*.

INGRAM, M. 1955 *An Introduction to the Biology of Yeasts*. London: Pitman.

JAY, J. M. 1970 *Modern Food Microbiology*. New York & London: Van Nostrand Reinhold.

KICHIRO, K., KURIMURA, Y., MAKIGUCHI, N. & ASI, Y. 1974 Determination of methanol strongly assimilating yeasts. *Journal of General and Applied Microbiology* **20**, 123–127.

KING, D. S. & JONG, S. C. 1975 *Sarcinosporon*: a new genus to accommodate *Trichosporon inkin* and *Prototheca filamenta*. *Mycotaxon* **3**, 89–94.

KING, D. S. & JONG, S. C. 1976 *Aciculoconidium*: a new hyphomycetous genus to accommodate *Trichosporon aculeatum*. *Mycotaxon* **3**, 401–408.

KODAMA, K. 1970 Saké yeasts. In *The Yeasts* Vol. 3, ed. Rose, A. H. & Harrison, J. C. pp. 225–284. London & New York: Academic Press.

KWON-CHUNG, K. J. 1977 Perfect state of *Cryptococcus uniguttulatus*. *International Journal of Systematic Bacteriology* **27**, 293–299.

KWON-CHUNG, K. J. 1978 Perfect and imperfect states of *Cryptococcus neoformans*. In *Abstracts of the XIIth International Congress of Microbiology, München* p. 48.

LANJOUW, J. *et al.* 1966 Cited by LODDER, J. (1970).

LODDER, J. ed. 1970 *The Yeasts — A Taxonomic Study* 2nd edn. Amsterdam: North-Holland.

LODDER, J. & KREGER-VAN RIJ, N. J. W. 1952 *The Yeasts — A Taxonomic Study*. Amsterdam: North-Holland.

MACKENZIE, D. A. 1925a The exquisite artistry of Ancient Egypt. In *Wonders of the Past* Vol. 1, ed. Hammerton, J. A. pp. 253–268. London: Fleetway House.

MACKENZIE, D. A. 1925*b* The soul's journey to Paradise. In *Wonders of the Past* Vol. 1, ed. Hammerton, J. A. pp. 339-351. London: Fleetway House.

MENDOZA, A. G. & PATINO, S. L. 1974 Criptococcosis generalizada asociada a leucemia granulocitica cronica. *Boletín de la Sociedad mexicana de microbiologia* 8, 11-15.

MILLER, J. D. A. & KING, R. A. 1975 Biodeterioration of metals. In *Microbial Aspects of the Deterioration of Materials* ed. Lovelock, D. W. & Gilbert, R. J. pp. 83-103. Society for Applied Bacteriology Technical Series No. 9. London: Academic Press.

MOORE, R. T. 1979 Septal ultrastructure in *Sirobasidium magnum* and its taxonomic implications. *Antonie van Leeuwenhoek* 45, 113-118.

NAKASE, T. & KOMAGATA, K. 1971 Significance of DNA base composition in the classification of yeast genus *Candida*. *Journal of General and Applied Microbiology* 17, 77-84.

NEUJAHR, H. Y. 1978 Degradation of phenols by yeast. *Process Biochemistry* 13, 3-7.

OLIVE, L. S. 1968 An unusual new heterobasidiomycete with *Tilletia*-like basidia. *Journal of the Elisha Mitchell Scientific Society* 84, 261-266.

OSMAN, H. G. & RAGAB, A. M. 1978 Biochemical and physiological studies on Riboflavin (R) and Flavine Adenine Dinocleotide (FAD) production by *Eremothecium ashbyii*. In *Abstracts of the XIIth International Congress of Microbiology, München* p. 129.

PARK, P. B. 1975 Biodeterioration in aircraft fuel systems. In *Microbial Aspects of the Deterioration of Materials* ed. Lovelock, D. W. & Gilbert, R. J. pp. 105-126. Society for Applied Bacteriology Technical Series No. 9. London: Academic Press.

PHAFF, H. J., MILLER, M. W. & MRAK, E. M. 1966 *The Life of Yeasts*. Cambridge, Massachusetts: Harvard University Press.

PHILLISKIRK, G. & YOUNG, T. W. 1975 The occurrence of killer character in yeasts of various genera. *Antonie van Leeuwenhoek* 41, 147-151.

RAINBOW, C. 1970 Brewer's yeasts. In *The Yeasts* Vol. 3, ed. Rose, A. H. & Harrison, J. C. pp. 147-224. London & New York: Academic Press.

REESEN, L. & STRUBE, R. 1978 Complete utilization of whey for alcohol and methane production. *Process Biochemistry* 13, 21-24.

RICHMOND, D. V. 1975 Effects of toxicants on the morphology and fine structure of fungi. *Advances in Applied Microbiology* 19, 289-319.

RODRIGUES DE MIRANDA, L. 1972 *Filobasidium capsuligenum* nov. comb. *Antonie van Leeuwenhoek* 38, 91-99.

RODRIGUES DE MIRANDA, L. 1978 A new yeast genus *Sporopachydermia*. *Antonie van Leeuwenhoek* 44, 439-450.

RODRIGUES DE MIRANDA, L. 1979 A new yeast genus of the Saccharomycetales, *Clavispora*. *Antonie van Leeuwenhoek* 45, 479-483.

ROMANO, A. H. 1966 In *The Fungi* Vol. 2, ed. Ainsworth, G. C. & Sussman, A. S. pp. 181-209. New York: Academic Press.

ROSE, A. H. 1977 *Economic Microbiology* Vol. 1. *Alcoholic Beverages*. London: Academic Press.

RYCHTERA, M., BARTA, J., FIECHTER, A. & EINSELE, A. A. 1977 Several aspects of the yeast cultivation on sulphite waste. *Process Biochemistry* 12, 26-30.

SAND, F. E. M. J., KOLFSCHOTEN, G. A. & VAN GRINSVEN, A. M. 1976 Yeasts isolated from proportioning pumps employed in soft drink plants. *Sonderdruck aus Brauwissenschaft, Jahrgang 29* 10, S. 3-7.

SAND, F. E. M. J. & VAN GRINSVEN, A. M. 1976 Investigation of yeast strains isolated from Scandinavian soft drinks. *Sonderdruck aus Brauwissenschaft Jahrgang 29* 11, S. 3-4

SANTA MARIA, J. 1959 Poliploidia en *Saccharomyces*. *Anales. Instituto nacional de investi gaciones agronómicas, Madrid* 8, 679-712.

SANTA MARIA, J. 1966 Characteristics delimiting the genus *Pichia* Hansen. *Antonie van Leeuwenhoek* 32, 197-201.

SCHEFFERS, W. A. 1966 Stimulation of fermentation in yeasts by acetoin and oxygen. *Nature, London* 210, 533-534.

SCHEFFERS, W. A. 1967 Effects of oxygen and acetoin on fermentation and growth in *Brettanomyces* and some other yeast genera. In *Atti del XIVth Congresso della Società Italiana di Microbiolgia, Messina-Taormina*.

SCHEFFERS, W. A. 1979 Anaerobic inhibition in yeasts (Custer's effect). *Antonie van Leeuwenhoek* **45**, 150.

SNEATH, P. H. & SOKAL, R. R. 1973 *Numerical Taxonomy. The Principles and Practice of Numerical Classification.* San Francisco: Freeman.

SOUW, P., SCHULTE, E. & REHM, H. J. 1978 Formation of l-tetradecane from n-tetradecane by *Candida parapsilosis*. In *Abstracts of the XIIth International Congress of Microbiology, München* p. 145.

STEVENS, B. J. & PAYNE, J. 1977 Cellulose and xylanase production by yeasts of the genus *Trichosporon. Journal of General Microbiology* **100**, 329.

STOKES, J. L. 1971 Influence of temperature on the growth and metabolism of yeasts. In *The Yeasts* Vol. 2, ed. Rose, A. H. & Harrison, J. S. New York & London: Academic Press.

STUMM, C., HERMANS, J. M. H., MIDDELBEEK, E. J., CROES, A. F. & DE VRIES, G. J. M. L. 1977 Killer-sensitive relationships in yeasts from natural habitats. *Antonie van Leeuwenhoek* **43**, 125-128.

TSUCHIYA, T., FUKAZAWA, Y. & KAWAKITA, S. 1965 Significance of serological studies on yeasts. *Mycopathologia et mycologia applicata* **26**, 1-15.

TSUCHIYA, T., FUKAZAWA, Y., TAGUCHI, M., NAKASE, T. & SHINODA, T. 1974 Serologic aspects on yeast classification. *Mycopathologia et mycologia applicata* **53**, 77-91.

ULLOA, M. & KURTZMANN, C. P. 1975 Occurrence of *Candida parapsilosis, C. tropicalis* and *Saccharomyces cerevisiae* in pozol from tabasco, Mexico. *Boletín de la Sociedad mexicana de microbologia* **9**, 7-12.

VAN DER WALT, J. P. 1970a The genus *Syringospora* quinquad emend. *Mycopathologia et mycologia applicata* **40**, 231-243.

VAN DER WALT, J. P. 1970b The perfect and imperfect states of *Sporobolomyces salmonicolor. Antonie van Leeuwenhoek* **36**, 49-55.

VAN DER WALT, J. P. 1970c Criteria and methods used in classification. In *The Yeasts — A Taxonomic Study* 2nd edn, ed. Lodder, J. pp. 34-113. Amsterdam: North-Holland.

VAN DER WALT, J. P. & SCOTT, D. B. 1971 The genus *Saccharomycopsis* Schiönning. *Mycopathologia et mycologia applicata* **43**, 279-288.

VAN DER WALT, J. P. & LIEBENBERG, N. V. D. W. 1973 The yeast genus *Wickerhamiella* gen. nov. (Ascomycetes). *Antonie van Leeuwenhoek* **39**, 121-128.

VAN UDEN, N. & VIDAL-LEIRIA, M. 1970 *Torulopsis* Berlese. In *The Yeasts — A Taxonomic Study* 2nd edn, ed. Lodder, J. pp. 1235-1308. Amsterdam: North-Holland.

VON ARX, J. A. 1972 On *Endomyces, Endomycopsis* and related yeast-like fungi. *Antonie van Leeuwenhoek* **38**, 289-309.

VON ARX, J. A. 1974 The genera of fungi sporulating in pure culture. Vaduz: Gantner.

VON ARX, J. A. & VAN DER WALT, J. P. 1976 The ascigerous state of *Candida chodatii. Antonie van Leeuwenhoek* **42**, 309-314.

VON DECHAU, P. & EMEIS, C. C. 1975 Pektinabbau durch Hefen. *Monatsschrift für Brauerei* **28**, 125-131.

WALKER, H. W. & AYRES, J. S. 1970 Yeasts as spoilage organisms. In *The Yeasts* Vol. 3, ed. Rose, A. H. & Harrison, J. S. pp. 463-527. London & New York: Academic Press.

WEIJMAN, A. C. M. 1979 Carbohydrate composition and taxonomy of *Geotrichum, Trichosporon* and allied genera. *Antonie van Leeuwenhoek* **45**, 119-127.

WIKÉN, T. É. 1972 Utilization of agricultural and industrial wastes by cultivation of yeasts. In *Fermentation Technology To-day. Proceedings of the IVth International Fermentation Symposium, Kyoto, Japan* ed. Terui, G. pp. 569-573.

WIKÉN, T. É., SCHEFFERS, W. A. & VERHAAR, A. J. M. 1961 On the existence of a negative Pasteur effect in yeasts classified in the genus *Brettanomyces Kufferath et van Laer. Antonie van Leeuwenhoek* **27**, 401-433.

YAMADA, Y., ARIMOTO, M. & KONDO, K. 1976 Co-enzyme Q system in the classification of apiculate yeasts in the genera *Nadsonia, Saccharomycodes, Hanseniaspora, Kloeckera* and *Wickerhamia. Journal of General and Applied Microbiology* **22**, 293-299.

YARROW, D. 1972 Four new combinations in yeasts. *Antonie van Leeuwenhoek* **38**, 357-360.

Generic Differentiation in Yeasts

N. J. W. KREGER-VAN RIJ

*Laboratory for Medical Microbiology, State University, Groningen,
The Netherlands*

Contents

1. Criteria for Differentiation

A GENUS is a group of species considered to be related to each other because they share one or more characters. A character or combination of characters also distinguishes the species of one genus from species in other genera. Delimitation of a genus is arbitrary and depends on the choice of characters. The following features are frequently, if not consistently, used for generic differentiation in the yeasts: methods of bud formation; production of true mycelium, formation of arthrospores and structure of the hyphal septum; presence of carotenoid pigments; shape and structure of the ascospores; asci evanescent or persistent (i.e. the spores are liberated or not); assimilation of nitrate; assimilation of inositol; coenzyme-Q system.

This list is not complete and other properties used for generic classi-fication are mentioned in the discussion of the genera in this paper.

Three characters listed are relatively new: structure of the septum, structure of the ascospore wall and nature of the coenzyme-Q system; these will be discussed briefly.

The difference in the ultrastructure of the wall of ascomycetous and basidiomycetous yeasts (Kreger-van Rij & Veenhuis 1971*a*), which according to Donk (1973) may also hold for the primary wall of other ascomycetes and basidiomycetes, is apparently fundamental. The same appears to be the case with the ascomycete and basidiomycete pore structure of the septum (Moore & McAlear 1961). Examination of septa in yeast hyphae revealed several types. In ascomycetous yeasts these are: (1) a plugged dolipore, i.e. a thickening of the wall around the central pore and a plug through the pore (Kreger-van Rij & Veenhuis 1969); (2) plasmodesmata or micropores (Kreger-van Rij & Veenhuis 1973); (3) a central connection which may be open or closed (Kreger-van Rij & Veenhuis 1973). In basidiomycetous yeasts the types are: (1) a simple pore with a tapered wall around it (Kreger-van Rij & Veenhuis 1971a; Johnson-Reid & Moore 1972); (2) a dolipore without a cap or parenthesome (Kwon-Chung & Popkin 1976); (3) a dolipore with a simple cap (Kreger-van Rij & Veenhuis 1971b); (4) a dolipore with a pore cap of cone-shaped dilations of endoplasmic reticulum (Kreger-van Rij & Veenhuis 1971a).

The occurrence of a dolipore or plasmodesmata in the septa of hyphae in some ascomycetous yeasts has facilitated the distinction between these true hyphae and pseudomycelium. In the latter only a closure line of the centripetally grown cross wall is visible. The dolipore is typical of one genus, *Ambrosiozyma*, but plasmodesmata are found in several yeast genera and also in other hemiascomycetes. In the basidiomycetous yeasts differences in septal structure, e.g. a simple pore in teliospore-forming yeasts and a dolipore in the Filobasidiaceae, accompany other differences. Variations in the pore cap among the yeast species of the Filobasidiaceae are not considered adequate for generic differentiation.

Early comparative studies of the internal structure of the ascospore wall of yeasts with the electron microscope were made by Kawakami (1960), Besson (1966) and Kreger-van Rij (1966). Sections of fixed material may show layers of various density and thickness, a ledge, ridges or warts. There is considerable variation in these features among the yeasts and they are certainly not all of the same taxonomic value. The comparative study of the ultrastructure of yeast ascospores, including development of the wall and the changes during germination, is far from complete, but it is already possible to recognize distinct types and I think that a selection of these may be useful for taxonomy. Some types of spores are found in a single genus, others, for instance hat-shaped spores, occur in widely different genera. Knowledge about the chemical composition of the spore wall, which at present is practically nil, may support the structural data.

Yamada and co-workers (Yamada & Kondo 1972, 1973; Yamada et al. 1973a, b, 1976a, b, 1977) have determined the coenzyme-Q system, i.e. the number of isoprene units attached to the quinone nucleus of the strains of a great number of yeast species of many genera and this study has provided noteworthy results. The co-Q varies among the yeasts between 6 and 10, and it appears that in general the species of each genus have the same number. However, exceptions exist, in some cases perhaps confirming already suggested heterogeneity of these genera. Alternatively, species with the same number may rightly belong to very different genera. So, the co-Q number is exclusive rather than inclusive. Some authors rate the co-Q highly and quote the number in the generic definition.

Biochemical tests such as DNA-homology experiments and the establishment of %G + C of DNA, have opened up the opportunity for a better delimitation of species. However, interspecific relationships can seldom be deduced from DNA complementarity data because the homology percentage between different species in one genus is very low (Price *et al.* 1978). On the other hand, Bicknell & Douglas (1970) found that rRNA–DNA homology percentages between *Saccharomyces* and *Kluyveromyces* species were very high and therefore, in this case, not suitable for distinguishing either. As far as the author knows, no new data have become available from this type of research. Base composition of species in one genus may cover a certain range but there is no sharp distinction from the species in other genera. Meyer & Phaff (1970) in a review on DNA composition of yeasts, mentioned that ascomycetous yeasts had a %G + C below 50 and basidiomycetous yeasts higher than 50, although a few ascomycetous species have since been found with a %G + C > 50.

Other new methods will probably be tried to elucidate interspecific relationship, such as comparison of certain proteins (Price *et al.* 1978). A useful quantification of these relationships may be attainable thereby, perhaps revealing generic relatedness. At present, the delimitation of genera is still largely a question of insight.

Wickerham (1951; Wickerham & Burton 1962) has put forward phylogenetic theories involving the concept of a genus, taking the genus *Hansenula* as an example. The characteristics of this genus are the shape of the spores, hat- or saturn-shaped, and the ability to assimilate nitrate. Wickerham distinguished between primitive and more advanced species by regarding the change from haploid to diploid species as a fundamental evolutionary trend. He also considered ecological data, such as association with coniferous trees or free-living states as indicators of phylogenetic relationship. Wickerham (1951) stressed the importance of the shape of the ascospores for generic differentiation. This feature is the main characteristic of the ascomycetous genera in the present classification which is also conspicious by the high percentage (50%) of monospecific genera.

The characters at present used for the definition of species are generally established by simple tests. More sophisticated methods, such as electron microscopy, often yield more fundamental knowledge and this is certainly required for the realization of a satisfactory classification. However, the results obtained with simple methods may more or less accidentally correlate with essential differences and are then useful. These indicator features are part of the usual standard description of species. They generally suffice for differentiation and identification.

Tables 1–3 give a survey of the classification system of the genera. The latter are divided into three groups: genera of the ascosporogenous yeasts; genera of the basidiosporogenous yeasts; genera of the yeasts belonging to the Fungi imperfecti.

TABLE 1

The ascosporogenous yeasts

	Spermophthoraceae	*Coccidiascus* *Metschnikowia* *Nematospora*
	Saccharomycetaceae Schizosaccharomycoideae	*Schizosaccharomyces*
	Nadsonioideae	*Nadsonia* *Hanseniaspora* *Saccharomycodes* *Wickerhamia*
	Lipomycoideae	*Lipomyces*
Endomycetales	Saccharomycoideae	*Ambrosiozyma* *Arthroascus* *Hyphopichia* *Stephanoascus* *Guilliermondella* *Saccharomycopsis* *Hansenula* *Pachysolen* *Citeromyces* *Wickerhamiella* *Dekkera* *Clavispora* *Cyniclomyces* *Lodderomyces* *Schwanniomyces* *Wingea* *Pachytichospora* *Sporopachydermia* *Pichia* *Debaryomyces* *Issatchenkia* *Kluyveromyces* *Saccharomyces*

TABLE 2

The basidiosporogenous yeasts

Teliosporic yeasts	*Leucosporidium* *Rhodosporidium* *Sporidiobolus*
Filobasidiaceae	*Filobasidium* *Filobasidiella*

TABLE 3
Yeasts belonging to the Fungi imperfecti

Blastomycetes	Sporobolomycetaceae	*Sporobolomyces* *Bullera*
	Cryptococcaceae	*Kloeckera* *Schizoblastosporion* *Trigonopsis* *Sympodiomyces* *Brettanomyces* *Malassezia* *Oosporidium* *Rhodotorula* *Phaffia* *Cryptococcus* *Sterigmatomyces* *Trichosporon* *Candida* *Torulopsis*

The genera will be briefly discussed with the help of Tables 4–6. In these tables the principal generic characters are given. If necessary, some other features are mentioned in the text with emphasis on the differences between the genera. For genera described after the appearance of *The Yeasts — A Taxonomic Study* (Lodder 1970), reference to the description is given.

2. Genera of the Ascosporogenous Yeasts

Two families are recognized in this group of yeasts; the Spermophthoraceae with needle- or spindle-shaped ascospores, and the Saccharomycetaceae in which the ascospores have a different shape.

The three genera of the Spermophthoraceae have asci which are much larger than the vegetative cells. They differ in the shape of the ascospores and in the production of true mycelium.

The single species of *Coccidiascus, Coccidiascus legeri*, is a parasite observed in *Drosophila* spp. but not yet cultured. It was first described by Chatton (1913) and more recently by Lushbaugh *et al.* (1976). In each ascus the two ascospores are entwined in a helix and free ascospores are spindle-shaped.

Metschnikowia species have needle-shaped spores, 1–2 per ascus. The species are homo- or heterothallic. In some of them the asci develop from chlamydospores, cells containing large lipid globules (Pitt & Miller 1968). These authors also found that dilute media and reduced temperature favoured sporulation.

TABLE 4

Genera of the ascosporogenous yeasts

Genus	True mycelium	dol/pl*	Shape of the ascospores	Spores liberated	h/d†	Nitrate	Fermentation	co-Q
Ambrosiozyma	+	dol	hat-shaped	v		v	+	7[6]
Arthroascus	+	–	hat- or saturn-shaped, smooth or warty	+	h	–	–	8[6]
Citeromyces	–		spherical, warty	–	h/d	+	+	8[3]
Clavispora	–		clavate	+	h	–	+	8[1]
Coccidiascus	–		spindle-shaped	+				
Cyniclomyces	–		oval, cylindrical	v	d	–	+	
Debaryomyces	–		spherical, warty, oval with ridges	–	h	–	v	9[6]
Dekkera	–		hat-shaped	+	d	v	+	
Guilliermondella	+	pl	spherical, oval, reniform	+	h	–	+	8[6]
Hanseniaspora	–		hat-shaped, spherical with a ledge, smooth or warty, spherical and warty	v	?	–	+	6[5]
Hansenula	–(+)	–	hat- or saturn-shaped	+	h/d	+	v	7, 8[3]
Hyphopichia	+	–	hat-shaped	+	h(d)	–	+	8[6]
Issatchenkia	–		spherical, warty	–	d	–	+	7[3]
Kluyveromyces	–		spherical, reniform	+	h/d	–	+	6[6]
Lipomyces	–		oval, smooth, warty or with ridges	+	h	–	–	9[7]
Lodderomyces	–		oblong	–	d	–	+	9[7]
Metschnikowia	–		needle-shaped	v	h/d	–	v	9[7]

Genus								Co-Q
Nadsonia	—		spherical, warty	—	h	—	v	6[5]
Nematospora	+		spindle-shaped	+	h?	—	+	6[7]
Pachysolen	—		hat-shaped	—	h/d	+	+	8[3]
Pachytichospora	—(+)	—	ellipsoidal, thick-walled	—	d	—	+	
Pichia	—(+)		spherical, hat- or saturn-shaped	+(—)	h/d	—	v	7, 8, 9[3]
Saccharomyces	—		spherical, oval, smooth or warty	—	h/d	—	+	6[6]
Saccharomycodes	—		spherical with narrow ledge	—	d	—	+	6[5]
Saccharomycopsis	+	pl(—)	spherical, hat- or saturn-shaped with an irregular ledge	v	h(d)	—	v	8[1], 9[6]
Schizosaccharomyces	v		spherical, oval, reniform	+	h	—	+	9, 10[4]
Schwanniomyces	—		spherical and warty with a ledge	—	h	—	+	9[7]
Sporopachydermia	—		spherical or oval, thick-walled	+	h/d	—	—	
Stephanoascus	+	pl	hemispherical, thick-walled	+	h	—	—	
Wickerhamia	—		cap-shaped	+	d	—	+	9[5]
Wickerhamiella	—		oblong	+	h	+	—	
Wingea	+		lenticular	—	h	—	+	9[3]

*dol, dolipore; pl, plasmodesmata.
†h, haploid; d, diploid.
nitrate, assimilation of nitrate.
co-Q, coenzyme-Q number; superscript is number of the reference.
v, variable; (+), seldom +; (−), seldom −.

References:

1 Yamada & Kondo (1972).
2 Yamada & Kondo (1973).
3 Yamada et al. (1973a).
4 Yamada et al. (1973b).
5 Yamada et al. (1976a).
6 Yamada et al. (1976b).
7 Yamada et al. (1977).

TABLE 5
The genera of the basidiosporogenous yeasts

Genus	Ballistospores	Carotenoid pigments	Nitrate assimilation	Inositol assimilation	Starch production	Fermentation	co-Q
Filobasidiella	−	−	−	+	+	−	10^2
Filobasidium	−	−	v	+	+	v	9, $10^{1,2}$
Leucosporidium	−	−	+	v	v	v	8, 9, 10^1
Rhodosporidium	−	+	v	v	v	−	8, 9, 10^2
Sporidiobolus	+	+	+	−	−	−	

Symbols and references as in Table 4.

TABLE 6
Genera of the Fungi imperfecti

Genus	asc/bas	True mycelium	Arthrospores	Ballistospores	Carotenoid pigments	Nitrate assimilation	Inositol assimilation	Starch production	Fermentation	co-Q
Sporobolomycetaceae										
Bullera	bas	v		+	−	v	v	v	−	
Sporobolomyces	bas	v		+	+	v	v	v	−	10^2
Cryptococcaceae										
Brettanomyces	asc	−(+)			−	v	−		+	
Candida	asc/bas	v	−		−	v	v	v	v	6, 7, 8, 9, 10^1
Cryptococcus	bas(asc)	−			v	v	+	+(−)	−	8, 9, 10^2
Kloeckera	asc	−			−	−	−	−	+	6^5
Malassezia		v			−				−	
Oosporidium	asc	+	−		−	+	−		−	
Phaffia	bas	−			+	−	−	+	+	
Rhodotorula	bas	−(+)			+	v	−	−	−	9, 10^2
Schizoblastosporion	asc	−			−	−	−	−	−	
Sterigmatomyces	bas	v	−		v	v	v	v	−	
Sympodiomyces	asc	+	−		−	−	+	−	−	
Torulopsis	asc/bas	−			−	v	−		v	6, 7, 8, 9, 10^1
Trichosporon	asc/bas	+	+		−	v	v	v	v	
Trigonopsis	asc	−			−	−	−	−	−	

asc, ascomycetous; bas, basidiomycetous.
Other symbols and references as in Table 4.

The two extremities of the spindle-shaped ascospores of *Nematospora coryli* consist of wall material, one of them very thinly drawn out, often as a whip-like appendage. The eight spores lie in two bundles of four.

The second ascosporogenous family, the Saccharomycetaceae, comprises four subfamilies: Schizosaccharomycoideae, Nadsonioideae, Lipomycoideae and Saccharomycoideae, distinguished by features of the vegetative and sexual reproduction.

Schizosaccharomyces, the single genus of the Schizosaccharomycoideae, is characterized by fission of single cells and is thus exceptional among the yeasts, which all form buds. One of the *Schizosaccharomyces* species also produces true mycelium splitting up into arthrospores. Diploidization takes place by fusion of two vegetative cells. In contrast with this, species of the genus *Endomyces*, which also produce true mycelium and arthrospores, form gametangia, i.e. special cells on hyphae or arthrospores which fuse to give a zygote developing to an ascus. Moreover, cells of *Endomyces* species may be multinuclear, whereas *Schizosaccharomyces* cells have one nucleus (Slooff 1970*a*). *Schizosaccharomyces* species produce 4–8 spores per ascus.

The Nadsonioideae are distinguished by bipolar budding on a more or less broad base. The vegetative cells are lemon-shaped and multiple scars are present at the two poles of the cells. The four genera are distinguished by the shape of the spores and the structure of the spore wall; moreover, there are differences in the mode of ascus formation.

In *Nadsonia* species mother cell and bud fuse to form a zygote which changes into an ascus or produces another bud which becomes the ascus. The brown spores are spherical and have a warty wall; one or two are formed per ascus.

Ascospores vary considerably among the *Hanseniaspora* species. There are four types (Kreger-van Rij 1977): hat-shaped and liberated from the ascus; spherical with a subequatorial ledge and warty; spherical with a ledge and not warty; spherical and warty without a ledge. The asci with spherical spores are persistent. Meyer *et al.* (1978) made DNA homology studies of strains of *Hanseniaspora* and *Kloeckera* species. As a result, they recognized six *Hanseniaspora* species and six similar, but non-sporogenous, *Kloeckera* species. The rough *Hanseniaspora* ascospores are much smaller than those of *Nadsonia*. All *Hanseniaspora* species produce a vigorous fermentation.

The third genus with bipolar budding is *Saccharomycodes* with a single species *S'codes ludwigii*. The four spherical spores in each ascus, each with a narrow subequatorial ledge, conjugate in pairs directly after germination while still in the ascus.

The last genus of the Nadsonioideae is *Wickerhamia*. The single species, *Wickerhamia fluorescens*, has cap-shaped spores, i.e. the spores have a brim which on one side protrudes much farther than on the other side. Usually one spore is formed per ascus.

The third subfamily, Lipomycoideae, includes one genus *Lipomyces*. Each *Lipomyces* species generally has a sac-like ascus containing 1–16 or more amber-coloured spores. The methods for ascus formation vary and often involve an active bud (Slooff 1970*b*). Sporulation is generally good on nitrogen-deficient media (Starkey 1946). *Lipomyces* cells are mostly capsulated.

The fourth subfamily of the Saccharomycetaceae, Saccharomycoideae, includes all ascomycetous yeasts not mentioned above. The ascospores are of various shapes, but not needle- or spindle-shaped; budding is never bipolar and generally multilateral; true mycelium, occasionally splitting up into arthrospores, may be formed, but always in addition to loose budding cells; and a sac-like ascus with many typical spores, as in *Lipomyces*, does not occur. Features for the differentiation of the 22 genera are: the occurrence of true mycelium and the structure of the septum; the shape and structure of the ascospores; the method of ascus formation and persistence or evanescence of the ascus; nitrate assimilation; fermentative ability, especially vigorous fermentation; coenzyme-Q system.

Six genera have species forming abundant true mycelium on the usual solid media. Buds may be formed on the hyphae, the blastospores or blastoconidia and, under suitable conditions, loose budding cells are present.

The genus *Ambrosiozyma* (van der Walt 1972) is distinguished from the others by a special structure of the septum, a plugged dolipore, visible under the light microscope as a small dot in the middle of the septum. Von Arx (1972) classified one of the species, *Ambrosiozyma platypodis* in a separate genus *Hormoascus* because it assimilates nitrate in contrast with the other species. The author does not follow this classification.

In the genus *Arthroascus* (von Arx 1972) with the single species *Arthroascus javanensis*, loose vegetative cells are mostly elongate, formed either as buds on a broad base or as arthrospores from the hyphae. The species is homothallic. Hat- or saturn-shaped ascospores are formed directly after fusion of two cells in one of them. *Arthroascus javanensis* has several features which are also typical of *Pichia* species, but it differs in producing abundant true mycelium. The septa have a central connection which may be open or closed.

Of the four other mycelial genera *Hyphopichia* stands out by the formation of asci as loose cells directly after conjugation of two yeast cells or without conjugation. In contrast, asci in the genera *Saccharomycopsis, Guilliermondella* and *Stephanoascus* arise on the hyphae. Von Arx & van der Walt (1976) characterized the genus by the formation of conidia on denticles. This feature is not specific and, therefore, the genus is not generally accepted. However, *Hyphopichia burtonii*, the single species, has multinuclear hyphal cells which have not been found in other ascomycetous yeasts (unpublished). The hyphae split up into arthrospores. *Hyphopichia burtonii* is heterothallic.

The genera *Stephanoascus*, *Guilliermondella* and *Saccharomycopsis* are distinguished by the shape of the ascospores. In *Stephanoascus ciferrii* (Smith *et al.* 1976), the single species of the genus, the spores are hemispherical with a wide wall at the convex side (Fig. 1*d*). Two are formed per ascus. Very typically, an apical cell is present on the ascus. The species is heterothallic.

Guilliermondella selenospora, also a single species in the genus, has short-oval to lunate ascospores of which the wall is thickened at the poles. The species is homothallic. Hyphal cells fuse and are separated again by a wall; both cells may then turn into asci, or buds are formed on them which become asci (Kreger-van Rij & Veenhuis 1976*b*). Four spores are formed per ascus; they may vary considerably in size and shape.

The ascospores of *Saccharomycopsis* species all have a ledge, but its shape and the structure of the spore wall varies among them. For instance, the spore wall of the type species *S. capsularis*, in sections under the electron microscope, has a broad dark outer layer and a light inner layer; the ledge is narrow and may disappear during development of the spore (Kreger-van Rij & Veenhuis 1975). *Saccharomycopsis vini* has two ledges (Kreger-van Rij 1969; Kurtzman & Wickerham 1973). *Saccharomycopsis crataegensis* has ellipsoidal spores with a longitudinal ledge and a warty wall (Kurtzman & Wickerham 1973). *Saccharomycopsis crataegensis* and *S. lipolytica* are both heterothallic, the other species are homothallic. *Stephanoascus*, *Guilliermondella* and *Saccharomycopsis* species all have septa with plasmodesmata with the exception of *S. lipolytica* which has a central connection, open (Fig. 2*a*) or closed (Kreger-van Rij & Veenhuis 1973). This species has a co-Q of 9, different from the other *Saccharomycopsis* species which have 8 (Yamada *et al.* 1976*b*). These features and also the shape of the ascospore which is variable, often with an irregular ledge (Fig. 1*f*), distinguish *S. lipolytica* from the other species.

Four genera are primarily characterized by the ability to assimilate nitrate: they are distinguished by the shape of the ascus and the ascospores. The genus *Hansenula*, the oldest and largest genus of the four with about 35 species, was comprehensively studied and described by Wickerham (1970). The ascospores are liberated from the ascus and each has a ledge which makes it hat- or saturn-shaped. A few species may produce some true mycelium. The *Hansenula* species are homo- or heterothallic, they vary in ploidy, pseudomycelium formation, pellicle formation, fermentative ability

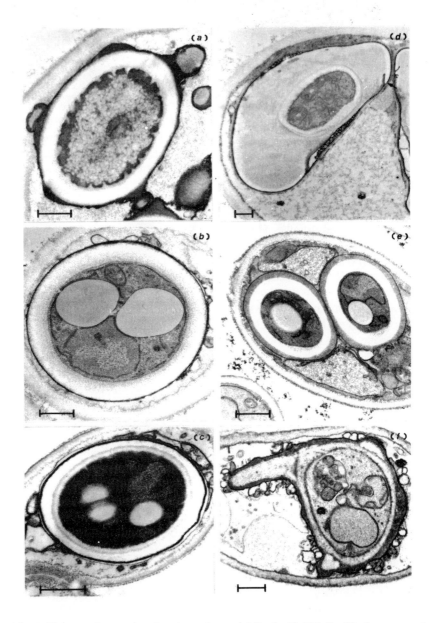

Fig. 1. Electron micrographs of sections of material fixed with $KMnO_4$. The bar represents 0·5 μm. *(a) Sacch. rouxii.* Ascospores with a warty wall. *(b) Sacch. bailii.* Ascospore with a smooth wall; the wall has a thin, dark outer layer and a broad, light inner layer. *(c) Sacch. florentinus.* The wall of the ascospore has a dark outer layer. The inner layer includes a greyish inner part and a lighter outer part. *(d) Stephanoascus ciferrii.* The wall of the asco-spore has a thin, dark outer layer. An inner light layer lies around the protoplast. In between, a wide greyish layer is present. Immature spores have a distinct ledge but during maturation the greyish layer extends and of the ledge only a small protrusion (arrow) is visible. *(e) K. africanus.* The two spores visible in the ascus have a wall with a light inner layer and a broad, dark outer layer. *(f) S. lipolytica.* The ascospore has an irregular ledge showing in sections as a small and a long protuberance. The wall has a light inner layer and a darkish, slightly warty outer layer.

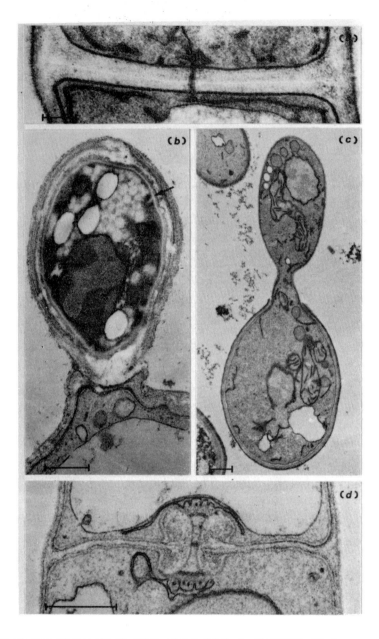

Fig. 2. Electron micrographs of sections of material fixed with KMnO$_4$. The bar represents 0·5 μm. *(a)* Septum of *S. lipolytica*. In the narrow central connection between the hyphal cells endoplasmic reticulum is visible. *(b)* *Filobasidiella neoformans*. Basidiospore still attached to the basidium. It has a thick, distinctly layered wall which seems to be partly continuous with the wall of the basidium. The inner part of the spore wall (arrow) is new. Older, detached spores were found to have shed the outer layers of the wall. *(c)* Budding cell of *St. nectairii*. The cell has a layered wall and the bud is formed on a short neck. This species forms orange-coloured cultures. *(d)* Septum in a hypha of *Filobasidiella neoformans* with a dolipore. The pore has at both sides a cap formed by endoplasmic reticulum.

and co-Q system number, most of them having 7 but *H. capsulata* and *H. holstii* have a co-Q of 8 (Yamada *et al.* 1973*a*).

The three other nitrate-positive genera are monospecific. The homothallic *Pachysolen tannophilus* has a typical ascus which consists of a cell with a long neck. Most of the wall is very thick with the exception of the tip of the neck where the four hat-shaped spores are situated.

Citeromyces matritensis produces one, seldom two, spherical, warty spores per ascus which are liberated. The species is heterothallic and cells of opposite mating type agglutinate.

Wickerhamiella domerquii (van der Walt & Liebenberg 1973) has oblong ascospores, one per ascus. The species is homothallic.

The genus *Dekkera* includes two species with both nitrate positive and negative strains. Characteristic of this genus is strong acid production and the short life of the cells. The species are homothallic and produce 1–4 hat-shaped spores per ascus.

The 12 nitrate-negative genera which follow are differentiated by the shape and structure of the ascospores and persistence or evanescence of the ascus; the co-Q is also considered. The monospecific genera are: *Clavispora, Cyniclomyces, Lodderomyces, Schwanniomyces, Wingea.*

Clavispora lusitaniae has 1–4 club-shaped spores per ascus. The species is heterothallic and the imperfect states are *Candida lusitaniae* and *C. obtusa* (Rodrigues de Miranda 1979).

Cyniclomyces guttulatus has 1–4 oval spores which may or may not be liberated from the ascus. This species is found in the stomach of rabbits. It requires an elevated CO_2 concentration for growth.

Lodderomyces elongisporus produces one, seldom two, oblong spores per ascus, which are not liberated. This species is homothallic and it can utilize higher alkanes.

Schwanniomyces alluvius has spherical, warty spores with an equatorial ledge, mostly one per ascus. Conjugation between mother cell and bud usually precedes ascus formation.

Wingea robertsii has lentiform, brown ascospores. Conjugation between mother cell and bud precedes ascus formation.

The ascospores of the genera *Pachytichospora* and *Sporopachydermia* are thick-walled, but of different structure (Kreger- van Rij 1978*b*).

Pachytichospora transvaalensis (van der Walt 1978) forms one, seldom two, ellipsoidal spores per ascus, which are not liberated. The species is heterothallic and was first classified in the genus *Saccharomyces.*

The two *Sporopachydermia* species (Rodrigues de Miranda 1978) produce 1–4 spores per ascus which are liberated. *Sporopachydermia lactativora* is heterothallic. Inositol is assimilated by both species which were first classified in *Cryptococcus.*

One of the large nitrate negative genera is *Pichia*, similar in many respects to the nitrate positive genus *Hansenula*, which includes a few species with true mycelium and also displaying homo- and heterothallism, variation in ploidy, pseudomycelium formation, pellicle formation, fermentative ability and co-Q number. Most species have ascospores with a ledge and evanescent asci. The ascospores of one species, *P. fluxuum*, have a ledge and a warty wall. A few species have spherical spores without a ledge and some authors, considering these species atypical of *Pichia*, have proposed another classification for them (van der Walt & Johannsen 1975).

The next genus, *Debaryomyces*, is smaller than *Pichia* and more homogeneous. The ascospores are warty and are not liberated from the ascus; usually 1–2 are formed per ascus. In one species, *Deb. marama*, the spores have ridges (Kurtzman *et al.* 1975). The vegetative cells are haploid and, frequently, conjugation between mother cell and bud precedes ascus formation. Fermentative ability varies, but is never vigorous. Price *et al.* (1978), by DNA homology studies, have corroborated the validity of some *Debaryomyces* species and reclassified others, including some *Pichia* species, in *Debaryomyces*.

Another genus with warty spores is *Issatchenkia*. In contrast with *Debaryomyces*, the vegetative cells are diploid and the species are heterothallic (Kurtzman & Smiley 1976). They form a pellicle on liquid media. Kurtzman (pers. comm.) considers the co-Q number of *Issatchenkia* (7) to be an important generic character which distinguishes the genus from *Debaryomyces* (9) and *Saccharomyces* (6).

The last two large ascomycetous genera are *Kluyveromyces* and *Saccharomyces*, both with strong fermentative ability and co-Q of 6.

Kluyveromyces is typified by spherical, reniform or sickle-shaped ascospores which are easily liberated. Most species produce 1–4 spores per ascus but *K. africanus* forms up to 16 and *K. polysporus*, 60 or more. Both homo- and heterothallic species occur. The internal structure of the spore wall is similar for most *Kluyveromyces* species and corroborates the homogeneity of the genus. However, a few species have aberrant spore walls, namely *K. waltii*, *K. thermotolerans* (Kreger-van Rij 1979) and *K. africanus* (unpublished) (Fig. 1e).

Finally, the genus *Saccharomyces* in the description of Lodder & Kreger-van Rij (1952) and van der Walt (1970) includes species with spherical or oval spores, with or without warts and with consistent asci. The genus thus defined includes several groups of species differing in the structure of the spore wall, in ploidy and methods of ascus formation. Some authors (van der Walt & Johannsen 1975; von Arx *et al.* 1977) proposed to separate the haploid species from *Saccharomyces* and classify them in the genera *Zygosaccharomyces* and *Torulaspora*. These genera are distinguished as follows: in *Zygosaccharomyces* conjugation precedes ascus formation and the ascospores are smooth; in *Torulaspora* usually no conjugation precedes ascus formation and the spores are mostly warty. However, Kreger-van Rij and Veenhuis (1976*a*) found that in the so-called *Torulaspora* group conjugation between mother cell and bud often precedes ascus formation; unconjugated asci with a protuberance were observed in all strains studied, but infrequently. Kreger-van Rij (1979) discovered that among the species of the proposed genus *Zygosaccharomyces* three types of ascospores exist (Fig. 1*a*, *b*, *c*) one of them similar to *Sacch. cerevisiae* spores (Fig. 1*c*). *Sacch. rouxii*, which according to von Arx *et al.* (1977) is synonymous with *Zygosacch. barkeri*, the type species of *Zygosaccharomyces*, has warty spores (Fig. 1*a*) similar to those of *Debaryomyces* and different from the warty spores of *Torulaspora* species. In view of the variation in spore wall structure, it seems advisable to delay any possible splitting of *Saccharomyces* until more data are available. Species of the genus *Saccharomyces* (van der Walt 1970) are homo- or heterothallic. Conjugation between separate cells and between mother cell and bud occur. In diploid species, germinating spores may fuse while still within the ascus (Kreger-van Rij 1978*a*).

3. Genera of the Basidiosporogenous Yeasts

Basidiomycetous yeasts are the haploid stages of various heterobasidiomycetes. Knowledge of these yeasts is now rapidly increasing from crossing experiments or from isolations revealing dikaryotic and sporulating states. However, for most basidiomycetous yeasts, the perfect form has not yet been found and they are still classified as Fungi imperfecti.

The basidiomycetous yeasts of which the perfect state is known have so far been classified in two groups: the teliospore-producing species and the family Filobasidiaceae.

The yeasts of the first group include three genera, *Sporidiobolus*, *Rhodosporidium* and *Leucosporidium*, in each of which two types of cell cycle exist. The first is a homothallic cycle in which a single yeast cell develops a uninucleate mycelium without clamp connections: teliospores are then formed on the hyphae which, during germination, produce metabasidia and basidiospores. The second type is a heterothallic cycle in which two yeast cells of different mating type fuse to give a dikaryotic mycelium with clamp connections. Teliospores are then formed in which karyogamy takes place and their germination produces basidia with haploid basidiospores. Fell (1970), who made valuable contributions to our knowledge of these organisms has described the two cell cycles in more detail.

The three genera of the teliosporic yeasts are easily distinguished: *Sporidiobolus* is the only one forming ballistospores; *Rhodosporidium* has carotenoid pigments which give the cultures a red, orange or yellow colour; *Leucosporidium* cultures do not have this colour. The septum of the dikaryotic mycelium in the species of these genera which have been examined has a narrow, simple pore.

In *Sporidiobolus* ballistospores are formed by the yeast cells and this phase closely resembles *Sporobolomyces*. Alternatively, crossing strains of *Sp. odorus* has produced the dikaryotic phase of *Sporidiobolus* (Bandoni *et al.* 1971). The two species *Sporid. johnsonii* and *Sporid. ruinenii* were both isolated in the dikaryotic phase.

The haploid states of some *Rhodosporidium* species are *Rhodotorula* strains. In contrast with *Rhodotorula* species, which do not assimilate inositol, assimilation of this compound varies among the *Rhodosporidium* species.

Leucosporidium species in the haploid stage are nitrate positive *Candida* species forming pseudo- and true mycelium. Most *Leucosporidium* species are homothallic, only *Leu. scottii* is both homo- and heterothallic (Fell & Phaff 1970).

The Filobasidiaceae, the second group of basidiomycetous yeasts, have slender, unseptate basidia with sessile basidiospores and no teliospores. Two of the genera, *Filobasidium* and *Filobasidiella* have a haploid state of yeast cells, but they differ in the method of basidiospore production. In *Filobasidium* (Olive 1968) the basidiospores are formed on top of the basidium in a petal-like whorl; in *Filobasidiella* (Kwon-Chung 1975) they are formed basipetally on the apex of the basidium from four spots giving four long chains (Fig. 2*b*).

The three *Filobasidium* species are heterothallic. They have a septum with a dolipore; in *Filobasidium floriforme* there is no pore cap, but in *F. capsuligenum* the cap consists of cupulate bodies. The latter species, in contrast with the others, has fermentative ability. Kwon-Chung (1977) discovered the perfect state of *Cryptococcus uniguttulatus* and named it *F. uniguttulatum*. The septum in this species also has a dolipore with a cupulate cap (unpublished).

The genus *Filobasidiella* includes the pathogenic species *Filobasidiella neoformans* and *Filobasidiella bacillispora*. Kwon-Chung (1975; Kwon-Chung *et al.* 1978) found both perfect states by crossing compatible strains of *Cr. neoformans* and *Cr. bacillisporus* respectively. The septum in hyphae of *Filobasidiella neoformans* has a dolipore, but no parenthesome (Kwon-Chung & Popkin 1976) or a simple one (Fig. 2*d*). Shadomy (1970) had previously found single strains of *Cr. neoformans* which produced aerial mycelium with clamp connections.

4. Genera of Yeasts Belonging to the Fungi Imperfecti

The imperfect yeasts are classified in two families: the Sporobolomycetaceae and the Cryptococcaceae. The species of the first family produce ballisto-spores, those of the second do not.

The Sporobolomycetaceae include two yeast genera: *Sporobolomyces* and *Bullera*. They differ in the shape of the ballistospores which are bean-shaped in *Sporobolomyces* and spheroidal to ovoidal in *Bullera*. In both genera, the spores are pointed at one pole, the site of attachment to the sterigma. *Sporobolomyces* species have orange to red carotenoid pigments. Cultures of *Bullera* species are cream-coloured.

The second family of the asporogenous yeasts, the Cryptococcaceae, includes a wide variety of genera. A natural classification is difficult since the important characters of sexual reproduction are lacking, hence most genera are characterized by typical features of the vegetative reproduction. This has led to five monospecific genera among the 15 genera. A hetero-geneous remainder is placed in the genera *Torulopsis* and *Candida*. The Cryptococcaceae include ascomycetous and basidiomycetous yeasts. In the ascomycetous yeasts spores have not been observed because the conditions for sporulation are not yet known or because the yeasts are heterothallic haploids and the compatible mating type has not yet been found. The asporogenous basidiomycetous yeasts are probably all heterothallic haploids. For some imperfect yeast species the perfect form is surmised because their morphological and physiological properties are similar to known perfect species, or a relationship has been demonstrated by DNA homology: this is however still a minority. Crossing experiments have brought to light new perfect forms, such as the basidiomycetous genus *Filobasidiella* and the ascomycetous genus *Sporopachydermia*.

It has become possible to distinguish with sufficient certainty between ascomycetous and basidiomycetous yeasts in the imperfect state by the structure of the cell wall. Also, basidiomycetous yeasts may be recognized by components of the extracellular heteropolysaccharides, production of extracellular deoxyribonuclease and the biosynthesis of tryptophan (Fell 1970). Moreover, there are other indicative features like a high mol %G + C, a positive urease test and starch formation. Some genera of the Cryptococcaceae are either ascomycetous or basidiomycetous, others are mixed.

Two ascomycetous imperfect genera are distinguished from the others by bipolar budding: *Kloeckera* and *Schizoblastosporion*. They differ in fermentative ability. *Kloeckera* species, the imperfect states of *Hansenia-spora* spp., can ferment sugars whereas the monospecific *Schizoblasto-sporion* cannot, and a perfect state is not known.

The monospecific genus *Trigonopsis* is characterized by the occurrence of triangular cells. Buds are oval and they are formed on short protrusions on the cell. Cells which have formed three buds have the typical triangular shape, but other shapes with more protuberances also occur.

In the genus *Sympodiomyces* (Fell & Statzell 1971) a bud is formed on a conidiophore developed from a yeast cell. After the first conidium, the conidiophore grows out sideways and produces a new bud and so on. The single species *Sympodiomyces parvus* has an ascomycetous wall and the hyphal septa have a closure line (unpublished).

The genus *Brettanomyces* is characterized primarily by a physiological feature, namely strong acid production from glucose: the cells are short-lived on malt agar. The perfect form of two species is known and classified in *Dekkera*.

The genus *Malassezia* has cells reproducing by monopolar budding, and short hyphae may also be formed. Originally, both forms had different names: *Malassezia* for the hyphae and *Pityrosporum* for the yeast cells, and in cultures, it was not possible to change one into the other. However, Keddie (1967) showed that the typical ultrastructure of the wall of the hyphae agreed with that of the yeast cells and Dorn & Roehnert (1977) induced *Pityrosporum* strains to form hyphae. The hyphae are present in a skin disease called pityriasis versicolor, the yeast cells are normally present on the skin and may be cultured on media with lipids; they have been described as *Pit. ovale* or *Pit. orbiculare*.

The monospecific genus *Oosporidium* produces true mycelium and buds formed laterally on a broad base. The cultures of *O. margaritiferum* are pink. The cell wall is ascomycetous and the septa have a simple pore (unpublished).

In two genera, *Rhodotorula* and *Phaffia*, all species have red or yellow cultures caused by the presence of carotenoid pigments. *Rhodotorula* lacks fermentative ability. Strains of *Rh. glutinis* appeared to be mating types of three different *Rhodosporidium* species (Fell 1970).

The monospecific genus *Phaffia* (Miller *et al.* 1976) has fermentative ability. No perfect state is known yet of the single species *Phaffia rhodozyma*.

The genus *Cryptococcus* is basidiomycetous with the exception of *Cr. melibiosum* for which no sporulating state has yet been found (Rodrigues de Miranda 1978). *Cryptococcus* species, in contrast with *Rhodotorula* and *Torulopsis*, assimilate inositol. A few *Cryptococcus* species have red or orange cultures. Imperfect forms of *Filobasidium*, *Filobasidiella* and *Tremella* species may be identified as *Cryptococcus* species. So far, the genera *Cryptococcus* and *Rhodotorula*, which are rather similar, have dissimilar perfect states: *Rhodotorula* in

the teliospore-forming genera, and *Cryptococcus* in the Filobasidiaceae. Most *Cryptococcus* and *Rhodotorula* species have capsulated
cells.

The genus *Sterigmatomyces* is characterized by the formation of buds on
stalks of variable length. Separation of mother cell and bud is generally in
the middle of the stalk. Most species have hyaline cultures; *St. nectairii*
(Fig. 2c) alone is orange. Perfect states of the basidiomycetous species are
not yet known.

The genus *Trichosporon* is characterized by the formation of loose
budding cells and true mycelium falling apart into arthrospores. It includes
both ascomycetous and basidiomycetous species. The type species,
Tr. cutaneum, has a basidiomycetous wall and septa with a dolipore
(Kreger-van Rij & Veenhuis 1971b). The mixed character of the genus has
induced some authors to propose a separation of basidiomycetous and
ascomycetous species in different genera (Kocková-Kratochvílová *et al.*
1977; Weijman 1979). So far, no perfect state has been found for the
basidiomycetous species. Ultrastructural studies have shown three types of
septa among the ascomycetous species, typical of at least three perfect
genera (unpublished). In view of this, the question arises whether it is
worthwhile to create temporarily new combinations for the imperfect
species. King & Jong (1975, 1976) have classified two *Trichosporon* species,
Tr. aculeatum and *Tr. inkin*, in two new genera: *Aciculoconidium* (1976)
and *Sarcinosporon* (1975), respectively. For the first species they stated that
arthrospores were not formed, which is not correct. *Trichosporon
aculeatum* has an ascomycetous wall and, like *Hyphopichia burtonii*, septa
with a central connection and several nuclei per hyphal cell (unpublished).
The genus *Sarcinosporon* was created because of a kind of cell, seemingly a
sporangium, which fell apart into spores. The author thinks this description
is confusing and agrees with do Carmo-Sousa (1970) who described
reproduction in *Tr. inkin* as cells dividing by septation in different planes.
This species has basidiomycetous cell walls and dolipores may be present in
the septa (unpublished). The difference between *Tr. cutaneum* and *Tr. inkin*
mainly lies in the mode of septation which is probably not sufficient for
describing a new genus.

The asporogenous yeasts not included in the genera mentioned above are
classified in the genera *Candida* and *Torulopsis*, the former producing
pseudomycelium, the latter without pseudomycelium. This distinction is of
practical importance to subdivide the large remaining group. However,
some species existing in a smooth and a rough form may be difficult
to place.

Van Uden & Buckley (1970) recognized 81 *Candida* species and 36
Torulopsis species, and these numbers have increased since then. *Candida*

and *Torulopsis* both include ascomycetous and basidiomycetous species, but relatively few of the latter. In *Candida*, species occur with true mycelium but without arthrospores, thus distinguishing the genus from *Trichosporon*, which has arthrospores. Both *Candida* and *Torulopsis* include imperfect forms of many genera, some surmised and many still unknown.

Van Uden & Buckley (1970) pointed to the artificial distinction between the two genera, but advised against a merging of them since this would necessitate a provisional change of name of many species which would be very confusing. Nevertheless, Yarrow & Meyer (1978) proposed abolition of the genus *Torulopsis*, arguing that the name *Torulopsis* was illegitimate. However, this nomenclatural argument for abolishing the genus may be met by retaining the name of *Torulopsis*, as indicated by Lodder & Kreger-van Rij (1952). Merging of *Torulopsis* with *Candida* results in a still more heterogeneous genus *Candida* without contributing to a clarification of natural relationships. To make the genera *Torulopsis* and *Candida* accessible for identification purposes subdivision into smaller, surveyable groups, not necessarily natural but easy to recognize, seems advisable.

The preparation of this paper was supported by the Netherlands Organization for the Advancement of Pure Science (ZWO). I thank Mr. J. Zagers for printing the photographs and Mr. M. Veenhuis for Fig. 2a.

5. References

BANDONI, R. J., LOBO, K. J. & BREZDEN, S. A. 1971 Conjugation and chlamydospores in *Sporobolomyces odorus*. *Canadian Journal of Botany* **49**, 683–686.

BESSON, M. 1966 Les membranes des ascospores de levures au microscope électronique. *Bulletin de la Société mycologique de France* **82**, 489–503.

BICKNELL, J. N. & DOUGLAS, H. C. 1970 Nucleic acid homologies among species of *Saccharomyces*. *Journal of Bacteriology* **101**, 505–512.

CHATTON, E. R. 1913 *Coccidiascus legeri n.g., n.sp.* levure ascosporée parasite des cellules intestinales de *Drosophila funebris* Fabr. *Compte rendu des séances de la Société de Biologie* **75**, 117–120.

DO CARMO-SOUSA, L. 1970 The genus *Trichosporon*. In *The Yeasts — A Taxonomic Study* 2nd edn, ed. Lodder, J. pp. 1309–1352. Amsterdam: North-Holland.

DONK, M. A. 1973 The Heterobasidiomycetes: a reconnaissance — IIIB. How to recognize a Basidiomycete? *Proceedings Koninklijke Akademie van Wetenschappen, Amsterdam, Series C* **76**, 14–22.

DORN, M. & ROEHNERT, M. 1977 Dimorphism of *Pityrosporum orbiculare* in a defined culture medium. *Journal of Investigative Dermatology* **69**, 244–248.

FELL, J. W. 1970 Yeasts with heterobasidiomycetous life cycles. In *Recent Trends in Yeast Research* Vol. 1, ed. Ahearn, D. G. pp. 49–66. Atlanta: Spectrum (Georgia State University).

FELL, J. W. & PHAFF, H. J. 1970 The genus *Leucosporidium*. In *The Yeasts — A Taxonomic Study* ed. Lodder, J. pp. 776–802. Amsterdam: North-Holland.

FELL, J. W. & STATZELL, A. C. 1971 *Sympodiomyces gen. n.*, a yeast-like organism from southern marine waters. *Antonie van Leeuwenhoek* **37**, 359–367.

JOHNSON-REID, J. A. & MOORE, R. T. 1972 Some ultrastructural features of *Rhodosporidium toruloides* Banno. *Antonie van Leeuwenhoek* **38**, 417–435.

KAWAKAMI, N. 1960 Electron microscopy on Ascomycetes. Investigation of fine structures and classification of sporogenous yeasts. *Memoirs of the Faculty of Engineering, Hiroshima University* **1**, 207–237.

KEDDIE, R. M. 1967 Tinea versicolor: the electron-microscope morphology of the genera *Malassezia* and *Pityrosporum*. In *XIII Congressus Internationalis Dermatologiae* Vol. 2, ed. Jadassohn, W. & Schirren, C. G. pp. 867–872. Berlin: Springer.

KING, D. S. & JONG, S. C. 1975 *Sarcinosporon*: a new genus to accommodate *Trichosporon inkin* and *Prototheca filamenta*. *Mycotaxon* **3**, 89–94.

KING, D. S. & JONG, S. C. 1976 *Aciculoconidium:* a new hyphomycetous genus to accommodate *Trichosporon aculeatum*. *Mycotaxon* **3**, 401–408.

KOCKOVÁ-KRATOCHVÍLOVÁ, A., SLAVIKOVÁ, E., ZEMEK, J. & KUŇIAK, L. 1977 The heterogeneity of the genus *Trichosporon*. *Proceedings of the 5th International Specialized Symposium on Yeasts, Kezthely, Hungary* p. 9.

KREGER-VAN RIJ, N. J. W. 1966 Some features of yeast ascospores observed under the electron microscope. *Proceedings, 2nd International Symposium on Yeasts, Bratislava* pp. 169–176.

KREGER-VAN RIJ, N. J. W. 1969 Taxonomy and systematics of yeasts. In *The Yeasts* ed. Rose, A. H. & Harrison, J. S. pp. 5–78. London & New York: Academic Press.

KREGER-VAN RIJ, N. J. W. 1977 Ultrastructure of *Hanseniaspora* ascospores. *Antonie van Leeuwenhoek* **43**, 225–232.

KREGER-VAN RIJ, N. J. W. 1978a Electron microscopy of germinating ascospores of *Saccharomyces cerevisiae*. *Archives of Microbiology* **117**, 73–77.

KREGER-VAN RIJ, N. J. W. 1978b Ultrastructure of the ascospores of the new yeast genus *Sporopachydermia* Rodrigues de Miranda. *Antonie van Leeuwenhoek* **44**, 451–456.

KREGER-VAN RIJ, N. J. W. 1979 A comparative ultrastructural study of the ascospores of some *Saccharomyces* and *Kluyveromyces* species. *Archives of Microbiology* **121**, 53–59.

KREGER-VAN RIJ, N. J. W. & VEENHUIS, M. 1969 Septal pores in *Endomycopsis platypodis* and *Endomycopsis monospora*. *Journal of General Microbiology* **57**, 91–96.

KREGER-VAN RIJ, N. J. W. & VEENHUIS, M. 1971a A comparative study of the cell wall structure of basidiomycetous and related yeasts. *Journal of General Microbiology* **68**, 87–95.

KREGER-VAN RIJ, N. J. W. & VEENHUIS, M. 1971b Septal pores in *Trichosporon cutaneum*. *Sabouraudia* **9**, 36–38.

KREGER-VAN RIJ, N. J. W. & VEENHUIS, M. 1973 Electron microscopy of septa in ascomycetous yeasts. *Antonie van Leeuwenhoek* **39**, 481–490.

KREGER-VAN RIJ, N. J. W. & VEENHUIS, M. 1975 Conjugation in the yeast *Saccharomycopsis capsularis* Schiönning. *Archives of Microbiology* **104**, 263–269.

KREGER-VAN RIJ, N. J. W. & VEENHUIS, M. 1976a Ultrastructure of the ascospores of some species of the *Torulaspora* group. *Antonie van Leeuwenhoek* **42**, 445–455.

KREGER-VAN RIJ, N. J. W. & VEENHUIS, M. 1976b Conjugation in the yeast *Guilliermondella selenospora* Nadson et Krassilnikov. *Canadian Journal of Microbiology* **22**, 960–966.

KURTZMAN, C. P. & SMILEY, M. J. 1976 Heterothallism in *Pichia kudriavzevii* and *Pichia terricola*. *Antonie van Leeuwenhoek* **42**, 355–363.

KURTZMAN, C. P. & WICKERHAM, L. J. 1973 *Saccharomycopsis crataegensis*, a new heterothallic-yeast. *Antonie van Leeuwenhoek* **39**, 81–87.

KURTZMAN, C. P., SMILEY, M. J. & BAKER, F. L. 1975 Scanning electron microscopy of ascospores of *Debaryomyces* and *Saccharomyces*. *Mycopathologia* **55**, 29–34.

KWON-CHUNG, K. J. 1975 A new genus, *Filobasidiella*, the perfect state of *Cryptococcus neoformans*. *Mycologia* **67**, 1197–1200.

KWON-CHUNG, K. J. 1977 Perfect state of *Cryptococcus uniguttulatus*. *International Journal of Systematic Bacteriology* **27**, 293–299.

KWON-CHUNG, K. J. & POPKIN, T. J. 1976 Ultrastructure of septal complex in *Filobasidiella neoformans (Cryptococcus neoformans)*. *Journal of Bacteriology* **126**, 524–528.

KWON-CHUNG, K. J., BENNETT, J. E. & THEODORE, TH. S. 1978 *Cryptococcus bacillisporus sp. nov.:* serotype B-C of *Cryptococcus neoformans. International Journal of Systematic Bacteriology* **28**, 616-620.

LODDER, J. ed. 1970 *The Yeasts — A Taxonomic Study* 2nd edn. Amsterdam: North-Holland.

LODDER, J. & KREGER-VAN RIJ, N. J. W. 1952 *The Yeasts — A Taxonomic Study.* Amsterdam: North-Holland.

LUSHBAUGH, W. B., ROWTON, E. D. & MCGHEE, R. B. 1976 Redescription of *Coccidiascus legeri* Chatton, 1913 (*Nematosporaceae*: Hemiascomycetidae), an intracellular, parasitic, yeast-like fungus from the intestinal epithelium of *Drosophila melanogaster. Journal of Invertebrate Pathology* **28**, 93-107.

MEYER, S. A. & PHAFF, H. J. 1970 Taxonomic significance of the DNA base composition in yeasts. In *Recent Trends in Yeast Research* Vol. 1, ed. Ahearn, D. G. pp. 1-29. Atlanta: Spectrum (Georgia State University).

MEYER, S. A., SMITH, M. T. & SIMIONE, F. P. 1978 Systematics of *Hanseniaspora* Zikes and *Kloeckera* Janke. *Antonie van Leeuwenhoek* **44**, 79-96.

MILLER, M. W., YONEYAMA, M. & SONEDA, M. 1976 *Phaffia* a new yeast genus in the Deuteromycotina (Blastomycetes). *International Journal of Systematic Bacteriology* **26**, 286-291.

MOORE, R. T. & MCALEAR, J. H. 1961 Fine structure of Mycota. Observations on septa of Ascomycetes and Basidiomycetes. *American Journal of Botany* **49**, 86-94.

OLIVE, L. S. 1968 An unusual new heterobasidiomycete with *Tilletia*-like basidia. *Journal of the Elisha Mitchell Scientific Society* **84**, 261-266.

PITT, J. S. & MILLER, M. W. 1968 Sporulation in *Candida pulcherrima, Candida reukaufii* and *Chlamydozyma* species: their relationships with *Metschnikowia. Mycologia* **60**, 663-685.

PRICE, C. W., FUSON, G. B. & PHAFF, H. J. 1978 Genome comparison in yeast systematics: delimitation of species within the genera *Schwanniomyces, Saccharomyces, Debaryomyces* and *Pichia. Microbiological Reviews* **42**, 161-193.

RODRIGUES DE MIRANDA, L. 1978 A new yeast genus: *Sporopachydermia. Antonie van Leeuwenhoek* **44**, 439-450.

RODRIGUES DE MIRANDA, L. 1979 *Clavispora*, a new yeast genus of the Saccharomycetales. *Antonie van Leeuwenhoek* **45**, 479-483.

SHADOMY, H. J. 1970 Clamp connections in two strains of *Cryptococcus neoformans.* In *Recent Trends in Yeast Research* Vol. 1, ed. Ahearn, D. G. pp. 67-72. Atlanta: Spectrum (Georgia State University).

SLOOFF, W. C. 1970*a* The genus *Schizosaccharomyces.* In *The Yeasts — A Taxonomic Study* 2nd edn, ed. Lodder, J. pp. 733-755. Amsterdam: North-Holland.

SLOOFF, W. C. 1970*b* The genus *Lipomyces.* In *The Yeasts — A Taxonomic Study* 2nd edn, ed. Lodder, J. pp. 379-402. Amsterdam: North-Holland.

SMITH, M. T., VAN DER WALT, J. P. & JOHANNSEN, E. 1976 The genus *Stephanoascus (Ascoideaceae). Antonie van Leeuwenhoek* **42**, 119-127.

STARKEY, R. L. 1946 Lipid production by a soil yeast. *Journal of Bacteriology* **51**, 33-50.

VAN UDEN, N. & BUCKLEY, H. 1970 The genus *Candida.* In *The Yeasts — A Taxonomic Study* 2nd edn, ed. Lodder, J. pp. 892-1087. Amsterdam: North-Holland.

VAN DER WALT, J. P. 1970 The genus *Saccharomyces.* In *The Yeasts — A Taxonomic Study* 2nd edn, ed. Lodder, J. pp. 555-718. Amsterdam: North-Holland.

VAN DER WALT, J. P. 1972 The yeast genus *Ambrosiozyma gen. nov.* (Ascomycetes). *Mycopathologia* **46**, 305-316.

VAN DER WALT, J. P. 1978 The genus *Pachytichospora gen. nov.* (Saccharomycetaceae). *Bothalia* **12**, 563-564.

VAN DER WALT, J. P. & LIEBENBERG, N. V. D. W. 1973 The yeast genus *Wickerhamiella. Antonie van Leeuwenhoek* **39**, 121-128.

VAN DER WALT, J. P. & JOHANNSEN, E. 1975 The genus *Torulaspora* Lindner. *CSIR Research Report No. 325* pp. 1-23.

VON ARX, J. A. 1972 On *Endomyces, Endomycopsis* and related yeast-like fungi. *Antonie van Leeuwenhoek* **38**, 289-309.

VON ARX, J. A. & VAN DER WALT, J. P. 1976 The ascigerous state of *Candida chodatii. Antonie van Leeuwenhoek* **42**, 309-314.

VON ARX, J. A., RODRIGUES DE MIRANDA, L., SMITH, M. T. & YARROW, D. 1977 The genera of yeasts and the yeast-like fungi. *Studies in Mycology No. 14* Centraalbureau voor Schimmelcultures, Baarn.

WEIJMAN, A. C. M. 1979 Carbohydrate composition and taxonomy of *Geotrichum, Trichosporon* and allied genera. *Antonie van Leeuwenhoek* **45**, 119–127.

WICKERHAM, L. J. 1951 Taxonomy of yeasts. 1. Techniques of classification. 2. A classification of the genus *Hansenula. U.S. Department of Agriculture Technical Bulletin No. 1029.*

WICKERHAM, L. J. 1970 The genus *Hansenula*. In *The Yeasts — A Taxonomic Study* 2nd edn, ed. Lodder, J. pp. 226–315. Amsterdam: North-Holland.

WICKERHAM, L. J. & BURTON, K. A. 1962 Phylogeny and biochemistry of the genus *Hansenula. Bacteriological Reviews* **26**, 382–397.

YAMADA, Y. & KONDO, K. 1972 Taxonomic significance of coenzyme Q system in yeasts and yeast-like fungi. In *Yeasts and Yeast-like Microorganisms in Medical Science. Proceedings of the Second International Specialized Symposium on Yeasts, Tokyo* pp. 63–69.

YAMADA, Y. & KONDO, K. 1973 Coenzyme Q system in the classification of the yeast genera *Rhodotorula* and *Cryptococcus*, and the yeast-like genera *Sporobolomyces* and *Rhodosporidium. Journal of General and Applied Microbiology* **19**, 59–77.

YAMADA, Y., OKADA, T., UESHIMA, O. & KONDO, K. 1973*a* Coenzyme Q system in the classification of the ascosporogenous yeast genera *Hansenula* and *Pichia. Journal of General and Applied Microbiology* **19**, 189–208.

YAMADA, Y., ARIMOTO, M. & KONDO, K. 1973*b* Coenzyme Q system in the classification of the ascosporogenous yeast genus *Schizosaccharomyces* and yeast-like genus *Endomyces. Journal of General and Applied Microbiology* **9**, 353–358.

YAMADA, Y., ARIMOTO, M. & KONDO, K. 1976*a* Coenzyme Q system in the classification of apiculate yeasts in the genera *Nadsonia, Saccharomycodes, Hanseniaspora, Kloeckera,* and *Wickerhamia. Journal of General and Applied Microbiology* **22**, 293–299.

YAMADA, Y., NOJIRI, M., MATSUYAMA, M. & KONDO, K. 1976*b* Coenzyme Q system in the classification of the ascosporogenous yeast genera *Debaryomyces, Saccharomyces, Kluyveromyces* and *Endomycopsis. Journal of General and Applied Microbiology* **22**, 325–337.

YAMADA, Y., ARIMOTO, M. & KONDO, K. 1977 Coenzyme Q system in the classification of some ascosporogenous yeast genera in the families *Saccharomycetaceae* and *Spermophthoraceae. Antonie van Leeuwenhoek* **43**, 65–71.

YARROW, D. & MEYER, S. A. 1978 Proposal for amendment of the diagnosis of the genus *Candida* Berkhout *nom. cons. International Journal of Systematic Bacteriology* **28**, 611–615.

A Mycologist's View of Yeasts

J. A. VON ARX

Centraalbureau voor Schimmelcultures, Baarn, The Netherlands

Contents

1. Introduction

MYCOLOGISTS recognize more than 50,000 fungus species of which only about 500 are regarded as yeasts; thus the yeasts represent only a small part of the fungal kingdom. The collections of the Centraalbureau voor Schimmelcultures (CBS) in Baarn and Delft comprise some 20,000 cultures of filamentous fungi representing nearly 8000 different species, and 4000 cultures of the 500 yeast species. About 170 of the yeast species are each represented by a single strain.

Taxonomists of the fungi are highly specialized. Even so, it is surprising to note that between no other groups of mycologists has there been such complete and consistent separation as between the workers on filamentous fungi and those on yeasts. Taxonomists of filamentous fungi are generally unable to identify yeast species; conversely, many yeast workers are frequently unable to recognize such common fungi as *Cladosporium herbarum* or even *Penicillium chrysogenum*, the producer of penicillin.

Why did the taxonomy of the yeasts develop along such separate lines? This cannot have happened because yeasts are able to ferment sugars, for numerous yeasts do not ferment them whereas many filamentous fungi do so; it cannot have been because of their ability to develop in liquid media, for fungi too occur in fruit juices and natural waters; nor can the absence of septate hyphae have been a factor because these are present in many yeasts, yet are absent in many fungi.

The reason for the separate development of yeast taxonomy is probably historical and was certainly influenced by the removal of the yeast cultures

from the CBS in Baarn to the microbiology laboratory in Delft by Professor
A. J. Kluyver in 1922. There, Kluyver applied the methods used in bacterial
taxonomy to these yeasts and his studies had great influence on yeast
taxonomy, especially in the Netherlands. Kluyver initiated and influenced
the three text books published between 1934 and 1952 (Stelling-Dekker
1931; Diddens & Lodder 1942; Lodder & Kreger-van Rij 1952). The
contributors to the last edition of *The Yeasts* (Lodder 1970) had worked
mostly under Kluyver. The delimitation of yeasts mainly on the basis of
their physiological characters, especially the ability to ferment and/or
assimilate a number of carbon compounds, is known today as work of the
'Dutch School'. Some morphological characters are employed in generic
delimitation, e.g. the shape of ascospores and the presence or absence of
hyphae or pseudohyphae. Emphasis is also placed on the formation of the
cells: taxonomists of yeasts distinguish between fission and unipolar,
bipolar and multilateral budding.

2. Conidiation in Yeasts

In mycology the delimitation of taxa is based primarily on morphological
characters, results of mating experiments, observations on submicroscopic
structure and chemical data (e.g. amyloidity or dextrinoidity of certain
structures). In the taxonomy of the Ascomycetes and their imperfect states,
conidiation (the mode of formation of new cells or conidia) has proved to
be a very useful character. Several kinds of conidiogenous cells and conidia
are distinguished. Most of these are also seen in yeasts and the terms
explained in Fig. 1 should be employed in yeast taxonomy in addition to
'fission' and 'budding'.

Phialidic (enteroblastic, basipetal) conidiation is observed in the red
yeasts, which should be classified in Sporobolomycetales, and also in the
yeast-like, vegetative states of *Taphrina*, *Symbiotaphrina*, *Protomyces*,
Microstroma and *Ustilago*. The species belonging to these latter genera are
mainly plant parasites that cause hypertrophies, swellings or witches'
broom growths. They have been classified partly in the Ascomycetes and
partly in the Basidiomycetes or Hyphomycetes, but may be closely related
to each other. A further common character is the presence of carotenoids.

Kreger-van Rij & Veenhuis (1971) published electron micrographs
showing the conidiation of *Rhodotorula glutinis*, with clear collars around
the conidiogenous opening and the formation of new inner walls.
No conclusions were drawn as to whether this was a form of phialidic
conidiation.

Enteroblastic conidiation or sporulation is also known in yeasts

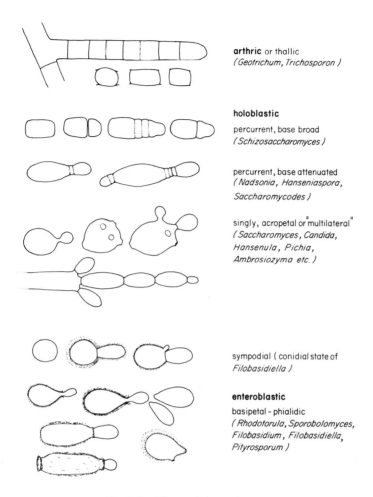

arthric or thallic
(*Geotrichum, Trichosporon*)

holoblastic

percurrent, base broad
(*Schizosaccharomyces*)

percurrent, base attenuated
(*Nadsonia, Hanseniaspora,
Saccharomycodes*)

singly, acropetal or "multilateral"
(*Saccharomyces, Candida,
Hansenula, Pichia,
Ambrosiozyma* etc.)

sympodial (conidial state of
Filobasidiella)

enteroblastic

basipetal - phialidic
(*Rhodotorula, Sporobolomyces,
Filobasidium, Filobasidiella,
Pityrosporum*)

Fig. 1. Conidiogenesis in yeasts.

belonging to the Basidiomycetes (Oberwinkler 1978). The basidiospores in *Filobasidiella neoformans* develop in basipetal succession on a swollen cell with several phialidic openings. The haploid yeast cells of *Filobasidium capsuligenum* and *Filobasidiella neoformans* also have either wide or narrow openings with basipetal conidiation.

In the yeasts now classified in the illegitimate genus *Cryptococcus* auct. non Kützing, budding is mostly enteroblastic–basipetal, but in *Cr. neoformans*, the anamorph of *Filobasidiella neoformans*, the author has also observed single budding from a broad base. New buds are formed sympodially, close to the preceding bud.

Basipetal conidiation is absent in all yeasts with asci. In this group two kinds of holoblastic budding are distinguished, referred to as multilateral and bipolar in the yeast literature. In most of the taxa, the buds are formed singly, leaving a small scar on release. Acropetal, often branched, chains are formed, especially in *Candida* species with pseudohyphae.

In bipolar, apiculate yeasts and in *Schizosaccharomyces* budding is restricted to the ends of the elongated cells. After release the cell elongates percurrently (through the scar) and forms a new bud. In the course of conidiation the conidiogenous parts of the cell become elongated and show annellations. In genera such as *Nadsonia* and *Hanseniaspora*, the conidiogenous locus is slightly attenuated, while in *Schizosaccharomyces* it is broad.

Arthric conidia, formed by fragmentation and separation of determinate, fertile hyphae into chains of conidia, are known mainly in the genera *Geotrichum* of the Dipodascaceae and *Trichosporon* of the Basidiomycetes and have no taxonomic significance. In some hyphal yeasts of other genera the hyphae may disintegrate into single cells.

3. Cell Chemistry

Other useful characters for the delimitation of homogeneous yeast taxa can be found in the chemistry of the cell wall or the whole cell. Bartnicki-Garcia (1970) proposed a new system for the fungal kingdom based on the chemical composition of the cell wall. He associated certain cell wall categories, defined by chemical properties, with particular taxonomic groups. In the 'chitin–glucan' category he placed the Chytridiomycetes, (Eu-) Ascomycetes and (Homo-) Basidiomycetes. He excluded the yeasts from the Ascomycetes and Basidiomycetes and classified them in two groups: the yeasts with asci were included in the 'mannan–glucan' category; the red yeasts in the 'chitin–mannan' category.

During the last six years my co-worker, Dr A. C. M. Weijman, has studied the chemistry of the cell wall or whole cell of a large number of fungi and yeasts. The Bartnicki–Garcia classification proved useful to some extent but many intermediate forms were found. Certain yeasts, most of which were mycelial, contained rather large amounts of chitin. In some basidiomycetous yeasts, e.g. in *Filobasidium capsuligenum*, only low amounts of chitin could be demonstrated.

Some other monosaccharides, such as rhamnose, galactose, xylose or fucose, could be detected in many species. The presence or absence of these monosaccharides often correlated with other characters, e.g. with the kind of conidiogenesis or with the % G + C of the DNA. Two examples of these observations will be discussed here: the first concerns the heterogeneity of the genera *Trichosporon* and *Geotrichum* (Table 1), the second the heterogeneity of the genus *Candida* (Table 2).

TABLE 1

Monosaccharide patterns and G + C values in Dipodascus, Geotrichum, Trichosporon *and related genera*

	G + C (%)	Chitin†	Pattern‡
Dipodascaceae (Saccharomycetes)			
Dipodascus aggregatus	42	moderate	
Dipodascus australiensis	41·5	moderate	
Geotrichum candidum	41·5	moderate	‡MAN, GLU, GAL
Trichosporon capitatum	31·5	low	
Trichosporon fermentans	42	moderate	
Trichosporon penicillatum	39	moderate	MAN, GLU
Trichosporon fennicum	35·5	moderate	
(= Hyphopichia burtonii)			
Filobasidiaceae (Basidiomycetes)			
Trichosporon beigelii	60–63	moderate	
Trichosporon dulcitum	54·5	moderate	
Trichosporon pullulans	57	moderate	
Geotrichum loubierii	59·5	moderate	GLU,MAN, XYL
Filobasidium capsuligenum	51	low	
Filobasidiella neoformans	49–52	low	
Sarcinosporon inkin		high	

*determined by GLC in 5N HCl in hydrolysates.
†determined by GLC in 1N HCl hydrolysates.
‡MAN, mannose; GLU, glucose; GAL, galactose; XYL, xylose.

The species of *Geotrichum* and *Trichosporon* were recently reclassified by von Arx *et al.* (1977) and by Weijman (1979). The classical character used to distinguish between the two genera, namely the presence of blastospores in *Trichosporon* and their absence in *Geotrichum*, has proved to have no taxonomic value. Cultural characters are, however, useful for delimiting the genera. *Geotrichum* forms aerial, septate hyphae which disintegrate into dry conidia whereas *Trichosporon* has mucoid, slimy colonies and lacks aerial hyphae. Blastoconidia may be present in both genera, but those in *Geotrichum* are holoblastic and those in *Trichosporon*, enteroblastic. Submicroscopic and chemical characters and the G + C content of the DNAs are appreciably different (Table 1). Consequently, some species of *Trichosporon* have been reclassified in *Geotrichum* and vice versa.

The heterogeneity of *Candida* (including *Torulopsis*, see Yarrow & Meyer 1978) has often been mentioned, but a revision of the genus has not yet been undertaken. *Candida* is the largest yeast genus and comprises *ca.* 200 species. We studied about 40 species, of which three really belonged to the genus *Apiotrichum* (in yeast literature, a synonym of *Candida*) which is close to *Trichosporon* and to the illegitimate genus *Cryptococcus.* Five more species should be transferred to *Rhodotorula.* These eight species show enteroblastic conidiation, a high G + C content of the DNA and a distinct cell chemical composition (Table 2) and should therefore be removed from *Candida.* The remaining species show holoblastic conidiation and a low G + C content, and can be retained in *Candida* which then includes only yeasts of ascomycetous affinity.

TABLE 2

Monosaccharide patterns and G + C values in Candida *(incl.* Torulopsis*) and related genera*

	G + C (%)	Chitin	Pattern
1. Candidaceae (ascomycetous)			
Candida albicans	36	low	
Candida boidinii	33	low	
Candida diddensii	40	trace	*MAN, GLU
Candida sake	40	low	
Candida tropicalis	35	low	
2. Sporobolomycetaceae (basidiomycetous)			
Candida bogoriensis	56	high	MAN, GLU, RHA, FUC
Candida javanica	59	high	
Candida muscorum		moderate	
Rhodotorula rubra	60	moderate	MAN, GLU, GAL, FUC
Sporobolomyces salmonicolor	64	moderate	
Sporobolomyces holsaticus	64	high	
Torulopsis fujisanensis	66	moderate	GLU, MAN, RHA, FUC
Torulopsis ingeniosa	56	moderate	MAN, GAL, GLU, FUC
3. Filobasidiaceae (basidiomycetous)			
Candida curvata	54·2	moderate	
Candida humicola	60·4	moderate	
Apiotrichum porosum	56·5	moderate	
Bullera alba	54·4	moderate	GLU, MAN, XYL
Filosbasidiella neoformans	51	low	
Filobasidium capsuligenum	51	low	
Trichosporon beigelii	60·0–62·5	moderate	

*MAN, mannose; GLU, glucose; GAL, galactose; XYL, xylose; RHA, rhamnose; FUC, fucose.

Among the yeasts of basidiomycetous affinity, two groups can be distinguished, the Filobasidiaceae and the Sporobolomycetaceae. Cells of the former group contain xylose (Xyl, Table 2) in addition to glucose (Glu) and mannose (Man). Xylose is absent in the Sporobolomycetaceae, but rhamnose (Rha, Table 2), galactose (Gal) and fucose (Fuc) may be present.

The Sporobolomycetaceae correspond to the Ustilaginales, and the Filobasidiaceae to the Cryptococcales, as defined by Oberwinkler (1978). The Sporobolomycetaceae contains mainly the red yeasts (*Rhodotorula, Sporobolomyces, Rhodosporidium*). The '*Candida*' species of this group are nearly colourless, but can be recognized by enteroblastic conidiogenesis and chemical composition of the cell. Typical members of the Filobasidiaceae (*Apiotrichum, Trichosporon, Bullera, Filobasidium, Filobasidiella*) can be recognized by the absence of carotenoids and the presence of xylose. Some yeasts with red carotenoid pigmentation also have cell walls containing xylose, for example, the species of the genera *Dioszegia* and *Phaffia*. Their relationship to either the Filobasidiaceae or the Sporobolomycetaceae is therefore still uncertain, and intermediates probably exist. The delimitation of the heterobasidiomycetous yeasts from the true Basidiomycetes has also become problematic since the discovery of the genera *Filobasidium* and *Filobasidiella*.

4. Delimitation of Ascomycetes and Saccharomycetes

If the asci-forming yeasts included genera like *Ascoidea, Cephaloascus, Dipodascus* and *Endomyces*, as in mycological text books, then their distinction from the Ascomycetes would be facilitated. The Ascomycetes should be restricted to fungi forming dikaryotic, ascogenous hyphae, ascomata (mostly perithecia and apothecia) and asci which usually contain four or eight ascospores. The genera *Saccharomycopsis* and *Dipodascus* are closely related, both having similar cultural characters, hyphae and hyphal septa with micropores (plasmodesmata). In text books on yeasts the genus *Saccharomycopsis* is included (as *Endomycopsis*), while *Dipodascus* is excluded. Another neglected genus is *Ascoidea*, which is also close to genera such as *Saccharomycopsis* or *Hyphopichia*. *Endomyces decipiens*, the type species of the genus *Endomyces*, is not known to develop in pure culture; it has been observed only on the natural substrate, that is gills of *Armillaria mellea* (von Arx 1977). It is morphologically similar to such species as *S. fibuligera*. *Cephaloascus fragrans*, the type species of the genus *Cephaloascus* is characterized by colonies pigmented due to the thick-walled, encrusted ascophores. In other respects it is similar to *Hormoascus platypodis*, an ambrosia yeast, treated in Lodder (1970) as both *Endomycopsis platypodis* and *Hansenula platypodis*.

Other genera which should be classified within the yeasts include *Eremothecium, Crebrothecium* and *Ashbya*, which are all close to *Nematospora* but which do not develop blastoconidia on agar media.

5. Conclusions

The treatment of the yeasts by Lodder (1970) is incomplete and schematic, nearly all the larger genera and many species are heterogeneous. Some key characters to the genera are incorrect, for example, the budding in *Rhodotorula*, *Cryptococcus* and other heterobasidiomycetous genera cannot be multilateral, but is enteroblastic–unipolar (the latter term is restricted to *Pityrosporum*). The generic names *Endomycopsis*, *Debaryomyces*, *Torulopsis* and *Cryptococcus* are illegitimate and the three first mentioned have since been rejected (van der Walt & Scott 1971; von Arx 1972, 1977; van der Walt & Johannsen 1978; Yarrow & Meyer 1978). Nearly all the larger genera, such as *Saccharomyces*, *Hansenula* and *Pichia* are heterogeneous, the delimitation of the latter two being arbitrary. More homogeneous genera have been proposed by von Arx *et al.* (1977). The classification of species based on the fermentation and assimilation of many carbon compounds resulted in heterogeneous taxa, as well as the description of identical species under different names. This has been shown by recent DNA–DNA homology studies and by mating experiments. The description of numerous superfluous 'species' distinguished mainly by the assimilation patterns, resulted in a classification which could only be applied by a few specialists. A more useful classification will have to be based on morphological characters, supplemented by mating experiments and certain physiological and ecological data. Physiological characters should be employed with discretion; their assessment is too complicated and time-consuming for quick identification, and above all their variability within a single species renders them unreliable.

6. References

BARTNICKI-GARCIA, S. 1970 Cell wall composition and other biochemical markers in fungal phylogeny. In *Phytochemical Phylogeny* ed. Harborne, J. B. pp. 81–103. London: Academic Press.

DIDDENS, H. A. & LODDER, J. 1942 *Die Hefesammlung des Centraalbureau voor Schimmelcultures.* II Teil. *Die Anaskosporogen Hefen, Zweite Hälfte.* Amsterdam: North-Holland.

KREGER-VAN RIJ, N. J. W. & VEENHUIS, M. 1971 A comparative study of the cell wall structure of basidiomycetous and related yeasts. *Journal of General Microbiology* **68**, 87–95.

LODDER, J. (ed.) 1970 *The Yeasts — A Taxonomic Study.* 2nd edn. Amsterdam: North-Holland.

LODDER, J. & KREGER-VAN RIJ, N. J. W. 1952 *The Yeasts — A Taxonomic Study.* 2nd edn. Amsterdam: North-Holland.

OBERWINKLER, F. 1978 Was ist ein Basidiomycet? *Zeitschrift für Mykologie* **44**, 13–29.

STELLING-DEKKER, N. M. 1931 *Die Hefesammlung des Centraalbureau voor Schimmelcultures.* I Teil. *Die Sporogenen Hefen.* Amsterdam: North Holland.

VAN DER WALT, J. P. & SCOTT, J. B. 1971 The yeast genus *Saccharomycopsis. Mycopathologia et mycologia applicata* **43**, 279–288.

VAN DER WALT, J. P. & JOHANNSEN, E. 1978 The genus *Debaryozyma* nom. nov. *Persoonia* **10**, 146–148.

VON ARX, J. A. 1972 On *Endomyces, Endomycopsis* and related yeast-like fungi. *Antonie van Leeuwenhoek* **38**, 289–309.

VON ARX, J. A. 1977 Notes on *Dipodascus, Endomyces* and *Geotrichum* with the description of two new species. *Antonie van Leeuwenhoek* **43**, 333–340.

VON ARX, J. A., RODRIGUES DE MIRANDA, L., SMITH, M. T. & YARROW, D. 1977 The genera of yeasts and the yeast-like fungi. *Studies in Mycology No. 14.* Centraalbureau voor Schimmelcultures, Baarn.

WEIJMAN, A. C. M. 1979 Carbohydrate composition and taxonomy of *Geotrichum, Trichosporon* and allied genera. *Antonie van Leeuwenhoek* **45**, 119–127.

YARROW, D. & MEYER, S. A. 1978 Proposal for amendment of the diagnosis of the genus *Candida*. *International Journal of Systematic Bacteriology* **28**, 611–615.

An Evaluation of Criteria for Yeast Speciation*

J. P. VAN DER WALT

Microbiology Research Group,
Council for Scientific and Industrial Research,
Pretoria, South Africa

Contents

1. Introduction

SOME 55 years ago the Dutch geneticist Lotsy (1925) remarked: "If it is true that everybody loves a lover because he personifies devotion, everybody should love a taxonomist for his devotion to the species-concept."

Despite this professed attachment to the species concept few of the many yeast taxonomists since the time of Guilliermond (1912) have stated explicitly what they believed the yeast species to be. Furthermore not one of these monographers has indicated any guidelines for the application of the so-called criteria they used for the demarcation of this taxon — what they did provide was a comprehensive list of phenotypic characters which could be used for descriptive purposes and which they have in one way or another over the years used for delimiting yeast species.

Consequently any evaluation of the vague, inferred criteria used, requires some inspection of the concept of the yeast species which has evolved in practice over the past 60–70 years.

*Throughout this chapter the term 'speciation' is used to mean 'species delimitation' — Author.

2. The Concept of the Yeast Species

A. The taxon based on phenotypic discontinuity

The first glimmering of the concept of yeast species emerged from Emil Christian Hansen's studies at the Carlsberg Laboratories almost a century ago when he introduced pure culture techniques. Hansen (1888) differentiated and classified the pure cultures he obtained on the basis of cellular and ascosporal morphology, characteristics of growth in liquid media, optimal growth temperatures and the ability to ferment different sugars. In so doing, he laid the basis for the phenotypic speciation which, until recently, has held sway in the yeast domain.

With the discovery of new taxa, subsequent taxonomists extended the list of phenotypic characters to include other features such as: mode of vegetative reproduction, mode of ascus formation, ploidy of the vegetative phase, pigmentation, the oxidative utilization of some 36 carbon sources, the utilization of nitrate and other nitrogenous compounds, tolerance to high sugar and salt concentrations and resistance to cycloheximide. As a full list of these phenotypic *differentiae* is to be found in the monograph edited by Lodder (1970) they require no further elucidation.

The taxon which is delimited exclusively on grounds of differences in morphological and physiological properties, or phenotypic discontinuity, agrees in essence with what Lotsy (1925) regarded as the 'linneon' and what Cain (1954) and Grant (1963) termed the 'taxonomic species'. This method of delimiting species is so rigid that its application leads to artificial systems and to taxonomic entities which come adrift when the properties applied for speciation are, for instance, subject to mutational changes. It must be emphasized that provided phenotypic characteristics only are employed in numerical taxonomy, the taxonomic species is still retained.

However, by 1970 the weaknesses inherent in speciation based exclusively on phenotypic differentiation had become apparent. Yeast taxonomists began to realize that phenotypic features, notably fermentative and assimilatory properties formerly employed for primary speciation, could be unstable (Scheda 1966; Scheda & Yarrow 1966, 1968) and coded by very small portions of the genome. Whilst it must be admitted that these features introduce serious limitations, it must not be inferred that phenotypic classification serves no taxonomic purpose.

B. The taxon delimited on the basis of interfertility

The second approach to the definition of the yeast species is yet again traceable to the Carlsberg Laboratory when the geneticist Öjvind Winge turned his attention to the brewing yeasts. Having elucidated the sexual cycle of *Saccharomyces cerevisiae* (Winge 1935) and having successfully hybridized phenotypically distinct strains, Winge & Lautsen (1939) concluded: ". . . that (1) two types generally ought to be reckoned to be the same species when hybridizing and their hybrid being fully fertile whereas (2) two types would be regarded as specifically different when they are unable to hybridize or their hybrid shows reduced fertility".

In delimiting the yeast species on the basis of interfertility Winge & Lautsen (1939) in fact followed Vandendries (1923) who, 16 years earlier, first proposed this approach to speciation in the Basidiomycetes. This approach is in keeping with the views of Nelson (1963) and some of the more contemporary mycologists (*see* Clémençon 1977) and subscribes to the tenets of Dobzhansky (1951) and Mayr (1967) who accept that the species is constituted by the sum total of interbreeding individuals or alternatively a reproductively isolated system of breeding populations. Speciation based on interfertility obviously maintains yeast systematics on a par with fungal taxonomy.

The yeast species, defined in terms of interfertility, reflects genetic or natural relationships and agrees with what Lotsy (1925) regarded as the 'syngameon' and Cain (1954) and Grant (1963) termed the 'biological species'. It is immediately apparent that whereas this definition makes provision for speciation among perfect taxa characterized by sexual cycles implicating the fusion of two independent, haploid cells, it cannot be applied to the so-called 'imperfect species' which do not represent haploid, heterothallic mating-types yet seemingly lack sexual stages.

Winge & Lautsen's (1939) directive found little application in the taxonomy of the perfect yeasts until Wickerham & Burton (1952, 1954) discovered that several of the so-called imperfect yeasts merely represented haploid mating-types of heterothallic species of *Kluyveromyces* and *Pichia*. Although the subsequent systematic application of interfertility as a criterion for speciation was rather limited in the case of the ascogenous taxa, it led to a more realistic demarcation of species in *Hansenula* (Wickerham 1970) and *Saccharomyces* (Neumann 1972; Yarrow 1972; Yarrow & Nakase 1975), and also to the discovery of numerous new heterothallic taxa.

The best examples of the successful application of interfertility as a criterion for speciation are probably to be found in the systematics of the heterothallic, basidiomycetous genera *Rhodosporidium* (Fell *et al.* 1970; Fell *et al.* 1973), *Leucosporidium* (Fell & Phaff 1970), *Filobasidium* (Rodrigues de Miranda 1972) and *Filobasidiella* (Kwon-Chung 1975, 1976).

C. The taxon delineated on the basis of nucleic acid analyses

It is only comparatively recently that yet a third approach to yeast speciation has been developed. It has its origin in modern developments in bacterial systematics resulting from comparison studies of the procaryotic genome (*cf.* Marmur *et al.* 1963) and hinges on an assessment of relatedness by means of nucleic acid analyses or a so-called comparison of genomes. It was only subsequent to this development in bacterial systematics and chiefly as a result of many initial studies (Belozersky & Spirin 1960; Storck 1966; Storck *et al.* 1969; Fell & Meyer 1967; Stenderup & Bak 1968; Nakase & Komagata 1968, 1969, 1970*a, b,* 1971*a, b, c, d, e*; Meyer 1970; Meyer & Phaff 1969, 1970; Bicknell & Douglas 1970; Ouchi *et al.* 1970) that genome comparison studies were found to have taxonomic significance for the yeasts.

At present, two particular features of the yeast genome are found useful for speciation purposes. The first involves the composition of the nuclear DNA bases. Differences of 1–2% in accurate determinations of the mean molar percentage of guanine plus cytosine (mol %G + C) present in two yeast strains are accepted as *prima facie* evidence that these strains are (i) precluded from sharing similar base sequences and (ii) indicative of their not being recent descendants from a common ancestor (Price *et al.* 1978).

The second feature hinges on the determination of nDNA complementarity between strains on the basis of the reassociation of their fragmented, denatured, single-stranded nDNA. Given the appropriate reassociation data, strains having *ca.* 80–100% of their base sequences in common may reasonably be considered to comprise the same species (Price *et al.* 1978).

Yeast DNA–RNA homology, although investigated (Bak & Stenderup 1969; Bicknell & Douglas 1970), has, as yet, found little or no application in species delineation.

D. The yeast species in terms of practical taxonomy

Although the application of the different criteria employed for yeast speciation, when applied to a given set of strains, may or may not result in the same taxonomic entity, these definitive criteria cannot, in practice, be separated one from another. This becomes apparent from the simple fact that the delimitation of species on the basis of either interfertility or genome comparison studies is, in practice, limited to phenotypically defined areas, i.e. to strains which have been assembled on the basis of their phenotypic properties. It must be emphasized that once the species has been delimited along either of these lines, its description still hinges on phenotypic properties which also serve as a basis for the construction of keys and the routine identification of strains.

In practice, a yeast species must ultimately be characterized in terms of its morphological, physiological, biochemical, sexual, genetic and ecological features.

3. nDNA Analyses or Genome Comparison Studies as Criteria
for Speciation

Price *et al.* (1978) maintain that: "The rationale for such genome comparison studies is accepted to be the restatement at molecular level of the evolutionary statement of the principle of common descent. If two organisms are related, they must retain in their genomes, base sequences that are descendant from a common ancestral base sequence; closely related organisms will have retained a greater portion of base sequences in common, than organisms that have highly diverged." It must nevertheless be emphasized that although this rationale may be tenable, it evades the issue as to whether it is possible, practically, to distinguish between ancestral and non-ancestral base sequences.

Moreover, it should also be realized that there are no formal or definitive guidelines for the construction of bacterial species based on nucleotide homology. Consequently, not only Marmur *et al.* (1963) but also Mandel (1969) and Moore (1974) have stressed the need to calibrate nucleic acid reassociation methods in terms of genetic tests, using transformation and conjugation. When applying these methods to the yeasts, such calibration must, of necessity, evaluate nDNA nucleotide homology between strains in terms of their interfertility.

DNA base compositions of strains are determined either by thermal denaturation procedures or by buoyant density determinations. It is important to note that as base composition determinations do not reflect base sequences, strains with identical % G + C values may still be representative of completely unrelated taxa.

Comparisons of polynucleotide sequence between strains hinge on the measurement of the re-annealment of their sheared, fragmented single-stranded nDNA. Several techniques have been developed for the estimation of the re-annealed, hybrid DNA. For the latest developments of these techniques the recent publications by Meyer *et al.* (1978) and Price *et al.* (1978) should be consulted. Results obtained by these techniques relate to the homology of base sequences, but they do not provide information as to whether the homologous base sequences represent actively transcribing genes which constitute only a limited portion of the DNA.

Nevertheless it is particularly as a result of the application of nDNA complementarity studies that the shortcomings of speciation based on morphological and physiological properties have become apparent and this has already resulted in a considerable remodelling of species. Of the numerous examples two may be cited: (i) In 1970 Phaff (1970*b*) accepted four species in *Schwanniomyces*. However, eight years later Price *et al.* (1978), on the basis of genome comparison studies, reduced the genus to a single species and a putative variety. (ii) Similarly in 1970, Phaff (1970*a*) recognized only three species in *Hanseniaspora* but Meyer *et al.* (1978) subsequently doubled the number of these species.

In these studies strains were assigned to the same species when degrees of nDNA complementarity were high. The setting of the lower limits of such high degrees of homology, however, has still to be agreed upon. Price *et al.* (1977) proposed 70% as the lower limit, but a year later (Price *et al.* 1978) reset this lower limit at "approximately 80%". When degrees of re-association fall below 80%, the interpretation of results becomes problematical. Apparently, degrees of reassociation between 20% and 80% are reputedly rarely encountered in the yeasts and many phenotypically similar yeast strains have been observed to show nDNA relatedness values of 10% or less (Price *et al.* 1978).

4. Interfertility as a Criterion for Speciation with Special Reference to the Genus *Kluyveromyces*

Ten years ago and on the basis of some 34 years of experience with the genus *Hansenula,* Wickerham (1970) concluded: "The usefulness of hybridization to yeast taxonomy has barely begun". Interfertility between yeast strains represents the virtual consummation of sexual reproduction and implies functional homology of their mating-type loci which are believed to comprise a cluster of genes, both regulatory and structural, which govern the expression of haploid and diploid functions in the yeast cells (Friis & Roman 1968; Fogel & Mortimer 1971; Mackay & Manney 1974; Crandall *et al.* 1977).

When dealing with heterothallic taxa, hybridization techniques hinge on either the availability or the recovery of compatible, haploid mating-types. These mating-types are then mass-mated with the strains under scrutiny and interfertility assessed by the formation of fertile, perfect states expressed as either asci, ustilospores or basidia. On the other hand, when dealing with homothallic taxa, the detection of interfertility requires recourse to the techniques developed for the hybridization of *Sacch. cerevisiae* and its biotypes. These techniques have been fully reviewed by Fowell (1969).

The first method, which involves micromanipulation, relies on the pairing of either single ascospores (Winge & Lautsen 1939) or of haploid vegetative cells (Chen 1950) and is a method readily applicable to strains with high ascosporal viability. The second method involves the application of the prototrophic-selection technique of Pomper & Burkholder (1949). In brief, the method relies on the mass-mating of either two comple-mentary, auxotrophic, mutant parents, or a prototroph and an auxotroph where the auxotroph utilizes a sugar not utilized by the prototroph. After mating and allowing for hybrid formation, the mated cultures are plated on a minimal medium which permits the development of recombinants only. Interfertility is assessed by the recovery of fertile hybrid progeny with combined parental phenotypes and by a comparison of recombination frequency data observed in intraspecific crosses (Johannsen 1978; Johannsen & van der Walt 1978). This method has the advantage that it obviates the problems arising from low ascosporal viability and permits the detection of recombination at very low frequencies.

Provided that the necessary precautions are taken in the choice of sexually-reactive strains as well as stable mutants, the prototrophic selection technique could find extensive application for speciation. The applicability of the prototrophic selection technique for the detection of interfertility became particularly evident from Johannsen's (1978) recent study of the genus *Kluyveromyces* which could serve as a model. By first examining crosses of mutants of the same species Johannsen observed the recombinant frequencies listed in Table 1. In such intraspecific crosses recombinant frequencies range from 10^3–10^6 per 10^8 cells of the mated populations (c.m.p.). Such frequencies are consequently taken to be indicative of relatedness at species level.

TABLE 1

Examples of recombinant formation between auxotrophic mutants of different strains of the same phenotypically-delimited Kluyveromyces *species (Johannsen 1978)*

Parental strains			Recovery medium (agar)	Recombinants /10^8 c.m.p.*
K. marxianus	×	*K. marxianus*		
[†]CBS 6923 Lys[§]		[‡]CSIR 293 Trp	Lactose	$1·7 \times 10^3$
CBS 6566 His		CSIR 293 A	Lactose	$1·4 \times 10^6$
K. lactis a	×	*K. lactis α*		
CBS 683 Lys		CBS 2359 Arg U	Lactose	$4·9 \times 10^5$
CBS 6315 Trp		CBS 2359 Arg U	Lactose	$3·3 \times 10^6$

*c.m.p., cells of the mated populations.
[†]CBS Centraalbureau voor Schimmelcultures, Baarn.
[‡]CSIR Council for Scientific & Industrial Research, Pretoria.
[§]Nutritional requirements: A, adenine; Arg, arginine; His, histidine; Lys, lysine; Trp, tryptophan; U, uracil.

In Table 2 the recombinant frequencies observed in crosses between phenotypically distinct strains are shown. All such crosses were characterized by the recovery of uninucleate hybrids which possessed combined parental phenotypes and formed viable ascospores. It is important to note that the cited recombinant frequencies fall well within the range observed in intraspecific crosses (see Table 1).

TABLE 2

Examples of recombinant formation between auxotrophic mutant strains of different, phenotypically-delimited Kluyveromyces *species (Johannsen 1978)*

Parental strains		Recovery medium (agar)	Recombinants /10^8 c.m.p.*
K. fragilis LAC[†]	× *K. dobzhanskii* MAL		
CBS 1556 Met[‡]	CBS 2104 A	Maltose	$9·6 \times 10^4$
CBS 1556	CBS 2104 A	Maltose	$2·6 \times 10^3$
K. lactis a MAL	× *K. marxianus*		
CBS 683 Trp	CBS 6566	Maltose	$5·0 \times 10^5$
CBS 683 Trp	CBS 6923	Maltose	$1·6 \times 10^6$
K. lactis α LAC	× *K. phaseolosporus*		
CBS 2359 Arg U	CBS 2103	Lactose	$1·6 \times 10^4$
K. vanudenii MAL	× *K. marxianus* LAC		
CBS 4372	CBS 6556 Leu Pro	Lactose	$8·4 \times 10^5$
K. vanudenii MAL	× *K. wickerhamii* LAC		
CBS 4372 His	CBS 2745 Trp	Maltose	$2·7 \times 10^3$

*c.m.p., cells of the mated populations.
[†]Sugar utilization: LAC, lactose; MAL, maltose.
[‡]Nutritional requirements: A, adenine; Arg, arginine; His, histidine; Leu, leucine; Met, methionine; Pro, proline; Trp, tryptophan; U, uracil.
Note that the recombinant frequencies fall within the range observed in crosses between mutant strains of the same species (Table 1).

Johannsen (1978), on the basis of some 900 crosses, involving all possible 190 combinations of the 20 described species and in each case taking a recovery of fertile hybrids at recombinant frequencies greater than $10^3/10^8$ c.m.p. as confirmation of interfertility, arrived at the interbreeding relationships shown in Fig. 1. In this diagram the 20 phenotypically delimited species of the genus *Kluyveromyces* are shown. The figures in brackets indicate the % G + C values reported for the type strains of these 20 taxa (Phaff, pers. comm.). The full lines represent fertile crosses in which recombinant frequencies characteristic of intraspecific crosses were observed. These lines link what are considered to be fully interfertile taxa. The broken lines represent fertile crosses in which recombinant frequencies were either less than $10^3/10^8$ c.m.p., or where higher frequencies were irregularly encountered. These lines consequently link taxa which display a limited degree of genetic relatedness.

Two syngameons or groups of interfertile taxa may be observed. The first and by far the largest is formed by taxa which hybridize with *K. marxianus* — the first validly published species of the genus. The second comprises taxa which hybridize with *K. wickerhamii*. It becomes apparent from an analysis of these results that ascosporal morphology cannot serve to

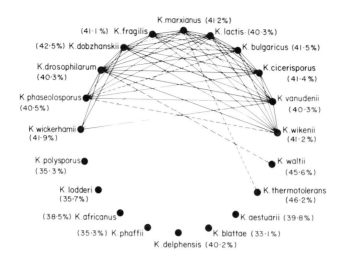

Fig. 1. Interbreeding relationships in *Kluyveromyces* (Johannsen 1978). Full lines represent fertile crosses in which recombination frequencies characteristic of intraspecific crosses, i.e. > $10^3/10^8$ c.m.p. were observed. Broken lines represent fertile crosses in which recombinant frequencies < $10^3/10^8$ c.m.p. were observed or where recombinant frequencies > $10^3/10^8$ c.m.p. were irregularly encountered.

differentiate species. This is evident from the fertile crosses between *K. marxianus* × *K. bulgaricus* and *K. cicerisporus* × *K. phaseolosporus* which respectively involve parents forming either crescentiform or spheroidal ascospores. Likewise the utilization of sugars such as maltose and lactose cannot serve to differentiate species. This is evident from the fertile crosses of *K. wikenii* × *K. lactis* and *K. fragilis* × *K. dobzhanskii*.

Figure 1 also illustrates the inadequacy of numerical taxonomy based exclusively on phenotypic characteristics. This is evident from the fact that *K. polysporus* and *K. lodderi* and also *K. phaffi* and *K. africanus* emerge as four distinct, genetically-isolated taxa, irrespective of the high-matching coefficients, reported for these taxa by Campbell (1972) and Kocková-Kratochvilová *et al.* (1972).

Presumed affinities between taxa on the basis of cell wall composition as reflected by proton magnetic resonance (p.m.r.) spectra of their cell wall mannans, also do not invariably correlate with relatedness determined by sexual recognition between taxa. This is evident from the fact that Gorin & Spencer (1970) assigned the cluster of interfertile taxa comprising *K. dobzhanskii*, *K. drosophilarum*, *K. fragilis*, *K. lactis* and *K. marxianus* to no less than four groups based on the p.m.r. spectra of their mannans. In view of these differences in the cell wall components of interbreeding strains and the fact that such components serve as immuno-determinants, it is not surprising that Sandula *et al.* (1974) failed to detect marked serological similarities between the interbreeding strains of *K. dobzhanskii* and *K. fragilis.*

Similarly, relatedness between strains as reflected by the immunological comparison of exo-β-glucanases, does not invariably correlate with relatedness based on interfertility. This is apparent from the recent report by Lachance & Phaff (1979) which suggests tht *K. phaseolosporus* is more closely related to *K. delphensis* — with which it does not hybridize — than to *K. drosophilarum* with which it is interfertile.

So much then for the limitations of phenotypic characters as criteria for species delineation in *Kluyveromyces.*

5. Comparison of Interfertility and Nucleic Acid Analyses as Criteria for Speciation in the Genus *Kluyveromyces*

Johannsen's (1978) study also permits an evaluation of interfertility and genome comparison data as fundamental criteria for speciation. From Fig. 1 it is apparent that strains with virtually identical mol % G + C values do not invariably represent the same species. This is evident from the fact that strains of *K. polysporus* and *K. phaffii* do not hybridize. It is also obvious that differences in nDNA base composition do not provide the expected exclusionary function in yeast taxonomy. This is evident from the fertile crosses of *K. lactis* × *K. dobzhanskii* and *K. cicerisporus* × *K. drosophilarum.* These interfertile parental taxa differ in their mol % G + C by 2·2% and 2·3%, respectively. If interfertility between strains is to be accepted as a criterion for conspecificity, differences of 1–2% in mol % G + C obviously cannot invariably be applied in the delineation of species in *Kluyveromyces.*

On comparing the nDNA reassociation data reported for *Kluyveromyces* by Martini & Phaff (1973), Martini (1973) and Phaff *et al.* (1978) with the results obtained by Johannsen (1978), two salient points emerge: (i) degrees of nDNA complementarity greater than 80% between strains invariably coincided with their interfertility as reflected by the recovery of fertile hybrid progeny and recombinant frequencies characteristic of intraspecific crosses; (ii) degrees of reassociation lower than 80% between strains did not invariably preclude their interfertility. No fewer than 12 such anomalies may be observed in the results submitted by Martini & Phaff (1973), Martini (1973) and Phaff *et al.* (1978), and three are shown in Table 3. The reassociation data presented derive from Martini's (1973) publication.

These observed discrepancies in relatedness as determined by DNA complementarity on the one hand and by interfertility on the other, are

TABLE 3

Comparison of interfertility and reported nDNA reassociation
between taxa in Kluyveromyces

Parental strains		Fertile hybrids* recovered on	Recombinant[†] frequencies/ 10^8 c.m.p.[‡]	DNA reassociation[§] %
K. fragilis LAC[ǀ] CBS 1556 Met**	× K. dobzhanskii MAL CBS 2104 A	Maltose agar	9.6×10^4	14
K. lactis MAL CBS 683 Trp	× K. marxianus CBS 6923	Maltose agar	1.6×10^6	12
K. marxianus LAC CBS 6923 His	× K. drosophilarum CBS 2896	Lactose agar	5.2×10^6	27

*Recovered hybrids possessed combined parental phenotypes.
[†]Data from Johannsen (1978).
[‡]c.m.p., cells of the mated populations.
[§]Data from Martini (1973).
[ǀ]Sugar utilization: LAC, lactose; MAL, maltose.
**Nutritional requirements: A, adenine; His, histidine; Met, methionine; Trp, tryptophan.
Note that the recombinant frequencies fall well within the range observed in crosses between mutant strains of the same species (Table 1).

difficult to explain. It is nevertheless apparent that genetic relatedness in terms of interfertility implies genic compatibility in respect of the entire organized yeast genome. In terms of nDNA complementarity it reflects high degrees of *in vitro* renaturation of fragments of denatured nDNA.

At this stage, the question of whether the application of nDNA reassociation data for yeast speciation has ever in fact been adequately evaluated in terms of interfertility, becomes pertinent. A survey of the literature shows that this evaluation has been inadequate and limited to selected observations on a single species, *Sacch. cerevisiae*.

Price *et al.* (1978), in their exposition of speciation based on nDNA re-association data, only cite the studies of Bicknell & Douglas (1970) and Ouchi *et al.* (1970) in which only some interfertile strains of *Sacch. cerevisiae* were reported to show degrees of DNA complementarity > 80%. They never-theless do not take into account the fact that not only Bicknell & Douglas but also Ouchi and his co-workers reported degrees of nDNA comple-mentarity as low as 40% between strains of *Sacch. cerevisiae* and the melibiose-fermenting biotypes cited as *Sacch. carlsbergensis* and *Sacch. uvarum*.

6. Conclusions and Future Prospects

The observed correlation between the interfertility of strains and their degrees of nDNA reassociation > 80% provides reliable grounds on which to accept such degrees of reassociation as criteria for conspecificity in *Kluyveromyces*. In view of the increasing reliance on nucleic acid analyses for speciation purposes, it is therefore obligatory that this correlation should also be confirmed in other perfect genera.

The fact that degrees of nDNA reassociation appreciably lower than 80% between strains of *Kluyveromyces* do not invariably preclude their inter-fertility, raises pertinent practical issues. Firstly, it is apparent that in *Kluyveromyces* the remodelling of taxa exclusively on the basis of degrees of nDNA reassociation > 80%, would obviously be too restrictive. Secondly, it raises the question as to whether the observed discrepancies in relatedness in terms of interfertility and nDNA reassociation data are limited to *Kluyveromyces* or whether they are possibly also manifested in other perfect genera.

In the event of this being so the application of degrees of nDNA reassociation > 80% might also be too restrictive a criterion for con-specificity among the imperfect yeasts, taxa where phenotypic speciation is inadequate and where there is little or no recourse to interfertility as a measure of genetic relatedness — an anomalous situation which emphasizes the difficulties of speciation based on the application of a single criterion. The failure to detect dangeardiens expressed either as asci, ustilospores or basidia in these so-called imperfect yeasts, has led to the tacit assumption that such species comprise strains that are either haploid or devoid of sexual stages. While this supposition may hold in the case of isolated, single mating-types of heterothallic taxa, this is by no means the general rule. This became apparent when an alternation of generations was detected in *Candida albicans* (van der Walt 1970), *C. tropicalis* (Sukroongrueng & Rodrigues de Miranda 1973) and *Torulopsis magnoliae* (van der Walt & Johannsen 1973) which implicated a dangeardien expressed as neither the ascus nor the basidium (van der Walt & Johannsen 1974). As these three taxa are nevertheless characterized by cell walls of the ascomycetous type, it has been inferred that while these species possessed the genes for meiosis, the genes for ascospore formation were either absent or not expressed (van der Walt & Johannsen 1974). The recent recovery by Rodrigues de Miranda & Török (1976) of an ascigeric phase from the presumed imperfect species *C. ingens* by treatment with the mutagen N-methyl-N'-nitro-nitrosoguanidine, could be interpreted in terms of de-repression of repressed regions of the genome which normally code for ascospore formation.

As the pattern or type of sexual cycle is, in general, highly characteristic of species, more incisive studies on the seemingly unmanifested or latent sexual stages of the imperfect yeasts may provide better guidelines for speciation among these taxonomically refractory, yet intriguing, organisms.

Maurice Ingram expressed similar views 11 years ago when he remarked: "At present, I think the most interesting new phenomena are the primitive forms of sexuality revealed among supposed asexual species, and the unexpected phylogenetic relationships which have consequently been indicated" (Ingram 1969).

7. References

BAK, A. L. & STENDERUP, A. 1969 Deoxyribonucleic acid homology in yeasts. Genetic relatedness within the genus *Candida. Journal of General Microbiology* **59**, 21-30.

BELOZERSKY, A. N. & SPIRIN, A. S. 1960 Chemistry of the nucleic acids of microorganisms. In *The Nucleic Acids* Vol. 3, ed. Chargraff, E. & Davidson, J. N. pp. 147-185. New York: Academic Press.

BICKNELL, J. N. & DOUGLAS, H. C. 1970 Nucleic acid homologies among species of *Saccharomyces. Journal of Bacteriology* **101**, 505-512.

CAIN, A. J. 1954 *Animal Species and their Evolution*. London: Hutchinson.

CAMPBELL, I. 1972 Numerical analysis of the genera *Saccharomyces* and *Kluyveromyces. Journal of General Microbiology* **77**, 279-301.

CHEN, S. Y. 1950 Sur une nouvelle technique de croisement des levures. *Comptes rendus hebdomadaire des séances de l'Académie des sciences* **230**, 1897-1899.

CLÉMENÇON, H. 1977 The species concept in Hymenomycetes. *Bibliotheca Mycologia. Vol. 61 Proceedings of a Herbette Symposium, University of Lausanne, Switzerland, 1976.*

CRANDALL, M., EGEL, R. & MACKAY, V. L. 1977 Physiology of mating in three yeasts. *Advances in Microbial Physiology* **15**, 307-398.

DOBZHANSKY, TH, 1951 *Genetics and the Origin of Species* 3rd edn, pp. 262-263. New York: Columbia University Press.

FELL, J. W. & MEYER, S. A. 1967 Systematics of yeast species in the *Candida parapsilosis* group. *Mycopathologia et mycologia applicata* **32**, 177-193.

FELL, J. W. & PHAFF, H. J. 1970 The genus *Leucosporidium* Fell, Statzell, Hunter & Phaff. In *The Yeasts — A Taxonomic Study* 2nd edn, ed. Lodder, J. pp. 776-802. Amsterdam: North-Holland.

FELL, J. W., PHAFF, H. J. & NEWELL, S. Y. 1970 The genus *Rhodosporidium* Banno. In *The Yeasts — A Taxonomic Study* 2nd edn, ed. Lodder, J. pp. 803-814. Amsterdam: North-Holland.

FELL, J. W., HUNTER, I. L. & TALLMAN, A. S. 1973 Marine basidiomycetous yeasts *(Rhodosporidium* spp. n.) with tetrapolar and multiple allelic bipolar mating systems. *Canadian Journal of Microbiology* **19**, 643-657.

FOGEL, S. & MORTIMER, R. K. 1971 Recombination in yeast. *Annual Review of Genetics* **5**, 219-236.

FOWELL, R. R. 1969 Sporulation and hybridization in yeasts. In *The Yeasts* Vol. 1, ed. Rose, A. H. & Harrison, J. S. pp. 301-383. London & New York: Academic Press.

FRIIS, J. & ROMAN, H. 1968 The effect of mating-type alleles on intragenic recombination in yeast. *Genetics* **59**, 33-36.

GORIN, P. A. J. & SPENCER, J. F. T. 1970 Proton magnetic resonance spectroscopy — an aid in identification and chemotaxonomy of yeasts. *Advances in Applied Microbiology* **13**, 25-89.

GRANT, V. 1963 *The Origin of Adaptations*. New York: Columbia University Press.

GUILLIERMOND, A. 1912 *Les Levures*. Paris: Octave Doin.

HANSEN, E. C. 1888 Undersogelser over Alkoholgjaersvampenes Fysiologi og Morfologi VII. Om Alkoholgjaersvampenes Forhold til Sukkerarterne. *Meddeleser fra Carlsberg Laboratoriet* **2**, 220-256.

INGRAM, M. 1969 Yeast science today and tomorrow. *Antonie van Leeuwenhoek* **35** Supplement: Yeast Symposium 1969, 7-29.

JOHANNSEN, E. 1978 *Hybridization studies within the genus* Kluyveromyces *van der Walt emend, van der Walt* Vols 1 & 2. Ph.D. Thesis, Rhodes University, Grahamstown, South Africa.

JOHANNSEN, E. & VAN DER WALT, J. P. 1978 Interfertility as basis for the delimitation of *Kluyveromyces marxianus*. *Archives of Microbiology* **118**, 45–48.

KOCKOVÁ-KRATOCHVILOVÁ, A., BLAGODATSKAYA, V. & HRONSKA, L. 1972 The grouping of the species within the genus *Kluyveromyces* van der Walt. In *Yeast-Models in Science and Tecnics*, ed. Kocková-Kratochvilová, A. & Minarik, E. pp. 375–387. Bratislava: Slovak Academy of Science.

KWON-CHUNG, K. J. 1975 A new genus *Filobasidiella*, perfect state of *Cryptococcus neoformans*. *Mycologia* **67**, 1197–1200.

KWON-CHUNG, K. J. 1976 A species of *Filobasidiella*, the sexual state of *Cryptococcus neoformans* B and C serotypes. *Mycologia* **68**, 942–946.

LACHANCE, M.-A. & PHAFF, H. J. 1979 Comparative study of molecular size and structure of exo-β-glucanases from *Kluyveromyces* and other yeast genera: evolutionary and taxonomic implications. *International Journal of Systematic Bacteriology* **29**, 70–78.

LODDER, J. ed. 1970 *The Yeasts — A Taxonomic Study* 2nd edn. Amsterdam: North-Holland.

LOTSY, J. P. 1925 Species or linneon? *Genetica* **7**, 487–506.

MACKAY, V. L. & MANNEY, T. R. 1974 Mutations affecting sexual conjugation and related processes in *Saccharomyces cerevisiae*. I. Isolation and phenotypic characterization of nonmating mutants. *Genetics* **76**, 225–271.

MANDEL, M. 1969 New approaches to bacterial taxonomy: perspective and prospects. *Annual Review of Microbiology* **23**, 239–274.

MARMUR, J., FALKOW, S. & MANDEL, M. 1963 New approaches to bacterial taxonomy. *Annual Review of Microbiology* **17**, 329–372.

MARTINI, A. 1973 Ibridazioni DNA/DNA tra species di lieviti del genere *Kluyveromyces*. *Annali della Facoltà di agraria della Università degli studi di Perugia* **28**, 1–15.

MARTINI, A. & PHAFF, H. J. 1973 Optical determination of DNA-DNA homologies in yeasts. *Annali di Microbiologia* **23**, 59–68.

MAYR, E. 1967 *Artbegriff und Evolution*. Hamburg–Berlin: Paul Parey.

MEYER, S. A. 1970 *DNA base composition homology in* Candida *species and related yeasts*. Dissertation, University of California, Davis.

MEYER, S. A., SMITH, M. TH. & SIMIONE, F. P. 1978 Systematics of *Hanseniaspora* Zikes and *Kloeckera* Janke. *Antonie van Leeuwenhoek* **44**, 79–96.

MEYER, S. A. & PHAFF, H. J. 1970 Taxonomic significance of DNA composition in yeasts. In *Recent Trends in Yeast Research* Vol. 1, ed. Ahearn, D. G. pp. 1–29. Spectrum, Monograph Series in the Arts and Sciences. Atlanta: Georgia State University.

MEYER, S. A., SMITH, M. TH. & SIMIONE, F. P. 1978 Systematics of *Hanseniaspora* Zikes and *Kloeckera* Janke. *Antonie van Leeuwenhoek* **44**, 79–96.

MOORE, R. L. 1974 Nucleic acid reassociation as a guide to genetic relatedness among bacteria. *Current Topics in Microbiology and Immunology* **64**, 105–128.

NAKASE, T. & KOMAGATA, K. 1968 Taxonomic significance of base composition of yeast DNA. *Journal of General and Applied Microbiology* **14**, 345–357.

NAKASE, T. & KOMAGATA, K. 1969 DNA base composition of the genus *Hansenula*. *Journal of General and Applied Microbiology* **15**, 85–95.

NAKASE, T. & KOMAGATA, K. 1970a Significance of DNA base composition in the classification of yeast genera *Hanseniaspora* and *Kloeckera*. *Journal of General and Applied Microbiology* **16**, 241–250.

NAKASE, T. & KOMAGATA, K. 1970b Significance of DNA base composition in the classification of yeast genus *Pichia*. *Journal of General and Applied Microbiology* **16**, 511–521.

NAKASE, T. & KOMAGATA, K. 1971a Significance of DNA base composition in the classification of yeast genus *Debaryomyces*. *Journal of General and Applied Microbiology* **17**, 43–50.

NAKASE, T. & KOMAGATA, K. 1971b Significance of DNA base composition in the classification of yeast genus *Cryptococcus* and *Rhodotorula*. *Journal of General and Applied Microbiology* **17**, 121–130.

NAKASE, T. & KOMAGATA, K. 1971c Significance of DNA base composition in the classification of yeast genus *Torulopsis*. *Journal of General and Applied Microbiology* **17**, 161–166.

NAKASE, T. & KOMAGATA, K. 1971d Significance of DNA base composition in the classification of yeast genus *Saccharomyces*. *Journal of General and Applied Microbiology* **17**, 227–238.

NAKASE, T. & KOMAGATA, K. 1971e Significance of DNA base composition in the classification of yeast genus *Candida*. *Journal of General and Applied Microbiology* **17**, 259–279.

NELSON, R. P. 1963 Interspecific hybridization in the fungi. *Annual Review of Microbiology* **17**, 31–48.

NEUMANN, I. 1972 Biotaxonomische und systematische Untersuchungen an einigen Hefen der Gattung *Saccharomyces*. *Beihefte zur Nova Hedwigia* **40**, 1–79.

OUCHI, K., SAITO, H. & IKEDA, Y. 1970 Genetic relatedness of yeast strains studied by the DNA-DNA hybridization method. *Agricultural and Biological Chemistry, Tokyo* **34**, 95–101.

PHAFF, H. J. 1970a The genus *Hanseniaspora* Zikes. In *The Yeasts — A Taxonomic Study* 2nd edn, ed. Lodder, J. pp. 209–225. Amsterdam: North-Holland.

PHAFF, H. J. 1970b The genus *Schwanniomyces* Klöcker. In *The Yeasts — A Taxonomic Study* 2nd edn, ed. Lodder, J. pp. 756–766. Amsterdam: North-Holland.

PHAFF, H. J., LACHANCE, M. A. & PRESLEY, H. L. 1978 Molecular approaches to the systematics of the yeast genus *Kluyveromyces*. In *Abstracts of the XIIth International Congress of Microbiology September 1978, München* p. 38.

POMPER, S. & BURKHOLDER. P. R. 1949 Studies on the biochemical genetics of yeast. *Proceedings of the National Academy of Sciences of the United States of America* **35**, 456–464.

PRICE, C. W., FUSON, G. B. & PHAFF, H. J. 1977 Trends in yeast systematics: molecular approaches. In *Proceedings of the 5th International Specialized Symposium on Yeasts, Keszthely, Hungary* pp. 1–35.

PRICE, C. W., FUSON, G. B. & PHAFF, H. J. 1978 Genome comparison in yeast systematics: Delimitation of species within the genera *Schwanniomyces, Saccharomyces, Debaryomyces* and *Pichia*. *Microbiological Reviews* **42**, 161–193.

RODRIGUES DE MIRANDA, L. 1972 *Filobasidium capsuligenum* nov. comb. *Antonie van Leeuwenhoek* **38**, 91–99.

RODRIGUES DE MIRANDA, L. & TÖRÖK, T. 1976 *Pichia humboldtii* sp. nov., the perfect state of *Candida ingens*. *Antonie van Leeuwenhoek* **42**, 343–348.

SANDULA, J., KOCKOVÁ-KRATOCHVILOVÁ, A. & SIKL, D. 1974 Immunochemical studies on mannans of the genera *Kluyveromyces* and *Saccharomyces*. *Journal of General Microbiology* **83**, 339–347.

SCHEDA, R. 1966 Merkmalsveränderungen bei Hefen der Gattung *Saccharomyces*. *Monatsschrift für Brauerei* **19**, 256–258.

SCHEDA, R. & YARROW, D. 1966 The instability of physiological properties used as criteria in the taxonomy of yeasts. *Archiv für Mikrobiologie* **55**, 209–225.

SCHEDA, R. & YARROW, D. 1968 Variation in the fermentative pattern of some *Saccharomyces* species. *Archiv für Mikrobiologie* **61**, 310–316.

STENDERUP, A. & BAK, A. L. 1968 Deoxyribonucleic acid base composition of some species within the genus *Candida*. *Journal of General Microbiology* **52**, 231–236.

STORCK, R. 1966 Nucleotide composition of nucleic acids of fungi. II. Deoxyribonucleic acids. *Journal of Bacteriology* **91**, 227–230.

STORCK, R., ALEXOPOULOS, C. J. & PHAFF, H. J. 1969 Nucleotide composition of deoxyribonucleic acid of some species of *Cryptococcus, Rhodotorula* and *Sporobolomyces*. *Journal of Bacteriology* **98**, 1069–1072.

SUKROONGRUENG, S. & RODRIGUES DE MIRANDA, L. 1973 A new aspect of the life-cycle of *Candida tropicalis*. *Antonie van Leeuwenhoek* **39**, 65–80.

VANDENDRIES, R. 1923 Nouvelles recherches sur la sexualité des Basidiomycetes. *Bulletin Sociéte royale de botanique de Belgique* **56**, 1–25.

VAN DER WALT, J. P. 1970 The genus *Syringospora* Quinquad emend. *Mycopathologia et mycologia applicata* **40**, 231–243.

VAN DER WALT, J. P. & JOHANNSEN, E. 1973 The perfect state of *Torulopsis magnoliae*. *Antonie van Leeuwenhoek* **39**, 635–647.

VAN DER WALT, J. P. & JOHANNSEN, E. 1974 The dangeardien and its significance in the taxonomy of the ascomycetous yeasts. *Antonie van Leeuwenhoek* **40**, 185–192.

WICKERHAM, L. J. 1970 The genus *Hansenula* H. & P. Sydow. In *The Yeasts — A Taxonomic Study* 2nd edn, ed. Lodder, J. pp. 226–315. Amsterdam: North Holland.

WICKERHAM, L. J. & BURTON, K. A. 1952 Occurrence of yeast mating types in nature. *Journal of Bacteriology* **63**, 449–451.

WICKERHAM, L. J. & BURTON, K. A. 1954 A classification of the relationship of *Candida guilliermondii* to other yeasts by a study of their mating types. *Journal of Bacteriology* **68**, 594–597.

WINGE, Ö. 1935 On haplophase and diplophase in some Saccharomycetes. *Comptes rendus des travaux du Laboratoire Carlsberg* **21**, 77–112.

WINGE, Ö. & LAUTSEN, O. 1939 On 14 new yeast types, produced by hybridization. *Comptes rendus des travaux du Laboratoire Carlsberg* **22**, 338–352.

YARROW, D. 1972 Centraalbureau voor Schimmelcultures, Baarn. Progress Report 1971 *Verhandelingen der Koninklijke Nederlandse Akademie van Wetenschappen*, Afdeling Natuurkunde, Sectie II **61**, 50–69.

YARROW, D. & NAKASE, T. 1975 DNA base composition of species of the genus *Saccharomyces*. *Antonie van Leeuwenhoek* **41**, 81–88.

Specificity of Natural Habitats for Yeasts and Yeast-Like Organisms

H. J. PHAFF AND W. T. STARMER*

*Department of Food Science and Technology,
University of California,
Davis, California, USA*

Contents

1. Introduction

THE FIELD of yeast ecology is concerned with the manner in which yeasts and yeast-like organisms live and propagate in nature, and where specific organisms can be found. Numerous surveys of substrates that were suspected to harbour yeasts, because of the availability of the necessary nutrients for their growth, have revealed that yeasts appear to be far less ubiquitous than many bacterial species. Specialization for habitat appears to be the rule rather than the exception. Early surveys of substrates were directed to a large extent to foods and food products and to the study of so-called wild yeasts that had invaded industrial processes (such as brewing, wine fermentation and baker's yeast production) where they caused undesirable changes or spoilage. Some early surveys of tree sap flowing from wounds on tree trunks (so-called slime fluxes or exudates) demonstrated the presence of quite unusual yeasts that were very different from those found associated with food products. Such findings stimulated further surveys into other substrates leading to the discovery of many new habitats

* Present address: Department of Biology, Syracuse University, Syracuse, NY 13210, USA.

for specific organisms. Although yeasts are often thought to be associated with fermenting, hexose sugar-containing substrates, the ability of yeasts to assimilate a variable number of nonfermentable organic compounds through the respiratory process, greatly expands the ability of specific yeasts to occupy an ecological niche containing compounds that can be metabolized by the necessary enzymes with which a yeast is endowed. Examples of such compounds are primary alcohols (ethanol, methanol), sugar alcohols (glucitol, mannitol, xylitol, erythritol), organic acids (citric, succinic, lactic, malic), pentose sugars (L-and D-arabinose, D-xylose), methyl pentoses (L-rhamnose, L-fucose), *meso*-inositol, hydrocarbons, and aromatic compounds (phenolics and derivatives). Many of these compounds that specific yeasts can assimilate are products of plant metabolism or products of other micro-organisms (fungi and bacteria) that produce them from precursors in plants.

The other principal nutrient requirement for yeast growth is a suitable nitrogen source. The best nitrogen source is a balanced mixture of amino acids, but this is rarely present in natural substrates. Yeasts vary in their ability to convert one amino acid into another by transamination. Ammonium nitrogen is suitable for nearly all yeasts, but an exception is *Cyniclomyces guttulatus* that lives and propagates in the intestinal tract of rabbits and cannot use ammonium ions for growth. Some yeasts (e.g. species of *Hansenula*) can assimilate nitrate and nitrite and a small number of species have lost the ability to reduce nitrate but can still utilize nitrite for growth (some strains of *Debaryomyces hansenii*). Also the ability to assimilate amines varies among yeast species. All of these variations in ability to assimilate various carbon and nitrogen sources undoubtedly contribute in an important way to habitat specificity. Mineral and vitamin requirements for yeast development in natural substrates are usually met adequately.

Specificity of yeasts in natural habitats is not only determined by the composition of nutrients present, but also by the presence of inhibitory compounds to which some yeasts are sensitive (and are therefore eliminated) and others resistant. For example, there are many antibiotic-producing strains of actinomycetes, especially in soil, that may produce sufficient concentrations of antibiotics to inhibit sensitive yeasts and thus affect the composition of the yeast community. Major metabolic products of bacteria can also inhibit yeast growth and in extreme cases kill yeast cells already present. For example, many yeasts are very sensitive to the undissociated form of acetic acid that may be produced from ethanol formed by yeast by the subsequent development of acetic acid bacteria. Some yeasts can even produce such high concentrations of acetic acid from ethanol that they kill themselves. This occurs in species of *Brettanomyces* and *Dekkera*.

Also contributing to the specificity of habitats for yeasts are the presence of natural compounds in plants that have a selective inhibitory capacity for the growth of yeasts. Two examples may be mentioned. Several species of *Hansenula* and *Pichia* that are associated with the bark beetles that attack pine trees grow vigorously in the sap of these trees in the presence of high concentrations of oleoterpenes that are inhibitory to the great majority of other yeasts. This has led to the highly specific yeast flora that is associated with these beetles and the trees that they parasitize. A second example concerns the inhibitory effect of triterpene glycosides in cactus species of the subtribe Stenocereinae on the growth of some cactus-specific yeasts (Starmer *et al.* 1979*b*). Of the two varieties of *P. amethinoina*, *P. amethionina* var. *amethionina*, which is found in cactus rots of the subtribe Stenocereinae, is tolerant to these glycosides. In contrast *P. amethionina* var. *pachycereana*, which is recovered from cacti of the subtribe Pachycereanae (which do not contain these glycosides), is sensitive to these compounds and the two varieties are therefore separated by habitat chemistry (Starmer *et al.* 1979*b*).

Other general environmental factors that influence the development of specific yeasts are oxygen concentration and temperature. Among the yeasts associated with cactus soft rots, some are strict aerobes and others also possess a fermentative form of metabolism. Thus, in the deeper layers of the thick decomposing plant tissue, the latter category of species has a selective advantage over the former. The same applies to other sugar-containing substrates, such as a fruit juice, where fermentative yeasts cause alcoholic fermentation in the bulk of the liquid and aerobic film yeasts grow on the surface. In some instances yeasts that have adapted to an anaerobic environment, such as the digestive tract of animals, may lose the respiratory ability altogether. This has been observed for *Torulopsis pintolopesii*, which occurs in the intestinal tract of small rodents (van Uden & Vidal-Leiria 1970).

There is definite evidence that the maximum temperature for growth is related to the natural environment in which species normally occur. The temperature range for yeast growth is from 0°C to about 47°C. *Leucosporidium scottii*, first isolated from chilled beef in Australia (where it was common) and later from Antarctic soils, has a maximum growth temperature of about 17°C. Later, some strains were isolated from ocean water with maximum growth temperatures up to 30°C. These may have originated in zones with a temperate climate (Fell & Phaff 1970). Yeasts associated with the intestinal tract of warm-blooded animals (*T. pintolopesii* and *Candida slooffii*) have minimal temperatures for growth of about 24–28°C and maxima greater than 40°C (van Uden & Buckley 1970; van Uden & Vidal-Leiria 1970). We also observed different maximum growth temperatures for the two varieties of *P. opuntiae* (Starmer *et al.* 1979*a*). *Pichia opuntiae* var. *opuntiae* grows in *Opuntia* cacti in a temperate climate of Australia and has a maximum growth temperature of *ca.* 33°C, whereas *P. opuntiae* var. *thermotolerans*, which occurs in columnar cacti of the hot North American Sonoran Desert, consistently grew at 37°C although not at 39°C. In general, however, most cactus-specific yeasts occurring in this desert have maximum growth temperatures of 42–45°C. Conversely, the great majority of species isolated from tree exudates in the Pacific Northwest of North America did not grow at 37°C and some had maxima below 30°C.

A. *Importance of correct taxonomic identity*

Studies of yeast ecology and the accuracy of assigning specific habitats to certain species depend heavily on our ability to identify correctly yeast isolates obtained from natural or domestic and industrial sources. Our systems of yeast identification are far from perfect, in part because of the relatively limited number of phenotypic properties that can be used to characterize species. In addition, recent molecular taxonomic studies have shown that the traditional fermentation and assimilation tests of single sugars (often determined by a single gene difference coding for a specific hydrolytic enzyme) are of little or no value in species differentiation (Price *et al.* 1978). Conversely, nuclear genome comparisons of yeast DNA from phenotypically similar species demonstrated that some of these have only negligible DNA sequences in common and therefore represent different species (Price *et al.* 1978). Clearly, an accurate picture of yeast habitats cannot be assembled if the species cannot be correctly identified. One example out of many concerns our own misidentification of a number of cactus-associated yeasts as *P. membranaefaciens* (Heed *et al.* 1976) to which they correspond closely by the system of Kreger-van Rij (1970*b*). Only after determining the nuclear DNA base composition of representative strains, did it become clear that none of these strains represented *P. membranae-faciens* (mol % G + C 44·3) but, instead, at least four new species of *Pichia*, i.e. *P. cactophila* (mol % G + C 36·9; Starmer *et al.* 1978*a*), *P. heedii* (mol % G + C 32·5; Phaff *et al.* 1978*a*), *P. amethionina* (mol % G + C 33·1; Starmer *et al.* 1978*b*) and *P. opuntiae* (mol % G + C 33·4; Starmer *et al.* 1979*a*). By the key of Kreger-van Rij (1970*b*) all would be identified as *P. membranaefaciens*. Subsequently we recognized less commonly used phenotypic properties to distinguish some from *P. membranaefaciens* (e.g. strong utilization of glucosamine by *P. cactophila* and an absolute requirement for L-methionine by *P. amethionina*), but for the others the habitat in specific cactus species constitutes the only presently known criterion for species separation. Similar examples could be cited for other yeasts where existing species have been shown by DNA analysis to be mixtures of several unrelated taxa (e.g. *C. sake*; Meyer *et al.* 1975). Obviously in cases where yeasts were misidentified, data on ecological habitats and niche specificity would have little meaning.

2. Methods of Isolation and Study

As an initial approach to studying the yeast flora of a natural substrate, samples of the material, in various dilutions, are streaked or plated directly on a complex agar medium that is likely to support the growth of the yeasts present. After yeast colonies develop fully they are carefully inspected with a dissection microscope and one colony of each morphological type is picked and purified by replating. Recognizing colonies with different morphology is the only approach to estimating the proportion of each type in the sample, yet the method has at least two significant drawbacks. The first is that certain species form identical or nearly identical colonies and can therefore not be recognized as separate entities. Also, some species of imperfect yeasts (notably *Candida* and *Torulopsis*) may show significant variation in colony morphology and could thus be picked as putative separate species. Such variations are thought to be caused by mutations of haploid clones that are expressed in different colony forms. A second problem is related to minority species that may be present in the sample. If a plate is not to be overcrowded with colonies (i.e. a maximum of 300), population densities of less than *ca.* 0·3% of the total would not be expected to occur on such plates. Fortunately, the total number of species in a natural substrate is usually rather small, ranging from one to four in a particular habitat.

If the yeast community of a certain habitat has been studied extensively and the prevalent species are known, it may become possible to use quantitative methods to enumerate their proportions including minority populations. This can be done by using selective synthetic media containing carbon and (or) nitrogen sources that are utilized by only one or two of the species present. After obtaining some experience with the necessary dilutions and the ability to recognize certain species by their colony morphology, quantitative counts of populations of the species present can be made. We have recently applied this technique to the yeast flora in cactus rots of Australian *Opuntia* species. Yeast Nitrogen Base (YNB; Difco) agar with *meso*-inositol as the carbon source enumerates *Cryptococcus cereanus* (Phaff *et al.* 1974), more recently known by its ascosporogenous state as *Sporopachydermia cereana* (Rodrigues de Miranda 1978). Occasionally slimy *Cryptococcus* species that are not cactus specific may occur on such plates, but they can be easily distinguished from *Sporopachydermia cereana*. YNB with methanol as the carbon source enumerates cactus-specific *T. sonorensis* (Miller *et al.* 1976) and occasional strains of the non-cactus specific *C. boidinii* that can be distinguished easily from *T. sonorensis*. YNB with glucosamine as the carbon source enumerates *P. cactophila* (Starmer *et al.* 1978a) one of the most common and widespread cactus-specific yeasts. This species has been isolated from all species of cacti tested and in areas that include the North American Sonoran Desert, southern Mexico, southern California, Hawaii, Spain, and Australia. *Pichia heedii* (Phaff *et al.* 1978a) which occurs predominantly in two major host plants [*Lophocereus schottii* (Engelm.) Britten & Rose (Senita) and *Carnegia gigantea* (Engelm.) Britten & Rose (Saguaro)] and only in the Sonoran Desert, can be enumerated in YNB plus D-xylose as the carbon source. *Cryptococcus cereanus* can also grow on this medium but the colony and cell morphologies of the two species are distinct.

A. Choice of media and incubation conditions

On solid media one of the problems that must be overcome is competition from bacteria and moulds. We find that some control of bacterial growth can be achieved by adjusting the pH of melted malt extract agar or yeast autolysate glucose agar (after autoclaving) with a precalculated volume of 1 N HCl to pH 3·5–3·8. Although many bacterial species are inhibited from growing at this pH, others that we find in plant material are surprisingly tolerant. However, with some practice it is not very difficult in most cases to differentiate yeast colonies from bacterial colonies with a wide-field microscope. Another, somewhat less convenient, technique is the incorporation of wide spectrum antibacterial agents in the melted agar media.

The control of moulds is more difficult. In each ecological survey, preliminary experiments usually reveal the seriousness of spreading mould colonies overgrowing the yeasts. Several antifungal agents (such as propionic acid at pH values of 3·5–4·0) also have an inhibitory action on some yeasts. Some dye solutions (e.g. 0·003% rose bengal) reduce the rate at which some fungi spread on plates. If the fungi are so abundant as to preclude yeast isolation by direct plating, the best approach in our experience is to inoculate samples of the yeast-containing substrate in acidified broth (pH *ca.* 3·5–3·7) and to place the flask on a shaker until the mould conidia have germinated into mycelial hyphae. The latter can then be filtered over sterilized glass wool and the yeasts in the filtrate can be plated directly on acidified media.

Incubation temperatures should generally be close to those under which the yeasts propagate in their natural environment, especially those from low temperature substrates (e.g. 10–12°C). Mesophilic yeasts are best incubated at 20–25°C, because many species are unable to grow at 30°C. Even those yeasts that can grow at 45–46°C generally grow well at 20°C. The psychrophobic species that occur specifically in the intestinal tract of warm-blooded animals (Travassos & Cury 1971; Travassos & Mendonça 1972) are an exception and incubation of yeasts from these sources is best done at 37°C.

B. Special growth requirements

Several yeasts have unusual growth requirements and as a consequence media and incubation conditions that are satisfactory for the vast majority of species must be adjusted. A few examples are given below.

Schizosaccharomyces octosporus strains grow poorly in synthetic media unless supplemented with about 15 mg/1 of adenine (Northam & Norris 1951).

Pichia amethionina is unable to convert hydrogen sulphite to hydrogen sulphide, a complex step in the metabolism of sulphate (Starmer *et al.* 1978*b*) and synthetic media must be supplemented with 10 mg/1 of L-methionine or L-cysteine for growth.

Pityrosporum ovale, which has its habitat on the scalp of humans or on the skin of certain animals, does not grow on ordinary complex media unless supplemented with a vegetable oil or with about 0·25% of Tween 80.

Brettanomyces or *Dekkera* species (the latter are the ascogenous forms of the former) occur as spoilage organisms in wine or in soft drinks. Their growth is extremely slow (colonies may not appear on complex media for 10–12 days) and lack of growth in the usual four or five days should not be interpreted as the absence of *Brettanomyces* in the samples tested. A raised level of thiamine to *ca.* 10 mg/1 of medium improves the growth rate somewhat. Two extra additions to isolation media are helpful in recognizing *Brettanomyces* species: calcium carbonate (0·5%) reveals the strongly acid-forming *Brettanomyces*, and supplementing the medium with 100 mg/1 of cycloheximide inhibits many other species, such as wine yeasts, that may also occur in bottled wines.

Cyniclomyces guttulatus (syn. *Saccharomycopsis guttulata*) occurs in the intestinal tract of domestic rabbits. The cells are so large *(ca.* 10 × 20 µm) that they were first observed microscopically in faecal matter or stomach contents in 1845. Culturing this yeast was not accomplished until *ca.* 1955, mainly because a requirement for high concentrations of gaseous CO_2 in the environment was not recognized until then. *Cyniclomyces guttulatus* does not assimilate ammonium nitrogen and must therefore be supplied with a balanced mixture of amino acids (e.g. from yeast autolysate), and its growth temperature ranges from 30–40°C. Buecher & Phaff (1970) studied the interaction between the gaseous environment, minimum growth temperature and longevity of the cells. Another interesting attribute of this species is that ascosporulation occurs only at temperatures that do not permit vegetative growth (optimum 18–20°C). Somewhat less fastidious in their growth requirements are the psychrophobic yeasts *T. pintolopesii* and *C. slooffii* (Travassos & Mendonça 1972). Both species occur in the intestinal tract of warm-blooded animals and are closely related by the criterion of DNA sequence complementarity (Mendonça-Hagler & Phaff 1975). Some strains of *T. pintolopesii* have an absolute requirement for choline (*ca.* 1 mg/1). *Candida slooffii* is less exacting; at 37°C choline inhibits growth, but at 43°C choline is essential in the medium, protecting it against thermal death. In addition vanadium (as $NaVO_3$, 10 $\mu g/l$) is required at 43°C. Another example of an environmentally imposed nutritional requirement relates to the growth of *Saccharomyces cerevisiae* under strictly anaerobic conditions. Continued growth under such conditions is possible only when the medium is supplemented with ergosterol plus oleic acid. Ergosterol can be replaced by oleanoic acid, a sterol-related compound that occurs in the whitish bloom of grapes.

Finally, a few words on the effect of isolation media on yeasts from environments high in salt or sugar content must be given. In general most yeasts that are associated with salt-preserved foods (e.g. species of *Debaryomyces* or *Pichia*) grow readily on salt-free media (e.g. malt extract agar) and may thus be regarded as halotolerant rather than halophilic. In natural environments, however, yeasts are occasionally isolated that have become dependent on salt. One example is *Metschnikowia bicuspidata* var. *australis*, strains of which were isolated in our laboratory from diseased brine shrimps grown in ponds with about 10–12% NaCl. These strains only grew in YNB-glucose media that were supplemented with 2% NaCl, although 10% malt extract media were also adequate for growth (Lachance *et al.* 1976). Because YNB media are used for identification purposes this modification proved to be essential.

In contrast to salt-tolerant yeasts, sugar-tolerant species generally thrive best in media of high sugar content and some grow poorly or not at all when transferred directly from high sugar media to ordinary isolation media. Similarly, on YNB synthetic media used for identification tests growth is very poor initially or altogether lacking. To isolate yeasts from such substrates it is best to use media with about 40% sugar, and when pure cultures have been obtained they can usually be adapted to media of lower osmotic pressure by transfers to gradually lower sugar concentrations (e.g. *Sacch. bisporus* var. *mellis* or *Sacch. rouxii*). *Eremascus albus*, on the other hand, is a yeast-like organism that does not seem to adapt to low sugar concentrations. It is a spoilage organism of foods with high osmotic pressure and seems to require a minimum of 40% sugar in media for it to grow properly. For these reasons it is advisable to use media of high sugar content (*ca.* 40% w/w) when isolating yeasts from food products preserved with sugar.

3. Specificity of Habitats

In the next section we describe certain habitats and hosts for yeasts that have been investigated in detail by yeast ecologists from many parts of the world. Numerous natural habitats have been discovered where yeasts occur in various population densities and with great specialization for habitat. Although many interesting yeast habitats have been discovered, information on the biochemical and environmental reasons for such high specificities is still very limited for most of these interactions between yeast and substrate. Because of the vast literature on yeast ecology a complete coverage is not possible in this space. For earlier general reviews the reader is referred to do Carmo-Sousa (1969) and Phaff *et al.* (1978*b*) and for marine yeasts to Fell (1976). Most of the reviews on the distribution of yeasts in nature cover the yeast flora by habitat or substrate, i.e. plant materials, animals, soil, water and air. In this paper we have followed the opposite approach and have selected a number of yeast species from ascomycetous, basidiomycetous and asporogenous genera with highly specific habitats. As will become evident, in some genera all species have a similar habitat. For the taxonomic arrangement of genera we have followed the system of Lodder (1970) with some important new species added where necessary.

A. Ascomycetous yeasts

(i) Citeromyces *Santa Maria*

This is a monotypic genus. *Citeromyces matritensis* was first isolated from a fruit preserved in syrup. Wickerham later reported that *T. globosa* isolated many years earlier from condensed milk was synonymous with *Cit. matritensis* (Wickerham 1970*a*). In our laboratory we have repeatedly isolated this species from slime fluxes of *Myoporum sandwicense* (D.C.) A. Gray, from trees on the island of Hawaii and these constitute the first isolations from natural sources (unpublished).

(ii) Coccidiascus *Chatton*

The only described species of this genus, *Coccidiascus legeri*, was observed in 1913 as vegetative cells and asci in the intestinal tract of *Drosophila funebris* but was not cultured. It was re-observed by Lushbaugh *et al.* (1976) in intestinal epithelial cells of *Drosophila melanogaster* and the authors provided a much more complete description. However, they too were unable to cultivate this species on artificial media.

(iii) Cyniclomyces *van der Walt et Scott*

Cyniclomyces guttulatus is the only species of the genus. It has been known for many years as *Saccharomycopsis guttulata*. The reader is referred to van der Walt & Scott (1971) for the arguments and justification of this change in nomenclature. This species occurs in the intestinal tract of domestic rabbits and is worldwide in distribution. Active budding takes place in the acidic environment of the stomach. No further vegetative growth occurs in the intestinal tract due to the high pH of the environment. Sporulation does not occur at body temperature but is thought to occur in faecal pellets at 18–20°C, the optimum temperature for sporulation. The yeast is introduced into weanling rabbits by coprophagy (Shifrine & Phaff 1958). The very demanding nutritional requirements of this yeast have been discussed under Section 2.*B.*

(iv) Debaryomyces *Lodder et Kreger-van Rij nom. cons.*

The best known species of the genus is *Deb. hansenii*, a highly salt-tolerant yeast that can be readily isolated from ocean water or from various salt-preserved food products (cheeses, bacon, pickled meats, dry salami, etc.). It tolerates salt concentrations up to 20% (w/v) although growth is slow at that concentration. Another salt-tolerant species is *Deb. tamarii* isolated from tamari-soya. Other species of *Debaryomyces* occur in soil. *Debaryomyces polymorpha*, with synonyms *Deb. phaffii* and *Deb. cantarellii* (Price & Phaff 1979), *Deb. pseudopolymorpha* and *Deb. castellii* have all been isolated exclusively from soils. The isolation of *Deb. vanriji* from both tree exudates and soils suggests that some soil yeasts may be washed into the soil by rain or be transmitted by insects feeding on tree exudates. For further information see Kreger-van Rij (1970*a*).

(v) Dekkera *van der Walt*

Dekkera is the ascosporogenous form of *Brettanomyces* (van der Walt 1970*a*) and the two genera are covered together because of their similar and

specific habitats. Species of these genera are important because of their occurrence as spoilage organisms in bottled wines and soft drinks (Kolfschoten & Yarrow 1970) causing turbidity and off-taste. Earlier in the century *Brettanomyces* species were used in the brewing industry to make certain types of ale (e.g. lambic beer in Belgium) but this practice is less common today. Little is known still on the natural origin of these yeasts and how they arrive in the bottled products. The best known species involved with wine spoilage is *D. intermedia.*

(vi) Hanseniaspora *Zikes*

Hanseniaspora is the ascosporogenous state of *Kloeckera* and the two genera are discussed together because of their similar habitat. They have in common the apiculate cell shape, caused by repeated bipolar budding. Physiologically the six species of *Hanseniaspora* and their imperfect counterparts in *Kloeckera* are rather similar but DNA base composition studies and DNA/DNA reassociation experiments revealed that the six species are distinct (Meyer *et al.* 1978). Their habitat is in decomposing fruit tissue or soil and the latter probably serves as an overwintering reservoir; it is quite easy to isolate *Hanseniaspora* or *Kloeckera* species from orchard or vineyard soil in the off-season. Undoubtedly, insects such as dried fruit beetles (*Carpophilus hemipterus*), drosophilas and bees act as vectors in the distribution of the yeasts. There is evidence that the various species of *Hanseniaspora* are to some extent geographically separated. In the Sacramento valley of California, for example, *H'spora uvarum* and *H'spora guilliermondii* are most common.

(vii) Hansenula *H. et P. Sydow*

This large genus contains a number of species with very specific habitats (Wickerham 1970*b*). All species of the genus have in common that they can assimilate nitrate and the ascospores are hat-, helmet- or saturn-shaped. Space does not permit a complete coverage of all of the 30 to 35 species but some examples will illustrate the diversity of habitats. A significant number of species is associated with various species of bark beetles (mainly belonging to the genera *Ips* and *Dendroctonus*) that attack *Pinus* and *Picea* species. The yeasts are introduced by adult beetles into the cambium and phloem. After ovipositing, the larvae that hatch distribute the actively budding yeasts in the galleries where they serve as food for the larval stage. Because of the girdling caused by the larvae, trees usually die rapidly and after pupation emerging adults leave the killed tree and attack other trees. In the insect frass of the galleries yeasts are present in very high concentration and can be readily isolated from this source. Common yeasts in various

Pinus species include *H. capsulata* and *H. holstii*. Less common species are *H. canadensis, H. bimundalis,* and *H. silvicola*. Fuson *et al.* (1979) provided evidence by DNA analysis that the two geographically separated varieties of *H. bimundalis* probably represent separate species. *Hansenula bimundalis* var. *bimundalis* is associated with *Pinus sylvestris* in Spain and *H. bimundalis* var. *americana* with *Pinus* species in western USA. Phenotypically the two varieties are nearly identical.

Other species of *Hansenula* have been isolated exclusively from soils (*H. californica, H. saturnus, H. dimennae*). The first two of these species appear to be world-wide in distribution and the latter came from soils in India and New Zealand. *Hansenula mrakii* has been isolated from both soils and swamp and creek water.

A remarkable species, *H. petersonii*, has been isolated repeatedly from human cadavers used for teaching and preserved with embalming fluid and stored in tanks with 3–4% phenol. Heavy growth of this species was reported in various internal organs (Wickerham 1970*b*).

Some species, such as *H. polymorpha* and a recently described species. *H. alni* (Phaff *et al.* 1979), are associated with exudates or insects in broad-leafed trees. Still others have been designated as free-living and have been isolated from a great many sources, including contaminants in industrial or food fermentations. The latter include *H. anomala, H. fabianii* and *H. subpelliculosa*. The latter is very tolerant to high sugar concentrations and has been isolated frequently from sugar-preserved foodstuffs.

(viii) Kluyveromyces *van der Walt emend v. d. Walt*

Several members of this genus were formerly included in *Saccharomyces*, but on the basis of the rapidly dehisced spores (kidney-shaped or spherical and smooth) were placed in a separate genus (van der Walt 1970*b*). This genus contains a number of lactose-fermenting species, *K. marxianus*, (syn. *K. fragilis*) and *K. lactis*. Both species have been isolated frequently from a variety of dairy products. A number of species of the genus have been isolated exclusively from soils (*K. polysporus, K. africanus, K. lodderi, K. phaffii*) and others from wild species of *Drosophila* (*K. drosophilarum, K. dobzhanskii*) that consumed these yeasts as food. *Kluyveromyces polysporus* and *K. africanus*, unusual in forming multi-spored asci, have been found only in South African soils thus far.

(ix) Lipomyces *Lodder et Kreger-van Rij*

Species of this genus have been isolated exclusively from soils and are world-wide in distribution. *Lipomyces* species can utilize a large number of

carbon compounds. This ability enables them to take advantage of the organic breakdown products of plant origin formed by other soil micro-organisms. Soil appears to be their true habitat. Five species are currently recognized. Species described since 1970 when Slooff reviewed the genus are *L. tetrasporus* (Nieuwdorp *et al.* 1974) and *L. anomalus* (Babjeva & Gorin 1975). The former appears to be world-wide in distribution but the latter has been isolated only from podzolic soil in northern USSR. In contrast to the other four species it has a low optimum temperature for growth (16–18°C).

(x) Metschnikowia *Kamienski*

The six species of this genus are characterized by needle-shaped asco-spores and have highly specialized aquatic or terrestrial habitats. Since Miller & van Uden (1970) reviewed the genus one new species, *M. lunata*, has been described (Golubev 1977). *Metschnikowia pulcherrima*, *M. reukaufii* and *M. lunata* are associated with nectar of flowers and appear to be distributed principally by bees. *Metschnikowia reukaufii* is also found frequently on overripe berry fruits from which it can be readily isolated. *Metschnikowia pulcherrima*, probably the most common of the three species, is often introduced to overripe grapes and orchard fruits when bees or wasps feed on them. The dried fruit beetle (*Carpophilus hemipterus*) is also involved in the distribution of *M. pulcherrima*. The aquatic species *M. zobellii* and *M. krissii* were cultivated for the first time by van Uden and co-workers in 1961 from Pacific Ocean water off the coast of California. Metschnikoff observed *M. bicuspidata* as early as 1884 as a parasite in the body cavity of the fresh water crustacean *Daphnia magna*, but he was unable to culture it. A similar yeast was seen by Kamienski in 1899 as a parasite in the brine shrimp *Artemia salina*. It is not clear why these investigators were unable to isolate these yeasts, because subsequent studies have not revealed any unusual growth requirements. *Metschnikowia bicuspidata*, because of its pathogenicity, can cause great losses to the brine shrimp industry where these crustaceans are grown in salt ponds for fish food in aquariums (Lachance *et al.* 1976).

(xi) Nadsonia *Sydow*

Until 1973 when Golubev described *N. commutata* only two species were known, *N. fulvescens* and *N. elongata*, isolated by Nadson and Konokotina from slime fluxes of oak and birch in Russia in the period 1911–1913 (Phaff 1970*a*). *Nadsonia fulvescens* has not been isolated again but *N. elongata* appears to be common in exudates of broad-leafed trees in certain

geographic regions. It has been isolated from such sources in the eastern USA but never in the Pacific Northwest in spite of extensive searches (Phaff *et al.* 1972). In contrast, these investigators found *N. elongata* common in exudates of many species of trees in Japan. This yeast is easily recognized because of the brown colonies formed on isolation plates due to the brown-coloured ascospores which are usually abundantly formed. We assume that the yeast is transmitted to the tree fluxes by a particular insect that utilizes the fluxes for breeding and that such insects are not present in the western USA. *Nadsonia commutata* has been isolated only once from field soil on East Folkland Island, USSR (Golubev 1973).

(xii) Nematospora *Peglion*

The single species of this genus, *Nem. coryli*, is characterized by forming eight spindle-shaped spores with whip-like appendages in large asci. This yeast is parasitic on a number of tropical or subtropical crop plants: cotton (the internal boll disease), lima beans and other legumes (yeast spot), coffee berries (coffee bean disease), tomatoes, pecans, hazelnuts and citrus fruit. The infection is introduced by hemipterous insects (bugs of the genus *Dysdercus* in particular). Spores or cells of *Nem. coryli* are carried on the mandibles or in stylet pouches of these insects and are introduced into the plant tissue during feeding. The cotton boll disease is of considerable economic importance because the growth of the yeast in immature, unopened bolls leads to a staining of the lint. A related yeast-like organism, *Ashbya gossypii*, is also associated with the cotton boll disease. For further details the reader is referred to Batra (1973).

(xiii) Pichia *Hansen*

This large genus (Kreger-van Rij 1970*b*) contains a divergent assemblage of species with varying degrees of habitat specificity. Several species are associated with bark beetles which infest coniferous trees. *Pichia pinus* has been isolated from the pine beetles *Dendroctonus* and *Ips* as well as from *Pinus* species infested by the beetles. *Pichia scolyti* is recovered from *Scolytus* beetles and from frass of these beetles in fir and Douglas fir. Many species of the genus *Pichia* are associated with *Drosophila* species, which utilize decomposing plant material. Decaying fruits such as tomatoes and citrus harbour *P. kluyveri* and *P. fermentans* (Vacek *et al.* 1979). The yeasts provide essential components in the diets of both the larval and adult stages of drosophilas. The exudates or slime-fluxes of broad-leafed deciduous trees also support yeasts utilized by the *Drosophila* fauna of temperate regions. Examples of tree-dependent *Pichia* species are *P. pastoris, P. salictaria, P. trehalophila, P. fluxuum, P. quercuum* and the recently

described *P. scutulata* (Phaff *et al.* 1976). Of these species *P. pastoris* has been found in Europe and is very common in fluxes of native *Populus* and *Salix* species as well as exotic species of *Ulmus* (Phaff *et al.* 1964) in the Pacific Northwest of North America but was virtually absent in Japan (Phaff *et al.* 1972). This could be due to a specific association of *P. pastoris* with fluxes of the above trees which are not found in Japanese forests. The two varieties of *P. scutulata* are geographically separated (Hawaii versus the Pacific coast area) and according to Kurtzman they represent species *in statu nascendi* (pers. comm.).

We have found a highly specific yeast community in the decaying stems of cacti, which are utilized by desert-adapted drosophilas for feeding and breeding. *Pichia cactophila* (Starmer *et al.* 1978*a*) is widespread among cactus necroses. This species has been found in the columnar cacti throughout Mexico as well as in *Opuntia* cacti recently introduced to Australia, Spain and Hawaii. Several cactus yeasts demonstrate variable ranges of their host plant specialization. *Pichia heedii* (Phaff *et al.* 1978*a*) is recovered mainly from senita and saguaro cactus in the North American Sonoran Desert. These cacti are members of the same cactus subtribe (Pachycereinae) and thus illustrate the extreme host specialization of the yeast. *Pichia amethionina* (Starmer *et al.* 1978*b*) is found in the same region as *P. heedii* but is isolated from two related subtribes of columnar cacti. *Pichia amethionina* var. *amethionina* is recovered from cacti of the subtribe Stenocereinae while the variety *pachycereana* is restricted to cacti of the subtribe Pachycereinae. The habitat specificity of the varieties is determined in part by the secondary metabolites formed by the plants. Cacti of the Stenocereinae contain copious amounts of triterpene glycosides, while cacti of the other subtribe do not contain these compounds (Gibson & Horak 1978). The triterpene glycosides inhibit growth of *P. amethionina* var. *pachycereana* but do not affect var. *amethionina*. The genetic basis of this difference resides in alleles at a single temperature sensitive locus (Starmer *et al.* 1979*b*). *Pichia opuntiae* (Starmer *et al.* 1979*a*) has two varieties each of which inhabits cacti of different tribes. The var. *opuntiae* is associated with the decaying cladodes of *Opuntia inermis* in Australia, while var. *thermotolerans* is recovered from rotting stems of giant columnar cacti of the Sonoran Desert.

Some *Pichia* species occur in the spoilage of food products. For example *P. membranaefaciens* is found in spoiled wine as a pellicle; *P. farinosa* is isolated as a contaminant of miso and saké in Japan and *P. etchellsii* and *P. ohmeri* have been isolated from fermenting cucumber brine.

(xiv) Saccharomyces *Meyen emend. Reess*

The species of this large genus show considerable diversity in habitat, but

most are associated with food products or fermentation processes. The classification of van der Walt (1970c) divides the species into four groups. Those belonging to Group I are diploid and include those used in the fermentation industry and some are contaminants or food spoilage organisms. In our experience *Sacch. cerevisiae* strains, typical of those used in the fermentation industry, are rarely if ever found in natural substrates. *Saccharomyces cerevisiae* var. *tetrasporus* of which many strains were isolated from the crops of wild species of *Drosophila* in the California mountains (Phaff *et al.* 1956) may represent the naturally occurring form from which the industrial strains are derived. Species of Group II are haploid (*Zygosaccharomyces*) and they have generally adapted to substrates with higher sugar contents. Examples are *Sacch. rouxii*, *Sacch. bisporus* and *Sacch. bailii*, commonly found in food products preserved with high concentrations of sugar. Species of Group III, sometimes referred to as the *Torulaspora* group, are also haploid but conjugation is usually restricted to a mother cell with a bud. Many of the species in this group that were separated by sugar assimilation reactions have been reduced to synonymy with *Sacch. delbrueckii* based on nuclear DNA reassociation experiments (Price *et al.* 1978). *Saccharomyces delbrueckii* and its synonyms are rather common in fermenting fruits, juices, wines and similar products. Most strains have a somewhat lower sugar tolerance than those of Group II. Van der Walt (1970c) placed some *Saccharomyces* species of uncertain affinity in Group IV. Two of the species have specific habitats. *Saccharomyces telluris* is the ascogenous state of *T. pintolopesii* and of several closely related psychrophobic species found in the intestinal tract of rodents and of other warm-blooded animals (Mendonça-Hagler & Phaff 1975). *Saccharomyces montanus* was the most common species isolated from the crops of wild species of *Drosophila* in the California mountains (Phaff *et al.* 1956) but information on the natural habitat of that species in the environment where the flies occur is lacking.

(xv) Saccharomycodes *Hansen*

This monotypic genus is represented by *S'codes ludwigii*, a fermentative yeasts reproducing by bipolar budding and forming four spheroidal spores each with an indistinct ledge (Phaff 1970b). This species is relatively rare and its natural habitat appears to be restricted to slime fluxes of *Quercus*. From there it may have been transmitted by insects to grape vineyards and then to early stages of wine fermentation from which sources it is isolated periodically.

(xvi) Schizosaccharomyces *Lindner*

Three of the four species of the fission yeasts, *Schiz. pombe*, *Schiz. japonicus* and *Schiz. malidevorans* are most commonly isolated from fermenting fruit juices or wine. Their ability to ferment L-malic acid may give them a competitive advantage over other yeasts in this habitat, unable to degrade malate. *Schizosaccharomyces octosporus* is highly osmotolerant and all strains studied by Slooff (1970) came from substrates high in sugar content. In our laboratory we have isolated this species repeatedly from sun-dried prunes in storage.

(xvii) Schwanniomyces *Kloecker*

The four species of this genus were reduced to synonymy with *Schw. occidentalis* by Price *et al.* (1978) on the basis of DNA base composition and homology. All strains studied were isolated from various soils. They assimilate a rather large variety of carbon compounds (including inulin and starch), a trait that is commonly observed in soil-inhabiting yeasts. Isolates of this genus are relatively rare.

B. Basidiomycetous yeasts

(i) Leucosporidium *Fell, Statzel, Hunter et Phaff*

Most of the species of this genus are psychrophilic and have maximum growth temperatures below 20°C. The reader is referred to Fell & Phaff (1970) for details on the life-cycle of these species. Because their separation is based on the assimilation and fermentation of sugars it is possible that future studies will lead to a reduction in number of species. Most of the strains have been isolated in the Antarctic from sea water, snow samples and soil. *Leucosporidium scottii* was first recognized as *C. scottii* (Lodder & Kreger-van Rij 1952). It had been isolated from chilled beef (where it was quite common) but also from substrates at higher temperatures. It was later recognized that the maximum growth temperature of this species is strain dependent and varies from 15–30°C.

(ii) Rhodosporidium *Banno*

Rhodosporidium species represent the sexual states of strains of *Rhodotorula*. The type species of the genus, *Rhodosp. toruloides* (Banno

1967), was based on mixing a number of *Rh. glutinis* strains from a variety of non-specific terrestrial sources. Later Fell *et al.* (1973) reported a number of additional species that were prevalent in marine habitats. Among these, *Rhodosp. diobovatum* was common in coastal waters and mangrove detritus in Florida. This species was also common in exudates of broad-leafed trees in Japan (Phaff *et al.* 1972) and it is possible therefore that the marine isolates may have been of terrestrial origin carried by run-off or by river water. The remaining species of the genus (*malvinellum, sphaero-carpum, bisporidiis, capitatum, dacryoidum, infirmo-miniatum*) are all of marine origin and were isolated in various locations of the southern Pacific and Indian Oceans, including the Antarctic Ocean, and at water depths ranging from the surface to over 4000 m deep.

(iii) Filobasidiella *Kwon-Chung*

Species of this genus form sessile basidiospores on basidia rather than teliospores as in the previous two genera. *Filobasidiella neoformans* (Kwon-Chung 1976*a*) represents the sexual state of *Cr. neoformans* serotypes A and D, whereas *F. bacillispora* (Kwon-Chung 1976*b*) is the sexual state of *Cr. neoformans* B and C serotypes. Both species are important human pathogens. Aside from numerous isolates from human patients afflicted with cryptococcosis the ecological niche of these species is not completely understood. Although isolations from pigeon droppings and nest areas in many parts of the world are common, the yeasts have also been isolated from soil, air and decaying plant materials. This suggests that pigeons may acquire the yeast from food contaminated with this organism (Kwon-Chung 1976*b*). The two species of *Filobasidiella* appear to be geographically separated (Wilson *et al.* 1968). *Filobasidiella bacillispora* (serotypes B and C) have been isolated only in California, whereas *Filobasidiella neoformans* (serotypes A and D) is distributed world-wide, except in California.

C. Ballistosporogenous yeasts

(i) Sporobolomyces *Kluyver et van Niel*

Phaff (1970*c*) accepted nine species in his review of the genus. Nearly all strains of these species have been isolated from plant leaves, stems or bark which appear to be their actual habitat. Because of their ability to discharge ballistoconidia with force and the possibility of their distribution by air currents it is not surprising that a number of *Sporobolomyces* strains have been isolated from the atmosphere or as air contaminants in the laboratory.

(ii) Sporidiobolus *Nyland*

The two species of this genus are similarly associated with plant leaves (Phaff 1970*d*). The number of strains that have been isolated is small. There is limited evidence for a sexual life cycle in the species of this genus.

(iii) Bullera *Derx*

The genus now contains five species. In addition to the three species reviewed by Phaff (1970*e*), *B. dendrophila* was described by van der Walt & Scott (1970) and *B. piricola* by Stadelmann (1975). *Bullera tsugae* and *B. dendrophila* are associated with insect frass from the bark of diseased trees and the other three species of the genus came from healthy leaves or leaves attacked by other fungi or insects.

D. *Asporogenous yeasts*

(i) Candida *Berkhout*

Most species of this large genus do not appear to have highly specific habitats; the origins of the strains studied by van Uden & Buckley (1970) are generally rather variable. This may be due, at least in part, to the incorrect placement of strains in certain species. This is not uncommon among imperfect yeasts because the lack of a sexual cycle eliminates an important classificatory parameter. Only a few examples will therefore be given of species that can be reproducibly isolated from certain habitats.

Candida albicans is well known as an opportunistic pathogen in humans, but it is also found as part of the normal intestinal flora in healthy individuals. It is only rarely isolated from natural sources, including sewage, suggesting that its survival outside the human body is poor. Other species that are periodically isolated from clinical material, e.g. *C. parapsilosis* and *C. tropicalis*, have been isolated repeatedly from non-clinical sources as well. *Candida boidinii*, a methanol-utilizing yeast, is associated with certain fleshy plant material that is rotting or is crushed. Originally isolated from tanning fluid prepared from tree bark, later isolates have been obtained from crushed olive pulp and more recently from *Opuntia* cactus rots in Hawaii and Australia (unpublished). The natural habitat of *C. tenuis* is associated with bark beetles that attack pine or spruce trees. Besides the strains listed by van Uden & Buckley (1970) three strains have been isolated in our laboratory from frass in *Picea* (Phaff *et al.* 1972 and unpublished) located in Oregon and in the Yukon Territory. We have obtained preliminary evidence (unpublished) that strains of *C. tenuis* isolated from sources other

than coniferous trees may represent different species. We have confirmed this for strains reported as *C. tenuis* from cacti in the Sonoran Desert (Heed *et al.* 1976).

Phaff & Miller (1961) have reported a highly specific association between the fig wasp, *Blastophaga psenes*, *C. guilliermondii* var. *carpophila* and the bacterium *Serratia plymuthica*. The wasp, which is involved in the pollination of figs, has its habitat, together with the two micro-organisms, in the flower ovaries (galls) of the caprifig and throughout the three successive annual caprifig crops.

Although *C. ingens* was originally isolated from winery and brewery material in South Africa, its ability to utilize short-chain fatty acids for growth has led to specialized habitats where these compounds occur in significant concentration. Starmer *et al.* (1976) isolated this species from 60% of organ pipe cactus rots which contain unusual amounts of short-chain fatty acids. The inhibitory action of these acids to some yeasts would give *C. ingens* a selective advantage in such substrates. Natural enrichment of *C. ingens* on effluent that had been fermented in a tank at a commercial piggery (a substrate high in C_2–C_6 fatty acids) confirms the specialization of this yeast on substrates of this composition (Henry *et al.* 1976).

(ii) Torulopsis *Berlese*

Of the 36 species reviewed by van Uden & Vidal-Leiria (1970) and the species subsequently described, only a few with well-established habitats will be cited here. *Torulopsis pintolopesii* (syn. *T. bovina,* Mendonça-Hagler & Phaff 1975) is restricted to the intestinal tract of various warm-blooded animals, including mice, rats, horses, pigs and cows. Strains of this species grow well at the low pH of the stomach (pH 1·0–1·5) but not at temperatures below 21–24°C. They assimilate only glucose and occasionally ethanol and lactic acid weakly. *Torulopsis glabrata* is also associated with warm-blooded animals and humans and may be an opportunistic pathogen, especially of the urinary tract. Diabetic individuals appear to be predisposed to such infections. This species appears to survive better in nature than *C. albicans* and is occasionally isolated from non-clinical sources. Its minimum temperature for growth is 6°C.

Strains of *T. stellata* are often associated with over-ripe grapes or grape juice and occasionally with other fermenting fruits such as figs and citrus, and with *Drosophila* species feeding on these substrates (Miller & Phaff 1962; Vacek *et al.* 1979). *Torulopsis versatilis* has been repeatedly isolated from fermenting cucumber brines where it has a competitive advantage because of its fermentative ability and tolerance to high NaCl concentrations (10–13% w/v).

Torulopsis sonorensis (Miller *et al.* 1976) occurs in necrotic tissue of all

cactus species studied thus far and is worldwide in distribution. We have isolated hundreds of strains of this methanol-utilizing yeast from cacti in the USA, Mexico, Hawaii, Spain and Australia. It presumably has a competitive advantage among the cactophilic species of yeast by being able to utilize for growth the methanol liberated from cactus pectin during the soft rot process.

4. Outlook for Future Ecological Studies

We visualize the most fruitful further advances in ecological knowledge of yeasts through yeast habitat studies in greater depth and, if possible, by quantitative methods. Many newly described species are represented by single or only a few strains. It is not uncommon that such strains represent adventitious yeasts that are not typical of the habitat sampled. Isolating organisms repeatedly from specific substrates or habitats is therefore highly desirable (Phaff *et al.* 1964), preferably on a broad geographic basis (Phaff *et al.* 1972; Starmer *et al.* 1978*a*).

Additional advantages of isolating numerous strains of new species are not only that it allows a more comprehensive characterization of new species but also that it greatly enhances the possibility of finding compatible mating types for haploid asporogenous strains. In one of our as yet unpublished new species of cactophilic yeast the proportion of the two mating types was very unequal and we would not have discovered the sexual state without having available a large number of strains.

Adequate characterization of strains from new habitats is most essential. When only limited phenotypic properties are available for identification (especially among asporogenous yeasts), taxonomic keys often lead to species which the organism in question may resemble only superficially. This had led to the situation that a number of yeast species do not represent a single taxon but are composed of two or more unrelated species. Determination of the nuclear DNA base composition (mol % G + C) and in some instances of DNA/DNA complementarity are highly valuable tools for accurate characterization of species (Price *et al.* 1978; Starmer *et al.* 1978*a*; Fuson *et al.* 1979). Once a species is adequately defined by such techniques one may hope to discover additional phenotypic properties, such as the utilization of less conventional sources of carbon and nitrogen, to separate such species from others they resemble. In some cases habitat constitutes the only criterion for the separation of species.

Finally, fruitful information is likely to come from studies of habitat chemistry as an explanation of the highly specific habitats of many yeasts (Starmer *et al.* 1979*b*). Thus far this area of yeast ecology has received only little attention.

5. References

BABJEVA, I. P. & GORIN, S. E. 1975 *Lipomyces anomalus* sp. nov. *Antonie van Leeuwenhoek* **41**, 185–191.

BANNO, I. 1967 Studies on the sexuality of *Rhodotorula*. *Journal of General and Applied Microbiology* **13**, 167–196.

BATRA, L. R. 1973 Nematosporaceae (Hemiascomycetidae): Taxonomy, pathogenicity, distribution, and vector relations. pp. 1–71. *United States Department of Agriculture, Technical Bulletin No. 1469* Washington, DC.

BUECHER, E. J. & PHAFF, H. J. 1970 Growth of *Saccharomycopsis* Schiönning under continuous gassing. *Journal of Bacteriology* **104**, 133–137.

DO CARMO-SOUSA, L. 1969 Distribution of yeasts in nature. In *The Yeasts* Vol. 1, ed. Rose, A. H. & Harrison, J. S. London & New York: Academic Press.

FELL, J. W. 1976 Yeasts in ocean regions. In *Recent Advances in Aquatic Mycology* ed. Gareth-Jones, E. B. New York: Halsted Press.

FELL, J. W. & PHAFF, H. J. 1970 *Leucosporidium* Fell, Statzel, Hunter et Phaff. In *The Yeasts — A Taxonomic Study* 2nd edn, ed. Lodder, J. Amsterdam: North-Holland.

FELL, J. W., HUNTER, I. L. & TALLMAN, A. S. 1973 Marine basidiomycetous yeasts (*Rhodosporidium* spp. n.) with tetrapolar and multiple allelic bipolar mating systems. *Canadian Journal of Microbiology* **19**, 643–657.

FUSON, G. B., PRICE, C. W. & PHAFF, H. J. 1979 Deoxyribonucleic acid sequence relatedness among some members of the yeast genus *Hansenula*. *International Journal of Systematic Bacteriology* **29**, 64–69.

GIBSON, A. C. & HORAK, K. E. 1978 Systematic anatomy and phylogeny of Mexican columnar cacti. *Annals Missouri Botanical Garden* **65**, 999–1057.

GOLUBEV, W. I. 1973 *Nadsonia commutata* nov. sp. *Mikrobiologia* **42**, 1058–1061.

GOLUBEV, W. I. 1977 *Metschnikowia lunata* sp. nov. *Antonie van Leeuwenhoek* **43**, 317–322.

HEED, W. B., STARMER, W. T., MIRANDA, M., MILLER, M. W. & PHAFF, H. J. 1976 An analysis of the yeast flora associated with cactiphilic *Drosophila* and their host plants in the Sonoran desert and its relation to temperate and tropical associations. *Ecology* **57**, 151–160.

HENRY, D. P., THOMPSON, R. H., SIZEMORE, D. J. & O'LEARY, J. A. 1976 Study of *Candida ingens* grown on the supernatant derived from the anaerobic fermentation of monogastric animal wastes. *Applied and Environmental Microbiology* **31**, 813–818.

KOLFSCHOTEN, G. A. & YARROW, D. 1970 *Brettanomyces naardenensis*, a new yeast from soft drinks. *Antonie van Leeuwenhoek* **36**, 458–460.

KREGER-VAN RIJ, N. J. W. 1970*a Debaryomyces* Lodder et Kreger-van Rij nom. conserv. In *The Yeasts — A Taxonomic Study* 2nd edn, ed. Lodder, J. Amsterdam: North-Holland.

KREGER-VAN RIJ, N. J. W. 1970*b Pichia* Hansen. In *The Yeasts — A Taxonomic Study*, 2nd edn, ed. Lodder, J. Amsterdam: North-Holland.

KWON-CHUNG, K. J. 1976*a Morphogenesis of *Filobasidiella neoformans*, the sexual state of *Cryptococcus neoformans*. *Mycologia* **68**, 821–833.

KWON-CHUNG, K. J. 1976*b A new species of *Filobasidiella*, the sexual state of *Cryptococcus neoformans* B and C serotypes. *Mycologia* **68**, 942–946.

LACHANCE, M. A., MIRANDA, M., MILLER, M. W. & PHAFF, H. J. 1976 Dehiscence and active spore release in pathogenic strains of the yeast *Metschnikowia bicuspidata* var. *australis*: possible predatory implication. *Canadian Journal of Microbiology* **22**, 1756–1761.

LODDER, J. ed. 1970 *The Yeasts — A Taxonomic Study* 2nd edn. Amsterdam: North-Holland.

LODDER, J. & KREGER-VAN RIJ, N. J. W. 1952 In *The Yeasts — A Taxonomic Study* 2nd edn, ed. Lodder, J. Amsterdam: North-Holland.

LUSHBAUGH, W. B., ROWTON, E. D. & MCGHEE, R. B. 1976 Redescription of *Coccidiascus legeri* Chatton, 1913 (Nematosporaceae: Hemiascomycetidae) an intracellular, parasitic, yeast-like fungus from the intestinal epithelium of *Drosophila melanogaster*. *Journal of Invertebrate Pathology* **28**, 93–107.

MENDONÇA-HAGLER, L. C. & PHAFF, H. J. 1975 Deoxyribonucleic acid base composition and

deoxyribonucleic acid/deoxyribonucleic acid hybrid formation in psychrophobic and related yeasts. *International Journal of Systematic Bacteriology* **25**, 222–229.

MEYER, S. A., ANDERSON, K., BROWN, R. E., SMITH, M. TH., YARROW, D., MITCHELL, G. & AHEARN, D. G. 1975 Physiological and DNA characterization of *Candida maltosa*, a hydrocarbon-utilizing yeast. *Archives of Microbiology* **104**, 225–231.

MEYER, S. A., SMITH, M. T. & SIMIONE, F. P., JR. 1978 Systematics of *Hanseniaspora* Zikes and *Kloeckera* Janke. *Antonie van Leeuwenhoek* **44**, 79–96.

MILLER, M. W. & PHAFF, H. J. 1962 Successive microbial populations in Calimyrna figs. *Applied Microbiology* **10**, 394–400.

MILLER, M. W. & VAN UDEN N. 1970 *Metschnikowia* Kamienski. In *The Yeasts — A Taxonomic Study* 2nd edn, ed. Lodder, J. Amsterdam: North-Holland.

MILLER, M. W., PHAFF, H. J., MIRANDA, M., HEED, W. B. & STARMER, W. T. 1976 *Torulopsis sonorensis*, a new species of the genus *Torulopsis*. *International Journal of Systematic Bacteriology* **26**, 88–91.

NIEUWDORP, P. J., BOS, P. & SLOOFF, W. C. 1974 Classification of *Lipomyces*. *Antonie van Leeuwenhoek* **40**, 241–254.

NORTHAM, B. E. & NORRIS, F. W. 1951 Growth requirements of *Schizosaccharomyces octosporus*, a yeast exacting towards adenine. *Journal of General Microbiology* **5**, 502–507.

PHAFF, H. J. 1970a *Nadsonia* Sydow. In *The Yeasts — A Taxonomic Study* 2nd edn, ed. Lodder, J. Amsterdam: North-Holland.

PHAFF, H. J. 1970b *Saccharomycodes* Hansen. In *The Yeasts — A Taxonomic Study* 2nd edn, ed. Lodder, J. Amsterdam: North-Holland.

PHAFF, H. J. 1970c *Sporobolomyces* Kluyver et van Niel. In *The Yeasts — A Taxonomic Study* 2nd edn, ed. Lodder, J. Amsterdam: North-Holland.

PHAFF, H. J. 1970d *Sporidiobolus* Nyland. In *The Yeasts — A Taxonomic Study* 2nd edn, ed. Lodder, J. Amsterdam: North-Holland.

PHAFF, H. J. 1970e *Bullera* Derx. In *The Yeasts — A Taxonomic Study* 2nd edn, ed. Lodder, J. Amsterdam: North-Holland.

PHAFF, H. J., MILLER, M. W. & SHIFRINE, M. 1956 The taxonomy of yeasts isolated from *Drosophila* in the Yosemite region of California. *Antonie van Leeuwenhoek* **22**, 145–161.

PHAFF, H. J. & MILLER, M. W. 1961 A specific microflora associated with the fig wasp, *Blastophaga psenes* Linnaeus. *Journal of Insect Pathology* **3**, 233–243.

PHAFF, H. J., YONEYAMA, M. & DO CARMO-SOUSA, L. 1964 A one-year, quantitative study of the yeast flora in a single slime flux of *Ulmus carpinifolia* Gled. *Rivista di patologia vegetale* Ser. III **4**, 485–497.

PHAFF, H. J., MILLER, M. W., YONEYAMA, M. & SONEDA, M. 1972 A comparative study of the yeast florae associated with trees on the Japanese islands and on the west coast of North America. In *Fermentation Technology* ed. Terui, G. pp. 759–774. Osaka, Japan: Society of Fermentation Technology.

PHAFF, H. J., MILLER, M. W., MIRANDA, M., HEED, W. B. & STARMER, W. T. 1974 *Cryptococcus cereanus*, a new species of the genus *Cryptococcus*. *International Journal of Systematic Bacteriology* **24**, 486–490.

PHAFF, H. J., MILLER, M. W. & MIRANDA, M. 1976 *Pichia scutulata*, a new species from tree exudates. *International Journal of Systematic Bacteriology* **26**, 326–331.

PHAFF, H. J., STARMER, W. T., MIRANDA, M. & MILLER, M. W. 1978a *Pichia heedii*, a new species of yeast indigenous to necrotic cacti in the North American Sonoran Desert. *International Journal of Systematic Bacteriology* **28**, 326–331.

PHAFF, H. J., MILLER, M. W. & MRAK, E. M. 1978b *The Life of Yeasts* 2nd edn. Cambridge, Massachusetts: Harvard University Press.

PHAFF, H. J., MILLER, M. W. & MIRANDA, M. 1979 *Hansenula alni*, a new heterothallic species of yeast from exudates of alder trees. *International Journal of Systematic Bacteriology* **29**, 60–63.

PRICE, C. W., FUSON, G. B. & PHAFF, H. J. 1978 Genome comparison in yeast systematics: delimitation of species within the genera *Schwanniomyces, Saccharomyces, Debaryomyces* and *Pichia*. *Microbiological Reviews* **42**, 161–193.

PRICE, C. W. & PHAFF, H. J. 1979 *Debaryomyces polymorphus* and *D. pseudopolymorphus*, new taxonomic combinations. *Mycologia* **71**, 444–445.

RODRIGUES DE MIRANDA, L. 1978 A new genus: *Sporopachydermia. Antonie van Leeuwenhoek* **44**, 439–450.

SHIFRINE, M. & PHAFF, H. J. 1958 On the isolation, ecology and taxonomy of *Saccharomycopsis guttulata. Antonie van Leeuwenhoek* **24**, 193–209.

SLOOFF, W. C. 1970 *Schizosaccharomyces* Lindner. In *The Yeasts — A Taxonomic Study* 2nd edn, ed. Lodder, J. Amsterdam: North-Holland.

STADELMANN, F. 1975 A new species of the genus *Bullera* Derx. *Antonie van Leeuwenhoek* **41**, 575–582.

STARMER, W. T., HEED, W. B., MIRANDA, M., MILLER, M. W. & PHAFF, H. J. 1976 The ecology of yeast flora associated with cactiphilic Drosophila and their host plants in the Sonoran Desert. *Microbial Ecology* **3**, 11–30.

STARMER, W. T., PHAFF, H. J., MIRANDA, M. & MILLER, M. W. 1978*a Pichia cactophila*, a new species of yeast found in decaying tissue of cacti. *International Journal of Systematic Bacteriology* **28**, 318–325.

STARMER, W. T., PHAFF, H. J., MIRANDA, M. & MILLER, M. W. 1978*b Pichia amethionina*, a new heterothallic yeast associated with the decaying stems of cereoid cacti. *International Journal of Systematic Bacteriology* **28**, 433–441.

STARMER, W. T., PHAFF, H. J., MIRANDA, M., MILLER, M. W. & BARKER, J. S. F. 1979*a Pichia opuntiae*, a new heterothallic species of yeast found in decaying cladodes of *Opuntia inermis* and in necrotic tissue of cereoid cacti. *International Journal of Systematic Bacteriology* **29**, 159–167.

STARMER, W. T., KIRCHER, H. W. & PHAFF, H. J. 1979*b* Evolution and speciation of host plant specific yeasts. *Evolution* **34**, 137–146.

TRAVASSOS, L. R. & CURY, A. 1971 Thermophilic enteric yeasts. *Annual Review of Microbiology* **25**, 49–74.

TRAVASSOS, L. R. & MENDONÇA, L. C. 1972 Vitamin requirements and induced nutritional imbalances as criteria in speciating psychrophobic yeasts. *Antonie van Leeuwenhoek* **38**, 379–389.

VACEK, D. C., STARMER, W. T. & HEED, W. B. 1979 Relevance of the ecology of *Citrus* yeasts to the diet of *Drosophila. Microbial Ecology* **5**, 43–49.

VAN DER WALT, J. P. 1970*a Brettanomyces* Kufferath et van Laer. In *The Yeasts — A Taxonomic Study* 2nd edn, ed. Lodder, J. Amsterdam: North-Holland.

VAN DER WALT, J. P. 1970*b Kluyveromyces* van der Walt emend. v. d. Walt. In *The Yeasts — A Taxonomic Study* 2nd edn, ed. Lodder, J. Amsterdam: North-Holland.

VAN DER WALT, J. P. 1970*c Saccharomyces* Meyen emend. Reess. In *The Yeasts — A Taxonomic Study* 2nd edn, ed. Lodder, J. Amsterdam: North-Holland.

VAN DER WALT, J. P. & SCOTT, D. B. 1970 *Bullera dendrophila* sp. n. *Antonie van Leeuwenhoek* **36**, 383–387.

VAN DER WALT, J. P. & SCOTT, D. B. 1971 The yeast genus *Saccharomycopsis* Schiönning. *Mycopathologia et mycologia applicata* **43**, 279–288.

VAN UDEN, N. & BUCKLEY, H. 1970 *Candida* Berkhout. In *The Yeasts — A Taxonomic Study* 2nd edn, ed. Lodder, J. Amsterdam: North-Holland.

VAN UDEN, N. & VIDAL-LEIRIA, M. 1970 *Torulopsis* Berlese. In *The Yeasts — A Taxonomic Study* 2nd edn, ed. Lodder, J. Amsterdam: North-Holland.

WICKERHAM, L. J. 1970*a Citeromyces* Santa Maria. In *The Yeasts — A Taxonomic Study* 2nd edn, ed. Lodder, J. Amsterdam: North-Holland.

WICKERHAM, L. J. 1970*b Hansenula* H. et P. Sydow. In *The Yeasts — A Taxonomic Study* 2nd edn, ed. Lodder, J. Amsterdam: North-Holland.

WILSON, D. E., BENNETT, J. E. & BAILEY, J. W. 1968 Serologic grouping of *Cryptococcus neoformans. Proceedings of the Society for Experimental Biology* **127**, 820–823.

Recent Research on Industrially Important Strains of *Saccharomyces cerevisiae*

A. H. ROSE

*Zymology Laboratory, School of Biological Sciences,
Bath University, Claverton Down, Bath, Avon, UK*

Contents

1. Introduction

THERE ARE three main groups of industrial processes which employ strains of *Saccharomyces cerevisiae*. The first includes the age-old production of alcoholic beverages using a wide range of sugar-containing vegetable and fruit extracts. With these are associated processes which use strains of *Sacch. cerevisiae* to produce industrial ethanol, processes which some years ago were partially superseded by growth of the synthetic alcohol industry, but which have recently come back into their own (Miller 1975; Rose 1976; Jackman 1977; Goldemberg 1978; Bu'Lock 1979). Production of alcoholic beverages and industrial alcohol by *Sacch. cerevisiae* exploits the ability of strains of this yeast to grow anaerobically when it ferments sugars to ethanol and carbon dioxide.

The second group of processes which employ *Sacch. cerevisiae* are more recent, and are those that involve production and use of baker's yeast. Until the end of the nineteenth century, surplus brewer's yeast was used to leaven doughs in breadmaking; indeed, it is so used to this day in some parts of the world. But largely following the development of the Vienna process in the latter half of the nineteenth century (Frey 1930), manufacture of baker's

yeast rapidly became an independent process, one which world-wide now yields a product of consistently high quality and stability (Burrows 1970, 1979). This development depended on the fact that strains of *Sacch. cerevisiae* are facultative anaerobes, a property not shared with many other yeasts. Efficient large-scale production of baker's yeast is possible because, under aerobic conditions, *Sacch. cerevisiae* oxidizes sugars completely to carbon dioxide and water and, in so doing, utilizes sugars efficiently. Pasteur first observed this effect (Pasteur 1879), and it has come to be recognized as part of the classical Pasteur effect. When used to leaven doughs, however, *Sacch. cerevisiae* carries out a fermentative metabolism, and obtains energy from alcoholic fermentation of sugars. Carbon dioxide becomes entrapped in the dough thereby creating the honeycomb texture typical of a well baked and leavened loaf. Ethanol, produced in equimolar yields with carbon dioxide, invariably disappears up the baking-oven chimney. What a waste in these days of energy shortage!

The third use to which strains of *Sacch. cerevisiae* are put industrially is in production of bulk biomass to be used as animal (or possibly human) food, otherwise known as single-cell protein (SCP). There are several review texts and articles dealing with SCP (Mateles & Tannenbaum 1968; Tannenbaum & Wang 1975; Laskin 1977), and from these it is evident that *Sacch. cerevisiae* is just one of a wide range of microbes that have been considered, or are being used, as SCP. It has, however, the advantage that it is nutritionally an accepted organism, a decided benefit over most of the other microbes contemplated as sources of SCP.

The three groups of processes differ considerably in the extent to which they have been studied from a microbiological standpoint, and in the likelihood of their being the subject of extensive research in the foreseeable future. Processes using *Sacch. cerevisiae* as a source of SCP are low-cost ventures and hardly justify in economic terms an extensive microbiological research input.

However, during the first half of the present century, there was a great deal of research into the production and properties of baker's yeast, so much so that it is, particularly in Great Britain and the USA, a very efficient industry indeed. This fact is not always appreciated. Producers of baker's yeast are, at their most efficient, able to guarantee virtually identical fermentative activity in bulk biomass in every batch retailed. Micro-biologists who regularly cultivate batches of cells on the laboratory scale, and who fight a constant battle with the 'variability' of different batches of cells, will surely appreciate the level of sophistication achieved by producers of baker's yeast. Microbiologically, the principal aim of the baker's yeast manufacturer is to ensure that strains used in commercial production do not deteriorate and change in properties. If they do, then the

producer returns to the master culture. The past year or so has seen the beginning of exciting new opportunities for improving strains used in baker's yeast production. These have come about following the development of techniques for introducing new genetic information into *Sacch. cerevisiae* by protoplast fusion (van Solingen & van der Plaat 1977) and transformation (Hinnen *et al.* 1978; Jansen *et al.* 1978). There is a growing market, world-wide, for active-dried yeast rather than the traditional pressed yeast, because of its considerably longer shelf life. The newly developed techniques for producing improved strains of *Sacch. cerevisiae* could well be of greatest value in the production of active-dried yeast.

Processes for manufacturing alcoholic beverages are not only more numerous and varied than those for manufacturing baker's yeast, but they are also more frequently subjected to technological innovation which often creates fresh demands on the strains of yeast employed. A cautionary word about nomenclature is required at this juncture. The name *Saccharomyces cerevisiae*, first applied by Meyen (1838) to distinguish beer yeasts from those isolated from other alcoholic beverages, has traditionally been reserved for top-fermenting or ale yeasts, the appellation *Saccharomyces carlsbergensis* having later been proposed by Stelling-Dekker (1931) for lager or bottom-fermenting yeasts (Mendlik 1937). The two species were separated formally on the basis of the ability of the latter to ferment the dissaccharide melibiose. The taxonomic treatise edited by Lodder (1970) came out with the recommendation that *Sacch. carlsbergensis* be included in the species *Sacch. uvarum* (van der Walt 1970) a species originally established to accommodate a yeast isolated from currant juice (Beijerinck 1894). The recommendation has not been rapturously received by industrial users of yeasts; it is not adhered to in this contribution to the Symposium.

The remainder of this article includes a review of the properties required of strains of yeast used to produce alcoholic beverages or industrial ethanol, and this is followed by a more detailed discussion of two properties, namely flocculation and the ability to tolerate moderately high concentrations of ethanol. The article concludes with a comment on ways of obtaining improved strains of industrial yeasts.

2. Properties Required of Strains of *Saccharomyces cerevisiae* used in Production of Alcoholic Beverages and Industrial Ethanol

A. Ability to effect an efficient alcoholic fermentation of sugars

The need for this property is self evident, but it is one which is rarely given detailed consideration. From a microbiological standpoint, it is of interest

in that it is a property not possessed to anything like the same degree by other groups of micro-organisms which might conceivably, taking into consideration other requirements (such as lack of pathogenicity), be used to manufacture beverages. Certain bacteria produce ethanol as a fermentation end-product, but together with other fermentation products which would render any beverages produced by these bacteria totally unacceptable.

Despite the importance of the fermenting ability of strains of *Saccharomyces*, very few quantitative studies have been made on this property. The fermenting ability of strains of *Saccharomyces* cannot be assessed simply by comparing their behaviour during production of alcoholic beverages and liquids, because the conditions under which they are acting — conditions such as sugar concentration, availability of other nutrients, and temperature — differ even when considering just one type of beverage. It is, however, possible to compare the fermenting ability of strains when these are examined under strictly defined laboratory conditions. So far, this has only been attempted to any great extent for strains used in the brewing of beer. Thorne (1954) introduced the term 'fermentation velocity' which he defined as the number of millilitres of carbon dioxide at s.t.p. evolved by one gram wet weight of yeast in one hour, using a standard nitrogen-free substrate solution containing 10% (w/v) glucose under strictly defined conditions. The rate of gas production is measured in a fermentometer apparatus, shaken at constant speed and temperature. Thorne (1954) reported a wide variation in the fermentation velocities of a collection of brewing yeasts, ranging from as little as 5 to as much as 35 ml carbon dioxide/g of yeast/h. A later publication from Thorne (1961) related the fermentation velocity of a strain to the nitrogen content of the cells, and he defined the ratio of these two values as the 'fermentation efficiency' of a strain. Surveying the fermentation efficiencies of a collection of yeasts, which included some 250 strains of bottom-fermenting yeasts and 150 strains of top-fermenting yeasts, Thorne (1961) found that the former group had significantly higher fermentation efficiencies ($14 \cdot 1 \pm 0 \cdot 1$) compared with top yeasts ($12 \cdot 2 \pm 0 \cdot 2$). Distillery yeasts in the collection had higher fermentation efficiencies ($13 \cdot 8 \pm 0 \cdot 4$) compared with ale yeasts ($12 \cdot 6 \pm 0 \cdot 3$).

Despite these attempts by Thorne to bring some uniformity into measurements of the fermenting ability of strains of *Saccharomyces*, he has attracted few fellow travellers. This is probably because most industrial users of yeasts wish only to compare a range of strains for a particular fermentation process, and so use conditions for measuring fermenting ability that resemble those that the strains will encounter under production conditions. An example of this more pragmatic approach to comparing the fermenting abilities of yeasts came in a report of a study on hybridizing

brewer's yeasts to produce improved strains for beer production. In this study, Anderson & Martin (1975) used fermentation ability as one of their more important characteristics. They quoted values for parent and hybrid strains in the range 5–10 μl carbon dioxide produced/min/mg dry weight at pH 5·0 and 25°C. As the efficiency of strains of *Sacch. cerevisiae* assumes greater importance in the production of alcoholic beverages and industrial ethanol, more attention will undoubtedly be paid to assessing quantitatively the fermenting ability of strains.

B. Production of desirable flavour compounds

The taste, flavour and bouquet of an alcoholic beverage are attributable in only small part to the ethanol content of the drink. The main attraction of an alcoholic beverage to the palate and to the sense of smell, and importantly too the principal components which distinguish one beverage from another, are many compounds present in the drink at fairly low concentrations, some of which are produced by action of the yeast on components of the sugar-containing raw material. The Herculean task of identifying these compounds in various alcoholic beverages and potable spirits is as yet far from complete.

Over the ages, the brewer, wine maker and distiller have many times discovered quite empirically that some yeasts produce a beverage or spirit that is particularly acceptable and attractive. There can be no doubt that at least some of the desirable qualities conferred by a yeast are attributable to the ability of the strain or mixture of strains to excrete in sufficient quantity certain organic compounds that affect the taste and smell of the beverage or spirit, and as such this is a microbiologically desirable property in a strain of *Sacch. cerevisiae* used for making beverages or spirits. However not until more data are available on the chemical nature of these flavour compounds, and on factors which influence their secretion by yeasts, will it be possible to select strains of *Sacch. cerevisiae* on this basis for use in the production of specific beverages and spirits (Brown & Clapperton 1978).

C. Strain stability and viability

The fact that alcoholic beverages, and fermented liquids that are subsequently distilled, have been produced by yeasts for centuries shows that the strains of *Saccharomyces* which brought about these fermentations must have been reasonably stable micro-organisms. The knowledge of yeast physiology acquired over the past century, and the advances in fermentation technology that have occurred in more recent times, have permitted a detailed assessment to be made of strain stability among species of industrially important yeasts, and at the same time have paradoxically made

the problem less important because industrial microbiologists have been persuaded to pay more attention to it. Nevertheless, even today some alcoholic beverages (such as certain ciders and table wines) are produced under conditions where the source of the principal strain of yeast is poorly understood, with the strain emerging as a result of a complex sequence of ecological changes in the flora of the fermenting juice. Such strains of *Saccharomyces* must have a very pronounced survival capacity. In recent years, industrial microbiologists have paid more attention to the conditions under which yeast, which is to be used in the production of alcoholic beverages, is cultured or harvested, and maintained, so that problems of strain stability are less frequently encountered although they are now far better appreciated.

Instability in a yeast strain can range from a tendency to lose viability after being harvested from a fermentation and stored, to the loss of ability to produce in low concentration one or more compounds that contribute to the flavour of a beverage or spirit. Manifestation of any type of instability must reflect changes in the yeast genome. Unfortunately, almost nothing is known about the genetic basis of the more subtle properties of commercial yeast strains, especially the ability to excrete desirable flavour compounds.

However, a few data are available on the retention of viability in strains of *Sacch. cerevisiae*. Death of a micro-organism usually follows quite rapidly when the organism is deprived of exogenous nutrients. This is a situation which yeast used in industry frequently encounters, as, for example, when stored after being harvested from a brewery fermentation or when pressed into a cake to be used in the production of potable spirits.

Death of a cell can follow the development of any one of a great variety of metabolic lesions. However, at the centre of any discussion on retention of microbial viability is the ability of a cell to produce ATP. Even when growing and dividing, but much more so when not growing, there is a continual turnover of microbial cell constituents, that is a breakdown followed by an ATP-requiring resynthesis of cell components. The energy or ATP required to keep a cell alive, without leading to reproduction of the cell, is known as the maintenance energy, and values for this requirement have been calculated for a number of micro-organisms. When a micro-organism is deprived of nutrients, the ability of that cell to remain viable will depend on the speed at which lytic enzymes break down cell constituents, and the extent to which the organism can call upon intracellular energy reserves and mobilize these reserve compounds to produce ATP. Not much is known about the activities of lytic enzymes in strains of *Sacch. cerevisiae*, or of the manner in which these activities are manifested when a yeast cell is deprived of nutrients. It is known, however, that several lytic activities in actively growing cells of *Sacch. cerevisiae* are located in

intracellular structures often referred to as vacuoles but more correctly as vesicles (Cartledge *et al.* 1977). The first report of this intracellular location of lytic enzyme activities described the presence of nucleases and proteases in yeast vesicles and, when reporting the discovery, Matile & Wiemken (1967) made the prescient comment that yeast vesicles may function as the yeast lysosome. There have since been several other reports of lytic enzymes present in yeast vesicles, and the list now includes glucanases (Matile *et al.* 1971; Cortat *et al.* 1972) as well as mannanases and lipases (Cartledge *et al.* 1977). Moreover, it has been shown that, during the cell cycle in *Sacch. cerevisiae*, large vesicles fragment to give a population of smaller vesicles, which become located in the region of envelope growth and are thought to be concerned in the growth process (Sentandreu & Northcote 1969). Research at the Eidgenössische Technische Hochschule in Zurich (Cortat *et al.* 1972; Matile *et al.* 1971) and in my own laboratory (Cartledge *et al.* 1977) has led to the separation of fractions rich in small and large vesicles, and further characterization of their enzymic and lipid composition. Presumably, when a yeast cell encounters starvation conditions, the enzymes contained in large and small vesicles escape from these structures and catalyse breakdown of cell constituents which, when the cell is unable to synthesize ATP, cannot be resynthesized.

Strains of *Saccharomyces* are known to synthesize two carbohydrate energy reserves, namely the disaccharide trehalose and the polysaccharide glycogen, and to store appreciable amounts of both compounds (Manners 1971). Chester (1963, 1964) reported that the amounts of trehalose and glycogen present in several brewing strains of *Sacch. cerevisiae* differed quite widely, although in general the glycogen content accounted for about 10% of the cell dry weight while the amounts of trehalose present were much smaller (Stewart *et al.* 1950). When cells of *Saccharomyces* spp. encounter starvation conditions, ATP can be produced by catabolism of trehalose, which involves action of the enzyme trehalase to yield glucose, and of glycogen, breakdown of which is catalysed by a phosphorylase and an amylo-1,6-glucosidase.

Although strains of *Sacch. cerevisiae* accumulate appreciable quantities of triacylglycerols and to a lesser extent, sterol esters, both of which are stored mainly in intracellular vesicles (Hossack *et al.* 1977), curiously little research has been carried out on the utilization of these lipids as energy sources by starved populations. However, Wilson & McLeod (1976) recently reported that fatty acids are oxidized by starving populations of *Sacch. cerevisiae*. These workers suggested that the loss in viability in starved populations may indeed be caused by impaired membrane function resulting from lipid catabolism.

There is evidence, too, that RNA can be used as a source of ATP by

yeasts when they are subjected to starvation conditions (Day *et al.* 1975). Strains of *Sacch. cerevisiae* (Higuchi & Uemura 1959) degrade RNA when incubated under nutrient-free conditions, and excrete ultraviolet-absorbing compounds. Ribonucleic acid is most likely used as a source of ATP when the reserves of trehalose and glycogen have been exhausted.

Research by Atkinson and his colleagues has begun to clarify the extent to which micro-organisms can be deprived of ATP before they lose the ability to multiply. This work has introduced the concept of the 'energy charge' or 'adenylate charge' of a population of micro-organisms (Atkinson 1968), which is calculated from values for the intracellular concentrations of AMP, ADP and ATP. Values for the energy charge fall in the range $1\cdot0-0\cdot0$. Actively growing micro-organisms possess an energy charge in the range $0\cdot7-0\cdot8$. When the energy charge of a suspension of *Escherichia coli* falls below about $0\cdot5$, the bacteria lose their viability (Chapman *et al.* 1971). The situation is, however, different with *Sacch. cerevisiae*, the energy charge of which can fall to as low as $0\cdot15$ without any impairment to cell viability (Ball & Atkinson 1975). This discovery would seem to explain, to some extent at least, how strains of *Sacch. cerevisiae* have been successfully maintained over the centuries for production of alcoholic beverages.

3. Flocculation

Some strains of *Saccharomyces* have a tendency, in the later stages of a batch fermentation, to agglomerate and to form flocs which settle to the bottom of the fermentation vessel. This phenomenon, which is known as 'flocculation', has for long been recognized as a very important one in production of alcoholic beverages. In order to achieve maximum conversion of sugar into ethanol and carbon dioxide, it is essential for the yeast to remain suspended in the fermenting liquid and not to flocculate. At the same time, the ability of the yeast to flocculate when the fermentation has been completed, or reached the desired stage, is an advantage to the manufacturer of alcoholic beverages because it greatly assists in the removal of the yeast from the beverage. In other words, the yeast should ideally flocculate only at the desired stage in a fermentation, a stage that varies with different beverages and with different types of any one beverage. Much also depends on the type of vessel in which the fermentation is being carried out. This is well illustrated in the brewing of beer, where the flocculation characteristics of a yeast used in open fermenting vessels may be quite unsuitable for producing beer in the recently introduced cylindro-conical vessels. The required flocculation characteristics of a yeast vary, therefore, not only with the type of beverage being produced, but also

with the design of the vessel in which it is being produced. With fermentations from which the yeast is not removed, such as those used in the manufacture of distilled spirits, flocculation of the yeast is on the whole an undesirable property, since the aim is to keep the yeast in suspension until the fermentation liquid is distilled.

Although the importance of yeast flocculation in the manufacture of alcoholic beverages has been appreciated for almost a century, yeast physiologists have until quite recently been ignorant of the biochemical basis of the phenomenon (Rainbow 1970; Stewart 1975, 1979). For some years, the consensus of opinion has been that anionic groups already present, or produced in stationary-phase cells, in the surface layers of yeasts form salt bridges with counterpart groups in adjacent cells through calcium bridges. Evidence for this consensus can be found in the review articles on flocculation already referred to. Research in recent years has concentrated mainly on three aspects of the flocculation phenomenon: namely the nature of the cell-surface anionic groups involved; the manner whereby flocculation is normally manifested only by cells in stationary-phase cultures, or in cultures about to enter the stationary phase of growth; and finally the biochemistry and morphology of floc structure.

The two types of anionic group which could be involved in bridge formation are carboxyl groups in acidic wall proteins and phosphodiester linkages in cell-surface phosphomannan. The weight of evidence is in favour of the former groups being involved in floc formation. Earlier, somewhat speculative evidence for a role for carboxyl groups came from Harris (1959) and Mill (1964), and this has more recently been supplemented by three additional sets of data. Jayatissa & Rose (1976) used the hydrogen fluoride technique to excise specifically phosphodiester groups from the wall phosphomannan, a technique which leaves the wall protein intact. If phosphodiester groups were the anionic groups involved in floc formation, their removal would be expected to decrease the extent of floc formation. Excision of phosphodiester groups, on the contrary, led to an increase in the degree of floc formation (Jayatissa & Rose 1976) for reasons which will be given later. The second line of evidence for the involvement of carboxyl groups in floc formation came from studies on the electrophoretic mobility of yeast cells in relation to their ability to form flocs. Using measurements of the electrophoretic mobility of cells to assess the density of amino, phosphodiester and carboxyl groups in the surface layers (down to 4·5 nm) of the wall, Beavan *et al.* (1979) reported a correlation between the acquisition of floc-forming ability and an increase in the density of surface carboxyl groups. These changes were frequently accompanied by a decrease in the density of phosphodiester groups, a finding which disputes the claim by Lyons & Hough (1970*a, b*, 1971). Finally, Stewart *et al.* (1975) also

reported evidence for a correlation between an increase in the density of surface carboxyl groups and the acquisition of flocculating ability, their evidence coming from an increased ability of cells to bind the cationic dye Alcian blue.

Based on the reported differences in the density of surface carboxyl and phosphodiester groups on *Sacch. cerevisiae*, Beavan *et al.* (1979) proposed a model to explain the acquisition of floc-forming ability. During exponential growth, cells have both types of anionic group in the surface layers, and these repel cells and keep them dispersed. Excision of phosphodiester groups, as shown by Jayatissa & Rose (1976), causes increased flocculation as a result of the elimination of a repulsive activity. As cultures enter the stationary phase of growth, the density of carboxyl groups in the surface layers increases in those strains that are able to form flocs and, when this density reaches a critical value, calcium-mediated floc formation takes place. The relation between the increase in the density of surface carboxyl groups and the sudden acquisition of floc-forming ability, as described by Beavan *et al.* (1979), is shown in Fig. 1.

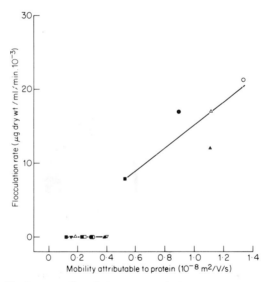

Fig. 1. Relationship between flocculation rate and electrophoretic mobility attributable to protein carboxyl groups for four strains of *Sacch. cerevisiae*. ●, ○ Strain 11; ▲, △ strain 13; ▼, ▽ strain 3; ■, □ strain 30. From Beavan *et al.* (1979). *Canadian Journal of Microbiology* 25.

Beavan *et al.* (1979) also set out to discover the biochemical basis of the changes in surface-charge density as cells of *Sacch. cerevisiae* acquire the ability to form flocs. Taking as their premise the view that alterations in the composition of wall-surface components in cells from stationary-phase

cultures probably arise as the result of the action of lytic enzymes, activities of which are known to increase when exponential growth in batch culture is arrested, they assayed cells of flocculent and non-flocculent strains for several enzyme activities including those that catalyse hydrolysis of proteins and mannan. In the strains studied, acquisition of flocculating ability was accompanied by a slight increase in amidase activity, and greater increases compared with non-flocculating populations in activities of leucine aminopeptidase, α-mannosidase and proteinase C. Activities of proteinases A and B showed no correlation with acquisition of flocculating ability. The increased activity of α-mannosidase could explain the decreased density of phosphodiester groups in populations of cells that develop flocculating ability. However, the principal problem is the need to explain how the greatly increased density of carboxyl groups is brought about, and it was not solved in this study, for it is not immediately apparent how increased activities of two peptidases (proteinase C is synonymous with carboxypeptidase Y) could cause such an increase. One alternative possibility is that the increased density of carboxyl groups arises as the result of the synthesis of new acidic wall proteins in flocculent cells that enter the stationary phase of growth.

Very little indeed is as yet known about the third of the currently researched aspects of yeast flocculation, namely the biochemistry and morphology of floc structure. One of the main puzzles is the nature of the calcium-mediated bridge. It is probably too simple to imagine that calcium ions form a direct bridge between pairs of carboxyl groups on adjacent cells. It is more likely that the bridge is longer, possibly involving connecting peptides. In this connection it is interesting to note that Holmberg (1978) has recently reported that a temperature-sensitive flocculating mutant, but not the non-flocculent wild-type strain, of *Sacch. cerevisiae* synthesizes a cell-surface peptide that could be involved in bridge formation. It is also widely held that floc formation in *Sacch. cerevisiae* involves not only calcium-mediated bridges between wall carboxyl groups on adjacent cells, but also secondary interactions involving other wall components. The most important of these secondary interactions are probably hydrogen bonds between mannose residues in the cell-surface mannans of adjacent cells. Evidence for this type of secondary interaction comes from the ability of sugars, and in particular mannose, to prevent floc formation (Eddy 1955; Jayatissa & Rose 1976). Nevertheless, much remains to be learned about the fine structures of flocs of *Sacch. cerevisiae*. It is likely too that there is not just one type of floc formation but several, possibly ones that can be distinguished in the scanning electron microscope.

A criticism often levelled at much of the recent research on flocculation in *Sacch. cerevisiae* is that it frequently involves comparisons of flocculent and genetically unrelated non-flocculent strains. This is a valid criticism, and the recent report by Holmberg & Kielland-Brandt (1978) of the isolation of a temperature-sensitive flocculating mutant of a non-flocculent wild-type strain of *Sacch. cerevisiae* strongly suggests that these wild-type and mutant organisms should be exploited in future work on flocculation.

4. Ethanol Tolerance

Ethanol is a well known cell poison and narcotic and, from the early days of research on yeast fermentations, the question as to how cells of *Sacch. cerevisiae* remain viable and metabolically active in media containing self-produced ethanol that can reach a concentration as high as 20% (v/v) has interested zymologists. Over half a century ago, Guilliermond & Tanner (1920) found that differences in the fermenting ability of species of *Saccharomyces*, as well as among different strains of any one species, were attributable to their capacity to tolerate the ethanol they produced. Since then, interest in the ethanol tolerance of strains of *Sacch. cerevisiae* has been fitful. Principally interested have been zymologists researching on saké yeasts (strains of *Sacch. cerevisiae*) which are among the most tolerant known. There has, however, been an upsurge of research activity during the past five years as interest has returned to industrial-scale production of ethanol, and as brewers in many countries have investigated the possibilities of high-gravity fermentations. High-gravity brewing of beers, followed by post-fermentation dilution to produce the marketed product, offers considerable economic advantages. Unfortunately, however, many production strains of *Sacch. cerevisiae* are unable to tolerate the concentration of ethanol (8–9% v/v) produced in such high-gravity beers, and cannot therefore be used to pitch subsequent fermentations.

A. Definition

Many workers in this field have not defined precisely what they mean by the term 'ethanol tolerance' of a strain of *Sacch. cerevisiae*. Suto and his colleagues (1951) measured ethanol tolerance in a strain of saké yeast (*Koyokai* No. 6) firstly by the extent to which the presence of ethanol (15% v/v final concentration) inhibited fermentative activity (Q_{CO_2} value), and secondly by the extent to which fermentative activity was depressed by previously steeping the yeast in 15% (v/v) ethanol at 30°C for 14 h. A similar method was published by Nojiro & Ouchi (1962) who expressed tolerance as the ratio of fermentative activities measured in the presence and absence of 18% (v/v) ethanol. These workers found that all of the yeast strains they examined had an ethanol tolerance in the range 20–30% (v/v); moreover, they failed to detect differences in tolerance between the saké and other strains of yeast included in the survey. This approach has the advantage that it relates tolerance to fermentative activity, thereby creating a relevance to performance of strains in batch semi-anaerobic fermentations. However, the methods do not quote ethanol tolerance directly in terms of a concentration of the alcohol.

Other workers have chosen to express ethanol tolerance of strains by the extent to which the alcohol inhibits growth. Inoue *et al.* (1962), for example, determined the concentration of ethanol which completely suppressed growth. The same approach was made by D. Susan Thomas, working in my laboratory. She grew strains of *Sacch. cerevisiae* (100 ml culture in 250 ml conical shake flasks; 250 rev/min) using a defined glucose-salts medium containing ethanol in 1% (v/v) step increases in concentration. Tolerance was defined as that concentration of ethanol which completely prevented growth after 72 hours incubation at 30°C. Some of her data are shown in Table 1. One of the strains listed in the Table (NCYC 366) is a British brewing yeast, and it clearly has a much lower ethanol tolerance than any of the other strains listed, all of which originated in distilleries or saké breweries.

TABLE 1

Ethanol tolerance of some strains of Sacch. cerevisiae

Strain	Lowest concentration of ethanol that just prevents growth during 72 h incubation in defined medium	
	v/v	mol/l
NCYC 366	7·0	1·2
NCYC 431	13·0	2·2
NCYC 478	12·0	2·0
NCYC 479	13·0	2·2
CBS 1198	12·0	2·0

Unpublished results of D. Susan Thomas and A. H. Rose.

Clearly, much more thought should be given to the problem of defining precisely what is meant by ethanol tolerance, if only because this will lead to a better defined basis from which all future comparisons can be made. In particular, it is essential that, in any agreed procedure, the method used for growing the strains of *Sacch. cerevisiae* must be described very precisely because growth under conditions that allow cells to produce ethanol could materially affect the ability of the cells to tolerate the catabolite.

B. Physiological basis

One of the first workers to sustain an interest in the physiology of ethanol tolerance in *Sacch. cerevisiae* was W. D. Gray. Having established that ethanol tolerance is not confined to members of any one genus or species of yeast (Gray 1941), he went on to report (Gray 1945) that induced tolerance of glucose in *Sacch. cerevisiae*, brought about by sequential transfer into media containing higher concentrations of the hexose, was accompanied by a decrease in ethanol tolerance. Examination of a range of yeast species subsequently revealed that those which tolerate high concentrations of ethanol store less lipid and carbohydrate compared with less tolerant strains (Gray 1948).

These earlier studies, while pioneering research on the physiological basis of ethanol tolerance in yeast, can be criticized on the grounds that they involved a comparison of different strains of *Sacch. cerevisiae*. Work in my own laboratory has sought to examine the physiological basis of ethanol tolerance in one strain of *Sacch. cerevisiae*, namely NCYC 366. The study formed part of an on-going programme aimed at establishing the relationship between lipid composition and function in the plasma membrane of this yeast. As a prelude to the study, methods were devised for effecting specific alterations to the lipid composition of the plasma membrane (Rose 1977). One of these techniques exploited the anaerobically-induced requirement in *Sacch. cerevisiae* for a sterol (Andreasen & Stier 1953) and an unsaturated fatty acid (Andreasen & Stier 1954), both of which requirements are fairly non-specific (Proudlock *et al.* 1968; Light *et al.* 1962). By growing cells of *Sacch. cerevisiae* NCYC 366 under strictly anaerobic conditions (Alterthum & Rose 1973) in media containing different sterols and unsaturated fatty acids, it is possible to obtain populations with plasma membranes enriched to about 70% with the exogenously supplied sterol (Hossack & Rose 1976) and to about 54% with the fatty-acyl residues supplied in the medium (Thomas *et al.* 1978). Arguing that the plasma membrane is the first sensitive organelle to make contact with ethanol when cells are suspended in a solution of the alcohol, and remembering that ethanol is an amphipathic compound, a study was made of the manner in which the sterol and fatty-acyl composition of the yeast plasma membrane affects the ability of populations to retain viability when suspended in buffer (pH 4·5) containing 1 mol/l ethanol. Populations retained viability to a greater extent when plasma membranes were enriched in linoleyl rather than oleyl residues, irrespective of the nature of the sterol in the membrane (Thomas *et al.* 1978). Moreover, populations of cells with membranes enriched in ergosterol or stigmasterol and linoleyl residues were more resistant to ethanol than populations of cells with membranes enriched in campesterol or cholesterol and linoleyl residues. When membranes were enriched in palmitoleyl residues, the protective effect was even greater than with membranes enriched in oleyl residues.

Further experiments (Thomas & Rose 1979) revealed that growth of cultures of *Sacch. cerevisiae* NCYC 366, under conditions which cause their membranes to become enriched in oleyl residues, is inhibited by ethanol (1·0 or 1·25 mol/l) to a greater extent than growth of the yeast in medium containing linoleic acid. Arguing that one of the major rate-limiting processes in microbial growth is transport of nutrients into cells, Thomas & Rose (1979) went on to show that accumulation of glucose, arginine and lysine is inhibited to a smaller extent in cells with linoleyl residue-enriched membranes compared with cells that have membranes enriched in oleyl residues. This was not true, however, for accumulation of phosphate.

TABLE 2

Effect of ethanol on rates of solute accumulation by Sacch. cerevisiae *NCYC 366 with plasma membranes enriched in ergosterol and either oleyl or linoleyl residues*

Fatty-acyl enrichment	Nature of solute	Concentration of solute (mM)	Rate of accumulation in absence of ethanol (nmol/min/mg dry wt)	Rate of accumulation in presence of ethanol (nmol/min/mg dry wt)	Percentage decrease in rate
Oleyl	Arginine	0·4	1·9	0·9	53
	Lysine	0·4	5·1	2·3	55
	Glucose	10·0	11·5	3·0	74
Linoleyl	Arginine	0·4	2·0	1·5	25
	Lysine	0·4	4·5	3·0	33
	Glucose	10·0	11·5	6·5	43

Values quoted are the means of three independent measurements for amino-acid accumulation and four independent measurements for glucose accumulation. Ethanol (0·5 mol/l final concentration) was added after 12 min incubation when measuring the rate of amino-acid accumulation, and after 8 min incubation when measuring the rate of glucose accumulation. From Thomas and Rose (1979).

While enrichment with the doubly unsaturated fatty-acyl residue clearly confers a measure of ethanol resistance on *Sacch. cerevisiae* NCYC 366, it is evident that the presence of these residues does not explain ethanol tolerance in all strains of *Sacch. cerevisiae*. A preliminary examination of some of the more tolerant strains listed in Table 1 (Thomas, unpublished observation) has shown that linoleyl residues are not present. Clearly, therefore, the genetically inherent resistance in these strains has a physiological basis different from that conferred experimentally on *Sacch. cerevisiae* NCYC 366.

5. Genetic Considerations

Although *Sacch. cerevisiae* has long been known to have a sexual cycle (Mortimer & Hawthorne 1969), apart from the breeding of baker's yeast (Fowell 1969), few attempts have been made to exploit this property in strain improvement programmes (Anderson & Martin 1975). However, the situation is, almost certainly, about to change.

There have, in the past two years, been reports of successful gene transfer in *Sacch. cerevisiae* using both protoplast fusion (van Solingen & van der Plaat 1977) and transformation (Hinnen *et al.* 1978; Jansen *et al.* 1978) as indicated in the Introduction to this review. These exciting developments offer the possibility of introducing into industrial strains of *Sacch. cerevisiae* genetic information that could confer additional desirable properties on these strains. In the future, this information could conceivably include the ability to produce wall lytic enzymes and to synthesize linoleyl residues, properties that might enhance the ability of strains to flocculate and to tolerate high concentrations of ethanol. At last, we see the prospect of the physiology and genetics of *Sacch. cerevisiae* marching hand in hand along the road to strain improvement in the yeast industries.

6. References

ALTERTHUM, F. & ROSE, A. H. 1973 Osmotic lysis of sphaeroplasts from *Saccharomyces cerevisiae* grown anaerobically in media containing different unsaturated fatty acids. *Journal of General Microbiology* **77**, 371–382.

ANDERSON, R. J. & MARTIN, P. A. 1975 The sporulation and mating of brewing yeasts. *Journal of the Institute of Brewing* **81**, 242–247.

ANDREASEN, A. A. & STIER, T. J. B. 1953 Anaerobic nutrition of *Saccharomyces cerevisiae*. I. Ergosterol requirement for growth in a defined medium. *Journal of Cellular and Comparative Physiology* **41**, 23–36.

ANDREASEN, A. A. & STIER, T. J. B. 1954 Anaerobic nutrition of *Saccharomyces cerevisiae*. II. Unsaturated fatty acid requirement for growth in a defined medium. *Journal of Cellular and Comparative Physiology* **43**, 271–281.

ATKINSON, D. E. 1968 The energy charge of the adenylate pool as a regulatory parameter. Interaction with feedback modifiers. *Biochemistry, New York* **7**, 4030–4034.

BALL, W. J. & ATKINSON, D. E. 1975 Adenylate energy charge in *Saccharomyces cerevisiae* during starvation. *Journal of Bacteriology* **121**, 975–982.

BEAVAN, M. J., BELK, D. M., STEWART, G. G. & ROSE, A. H. 1979 Changes in electrophorectic mobility and lytic enzyme activity associated with development of flocculating ability in *Saccharomyces cerevisiae*. *Canadian Journal of Microbiology* **25**, 888–895.

BEIJERINCK, M. W. 1894 *Schizosaccharomyces octosporus*, eine achtsporige Alkoholhefe. *Zentralbatt für Bakteriologie, Parasitenkunde, Infektionskrankheiten und Hygiene* **16**, 49–58.

BROWN, D. G. W. & CLAPPERTON, J. F. 1978 Discriminant analysis of sensory and instrumental data on beer. *Journal of the Institute of Brewing* **84**, 318–323.

BU'LOCK, J. D. 1979 Industrial alcohol. *Symposium of the Society for General Microbiology* **29**, 309–325.

BURROWS, S. 1970 Baker's yeast. In *The Yeasts* Vol. 3, ed. Rose, A. H. & Harrison, J. S. London: Academic Press.

BURROWS, S. 1979 Baker's yeast. In *Economic Microbiology* Vol. 4, ed. Rose, A. H. London: Academic Press.

CARTLEDGE, T. G., ROSE, A. H., BELK, D. M. & GOODALL, A. A. 1977 Isolation and properties of two classes of low-density vesicles from *Saccharomyces cerevisiae. Journal of Bacteriology* **132**, 426–433.

CHAPMAN, A. G., FALL, L. & ATKINSON, D. E. 1971 Adenylate energy charge in *Escherichia coli* during growth and starvation. *Journal of Bacteriology* **108**, 1072–1086.

CHESTER, V. E. 1963 The dissimilation of the carbohydrate reserves of a strain of *Saccharomyces cerevisiae. Biochemical Journal* **86**, 153–160.

CHESTER, V. E. 1964 Comparative studies in the dissimilation of reserve carbohydrate in four strains of *Saccharomyces cerevisiae. Biochemical Journal* **92**, 318–323.

CORTAT, A., MATILE, P. & WIEMKEN, A. 1972 Isolation of glucanase-containing vesicles from budding yeast. *Archiv für Mikrobiologie* **82**, 189–205.

DAY, A., ANDERSON, E. & MARTIN, P. A. 1975 Ethanol tolerance of brewing yeasts. *Proceedings of the European Brewing Convention, Nice* pp. 377–391.

EDDY, A. A. 1955 Flocculation characteristics of yeasts. II. Sugars as dispersing agents. *Journal of the Institute of Brewing* **61**, 313–318.

FOWELL, R. R. 1969 Sporulation and hybridization of yeasts. In *The Yeasts* Vol. 1, ed. Rose, A. H. & Harrison, J. S. London: Academic Press.

FREY, C. N. 1930 Production of baker's yeast. *Industrial and Engineering Chemistry* **22**, 1154–1162.

GOLDEMBERG, J. 1978 Brazil: energy options and current outlook. *Science, New York* **200**, 158–164.

GRAY, W. D. 1941 Studies on the alcohol tolerance of yeasts. *Journal of Bacteriology* **42**, 561–574.

GRAY, W. D. 1945 The sugar tolerance of four strains of distiller's yeast. *Journal of Bacteriology* **49**, 445–452.

GRAY, W. D. 1948 Further studies on the alcohol tolerance of yeast; its relationship to cell storage products. *Journal of Bacteriology* **55**, 53–59.

GUILLIERMOND, A. & TANNER, F. W. 1920 *The Yeasts.* New York: John Wiley.

HARRIS, J. O. 1959 Possible mechanism of yeast flocculation. *Journal of the Institute of Brewing* **65**, 5–6.

HIGUCHI, M. & UEMURA, T. 1959 Release of nucleotides from yeast cells. *Nature, London* **184**, 1381–1383.

HINNEN, A., HICKS, J. B. & FINK, G. R. 1978 Transformation of yeast. *Proceedings of the National Academy of Sciences of the United States of America* **75**, 1929–1933.

HOLMBERG, S. 1978 Isolation and characterization of a polypeptide absent from non-flocculation mutants of *Saccharomyces cerevisiae. Carlsberg Research Communications* **43**, 401–413.

HOLMBERG, S. & KIELLAND-BRANDT, M. C. 1978 A mutant of *Saccharomyces cerevisiae* temperature sensitive for flocculation. Influence of oxygen and respiratory deficiency on flocculence. *Carlsberg Research Communications* **43**, 37–47.

HOSSACK, J. A. & ROSE, A. H. 1976 Fragility of plasma membranes in *Saccharomyces cerevisiae* enriched with different sterols. *Journal of Bacteriology* **127**, 67–75.

HOSSACK, J. A., BELK, D. M. & ROSE, A. H. 1977 Environmentally-induced changes in the neutral lipids and intracellular vesicles of *Saccharomyces cerevisiae* and *Kluyveromyces fragilis. Archives of Microbiology* **114**, 137–142.

INOUE, T., TAKAOKA, Y. & HATA, S. 1962 Studies on *saké* yeast. V. On the conditions which effect the tolerance of yeast to alcohol. *Journal of Fermentation Technology, Japan* **40**, 511–521.

JACKMAN, E. A. 1977 Distillery effluent treatment in the Brazilian National Alcohol Programme. *The Chemical Engineer* 239–242.

JANSEN, G. P., BARNEY, M. C. & HELBERT, J. R. 1978 Genetic transformation of *Saccharomyces cerevisiae. Proceedings of the Annual Meeting of the American Society for Microbiology* A 37.

JAYATISSA, P. M. & ROSE, A. H. 1976 Role of wall phosphomannan in flocculation of *Saccharomyces cerevisiae. Journal of General Microbiology* **96**, 165–174.

LASKIN, A. I. 1977 Single cell protein. *Annual Reports on Fermentation Processes* **1**, 151–175.

LIGHT, R. J., LENNARZ, W. J. & BLOCH, K. 1962 The metabolism of hydroxystearic acids in yeast. *Journal of Biological Chemistry* **237**, 1793–1800.

LODDER, J. ed. 1970 *The Yeasts — A Taxonomic Study*, 2nd edn. Amsterdam: North-Holland.

LYONS, T. P. & HOUGH, J. S. 1970a Cation binding of yeast cell walls. *Biochemical Journal* **119**, 10P.

LYONS, T. P. & HOUGH, J. S. 1970b Flocculation of brewer's yeast. *Journal of the Institute of Brewing* **76**, 564–571.

LYONS, T. P. & HOUGH, J. S. 1971 Further evidence for the cross-bridging hypothesis for flocculation of brewer's yeast. *Journal of the Institute of Brewing* **77**, 300–305.

MANNERS, D. J. 1971 The structure and biosynthesis of storage carbohydrates in yeast. In *The Yeasts* Vol. 2, ed. Rose, A. H. & Harrison, J. S. London: Academic Press.

MATELES, R. I. & TANNENBAUM, S. eds 1968 *Single-Cell Protein*. Cambridge, Massachusetts: MIT Press.

MATILE, P. & WIEMKEN, A. 1967 The vacuole as the lysosome of the yeast cell. *Archiv für Mikrobiologie* **56**, 148–155.

MATILE, P., CORTAT, A., WIEMKEN, A. & FREY-WYSSLING, A. 1971 Isolation of glucanase-containing particles from *Saccharomyces cerevisiae. Proceedings of the National Academy of Sciences of the United States of America* **68**, 636–640.

MENDLIK, F. 1937 Some aspects of the scientific development of brewing in Holland. *Journal of the Institute of Brewing* **43**, 294–300.

MEYEN, J. 1838 *Wiegmann Archiv für Naturgeschichte* **4**, Band 2. 100.

MILL, P. J. 1964 The nature of the interactions between flocculent cells in the flocculation of *Saccharomyces cerevisiae. Journal of General Microbiology* **35**, 53–60.

MILLER, D. L. 1975 Ethanol production and potential. *Biotechnology and Bioengineering Symposia* **5**, 345–352.

MORTIMER, R. K. & HAWTHORNE, D. C. 1969 Yeast genetics. In *The Yeasts* Vol. 1. eds. Rose, A. H. & Harrison, J. S. London: Academic Press.

NOJIRO, K. & OUCHI, K. 1962 Fermenting capability and alcohol tolerance of yeast. I. Alcohol tolerance factors involved in the fermenting capability. *Journal of the Society of Brewing, Japan* **57**, 824–833.

PASTEUR, L. 1879 *Études sur la bière*, translated by Faulkner, F. & Robb, D. C. London: Macmillan.

PROUDLOCK, J. W., WHEELDON, L. W., JOLLOW, D. J. & LINNANE, A. W. 1968 Role of sterols in *Saccharomyces cerevisiae. Biochimica et Biophysica Acta* **152**, 434–437.

RAINBOW, C. 1970 Brewer's yeast. In *The Yeasts* Vol. 3, ed. Rose, A. H. & Harrison, J. S. London: Academic Press.

ROSE, A. H. 1976 History and scientific basis of alcoholic beverage production. In *Economic Microbiology* Vol. 1, ed. Rose, A. H. London: Academic Press.

ROSE, A. H. 1977 Dialling the composition of the yeast plasma membrane. In *Alcohol, Industry and Research* ed. Forsander, O., Eriksson, K., Oura, E. & Jounela-Eriksson, P. Helsinki: State Alcohol Monopoly.

SENTANDREU, R. & NORTHCOTE, D. H. 1969 The formation of buds in yeast. *Journal of General Microbiology* **55**, 393–398.

STELLING-DEKKER, N. M. 1931. *Die Sporogenen Hefen. Verhandelingen der Kon Academie van Wetenschappen, Section 11*, **28**, 1.

STEWART, G. G. 1975 Yeast flocculation. *Brewer's Digest* March, 42–57.

STEWART, G. G. 1979 Yeast flocculation. In *Brewing Science* ed. Pollock, J. R. A. & Pool, A. A. London: Academic Press.

STEWART, G. G., RUSSELL, I. & GARRISON, E. F. 1975 Some considerations of the flocculation characteristics of ale and lager yeast strains. *Journal of the Institute of Brewing* **81**, 248–257.

STEWART, L. C., RICHTMYER, N. K. & HUDSON, C. S. 1950 The preparation of trehalose from yeast. *Journal of the American Chemical Society* **72**, 2059–2061.

SUTÔ, T., TAGAKI, T. & UEMURA, T. 1951 Physiology of saké yeast. Fermenting capacity of the yeast in heavily concentrated mashes. *Journal of Fermentation Technology, Osaka* **29**, 447–452.

TANNENBAUM, S. R. & WANG, D. I. C. eds 1975 *Single-Cell Protein II.* Cambridge, Massachusetts: MIT Press.

THOMAS, D. S., HOSSACK, J. A. & ROSE, A. H. 1978 Plasma-membrane lipid composition and ethanol tolerance in *Saccharomyces cerevisiae. Archives of Microbiology* **117**, 239–245.

THOMAS, D. S. & ROSE, A. H. 1979 Inhibitory effect of ethanol on growth and solute accumulation by *Saccharomyces cerevisiae* as affected by plasma-membrane lipid composition. *Archives of Microbiology,* **122**, 49–55.

THORNE, R. S. W. 1954 Fermentation velocity of yeasts. I. Measurement of fermentation velocity and the effects of experimental conditions. *Journal of the Institute of Brewing* **60**, 227–237.

THORNE, R. S. W. 1961 Fermentation velocity of brewery yeasts. *Brewers' Digest* **36**, (7), 38.

VAN DER WALT, K. 1970 In *The Yeasts — A Taxonomic Study*, 2nd edn, ed. Lodder, J. Amsterdam: North-Holland.

VAN SOLINGEN, P. & VAN DER PLAAT, J. B. 1977 Fusion of yeast protoplasts. *Journal of Bacteriology* **130**, 946–947.

WILSON, K. & MCLEOD, B. J. 1976 The influence of conditions of growth on the endogenous metabolism of *Saccharomyces cerevisiae*: effect on protein, carbohydrate, sterol and fatty acid content and on viability. *Antonie van Leeuwenhoek* **42**, 397–410.

Yeast Spoilage of Fresh and Processed Fruits and Vegetables

C. DENNIS AND R. W. M. BUHAGIAR

ARC Food Research Institute, Colney Lane, Norwich, UK

Contents

1. Introduction

IN THEIR REVIEW of yeasts as spoilage organisms, Walker & Ayres (1970) stated that microbial spoilage of foods is commonly the result of combined activities of yeasts, moulds and bacteria; depending upon the environment however, one of these groups of microbes may prevail over the others. Ingram (1958) considered that the most important factors in determining the ability of yeasts to compete with moulds and bacteria on food were the numbers and types of contaminating yeasts, available nutrients, pH, redox potential, temperature during processing and storage and relative humidity or water activity of the surroundings of the food product. In addition, the presence of chemical preservatives also influences the relative importance of potential spoilage organisms.

Yeasts form part of the natural microflora of most fruits and vegetables although their relative proportion varies from commodity to commodity as well as being influenced by environmental, harvesting and storage conditions (Ingram 1958; Last & Price 1969; Kunkee & Amerine 1970; Beech & Davenport 1970; Buhagiar & Barnett 1971). Yeasts, however, are rarely encountered as spoilage organisms of fresh (unprocessed) fruits and vegetables but attain greater importance in processed products where competition with moulds and bacteria is often reduced by the action of preservatives, change in pH, and storage conditions.

Spoilage of fresh fruits and vegetables by yeasts usually results from their fermentative activity rather than degradation of the plant tissue by the action of cell wall degrading enzymes, although some yeasts are capable of producing pectolytic, cellulolytic, and xylanolytic enzymes (Luh & Phaff 1951, 1954*a*, *b*; Dennis 1972; Stevens & Payne 1977).

A wide variety of processed fruits and vegetables such as those preserved in salt or sulphite liquors, as well as acetic acid preserves have been reported to suffer yeast spoilage (Ingram 1958; Dakin & Day 1958; Walker & Ayres 1970; Mundt 1978; Splittstoesser 1978). Spoilage occurs as the production of off-odours and off-flavours, in the formation of surface films, excessive gas production or as changes in texture leading to softening of the product. The spoilage yeasts are often tolerant of high salt and/or acidic conditions (Bell & Etchells 1952; Walker & Ayres 1970). Mundt (1978) has recently reviewed the methods of production of brined vegetables and some of the spoilage problems encountered. In fermented vegetables there is competition between the desired fermentative organisms such as the lactic acid bacteria and the spoilage yeasts. The temperature during fermentation markedly influences this interaction, spoilage occurring at the higher temperatures which favour the development of yeasts.

2. Fresh Fruits and Vegetables

Lowings (1956) reported that *Kloeckera apiculata* caused strawberries to soften, exude fluid and become pale in colour when inoculated on injured fruits. The present authors have not observed *Kl. apiculata* to grow on naturally injured strawberries and although this was the only yeast isolated from blackcurrants picked at the end of the harvesting period, it did not spoil damaged fruits (Buhagiar, unpublished).

De Camargo & Phaff (1957) provided evidence that *Hanseniaspora uvarum* (and the imperfect form *Kl. apiculata*) and *Pichia kluyveri* caused spoilage of tomatoes in California. The essentially identical yeast flora of the alimentary canal of the fruit fly, *Drosophila melanogaster*, and fermenting tomatoes indicated the fruit fly as an important vector in transmitting the yeasts to cracked or damaged fruit in the field, thus increasing the incidence of spoilage.

Drosophila spp. (mainly *Drosophila melanogaster*) have also been shown to carry *H'spora valbyensis*, *H'spora uvarum (Kl. apiculata)* and *Torulopsis stellata*, the spoilage yeasts responsible for fermentative spoilage of calimyrna figs in California (Miller & Phaff 1962). Fermentative spoilage or 'souring' is one of the most common types of microbial spoilage in the fig industry in California (Mrak *et al.* 1942). Miller & Phaff (1962) showed that the apiculate yeasts and *T. stellata* usually comprise 90% of the spoilage flora of figs; this contrasts with the earlier finding of Mrak *et al.* (1942) and Miller & Mrak (1953) where *Saccharomyces cerevisiae* and

Candida krusei were listed as the most numerous yeasts of spoiled figs. Miller & Phaff (1962), however, considered that *Saccharomyces* spp. do not play a role in fruit spoilage.

Similar fermentative spoilage or 'souring' also occurs in dates (Fellows & Clague 1942; Zein *et al.* 1956; Salik *et al.* 1979). Although *Saccharomyces* spp. are associated with such spoilage of dry dates (Mrak *et al.* 1942), spoilage of soft dates is caused by *H'spora valbyensis* and *C. guilliermondii* (Zein *et al.* 1956).

Apart from the fermentative spoilage of the various fruits mentioned above certain yeasts are potentially capable of causing spoilage through the action of their cell wall degrading enzymes. Pectinolytic yeasts do not, however, usually infect fresh fruits and vegetables. A notable exception was recently found in relation to spoilage of forced rhubarb petioles (Buhagiar & Dennis, unpublished). Buhagiar (unpublished) found *Trichosporon cutaneum* to be present invariably on leaves, petioles, stems and roots (10^4–10^5/g) of forced rhubarb plants. In the early 1970s the introduction of pre-packed rhubarb required cutting petioles into short uniform lengths prior to packing into a cellophane-covered pack for retail sale. Where such packs were stored at ambient temperatures the damaged areas of the petioles, particularly the cut ends, provided sites for infection by micro-organisms, resulting in spoilage of the petioles (Robinson & Tomalin, unpublished). Isolations made from the leading edge of such rots frequently revealed the presence of large numbers of the yeast *Tr. cutaneum*, often in combination with *Pseudomonas* and *Penicillium* spp. Inoculation of cut petioles confirmed that *Tr. cutaneum* could initiate similar spoilage. Strains of this yeast together with *Tr. pullulans* were the first yeasts shown to be cellulolytic and xylanolytic (Dennis 1972; Stevens & Payne 1977).

During the preparation of commercial packs of rhubarb the contaminating spoilage organisms, including *Tr. cutaneum* were spread during cutting, and frequent cleaning of the knives markedly reduced the incidence of spoilage.

This report of *Tr. cutaneum* attacking cell walls of fresh rhubarb, together with the reports of Lodder & Kreger-van Rij (1952) and Ingram (1958) of *Nematospora* rotting fruits, appear to be rare instances of yeasts causing rotting of plant tissue.

3. Processed Fruits and Vegetables

A. Fruit and vegetable salad preparations

Over recent years there has been an increase in the production for both retail and catering outlets in the UK, of preparations containing peeled, sliced, shredded or diced pieces of root, green or salad vegetables. Many such preparations are ready for consumption and consist of chopped vegetables in mayonnaise (e.g. coleslaw, potato salad). Kurtzman *et al.* (1971) reported yeasts, in particular *Sacch. bailii*, to cause spoilage of mayonnaise. Recent observations on a range of mayonnaise-based salads by the present authors, indicates that yeasts also play an important role in the spoilage of these products, especially in preparations which contain a mixture of fruits, nuts and vegetables. Spoilage occurred as gas production, off-odours and the presence of films of yeast on the surface of the salads. *Saccharomyces* species appear to be the main cause of spoilage of coleslaw, coleslaw and fruit preparations and potato salad. The species concerned vary between batches of salad and also according to storage temperature. Coleslaw stored at 5°C in the laboratory was spoiled by *Sacch. dairensis* whereas *Sacch. exiguus* was the only yeast isolated from samples spoiled during commercial distribution. This may indicate an effect of storage temperature as further samples stored in the laboratory at 5°C, where *Sacch. dairensis* was not isolated in the freshly prepared coleslaw, did not spoil within 21 days storage. Samples from the latter batch were spoiled by *Sacch. exiguus* when stored at 10–20°C. *Saccharomyces dairensis* is the only micro-organism which has shown a marked increase in viable count during storage at 5°C (Fig. 1). *Saccharomyces exiguus* declines to undetectable levels during a 14–18 day storage period while other yeasts such as *Candida* sp., *Tr. cutaneum* and *P. etchellsii* decline more rapidly. The viable count of lactobacilli shows little change during storage at 5°C (Fig. 1). The main source of inoculum of the spoilage yeasts appears to be the cabbage. Steinbuch (1967) indicated that the incidence of yeasts (including *Sacch. exiguus*) tends to increase on cabbages stored for long periods. In relation to cabbage used for sauerkraut manufacture, he considered the machines used for stripping the cabbage to be a means of spreading contamination and emphasized the need for frequent cleaning. A similar situation could exist during the preparation of the cabbage for coleslaw.

Samples of potato salad stored at 10–20°C showed a rapid increase in the count of both yeasts (predominantly *Sacch. exiguus*) and lactobacilli. *Saccharomyces exiguus* multiplied in samples stored at 5°C although no spoilage was apparent during 21 days storage. The increase in yeast count at the higher temperatures coincided with the occurrence of gas production, off-odours and a film of yeasts on the surface. *Saccharomyces exiguus* was the predominant yeast isolated from the spoiled samples except on the surface of the salad, where *P. membranaefaciens* was prevalent.

B. Salt brined fruits and vegetables

In the case of sauerkraut, yeasts have been implicated in discolorations of the cabbage occurring during fermentation (Fred & Peterson 1922;

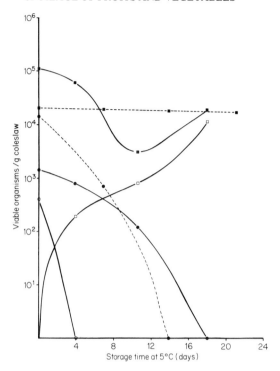

Fig. 1. Viable counts of yeasts and lactobacilli on coleslaw stored at 5°C. ■, lactobacilli; ●, *Sacch. exiguus*; □, *Sacch. dairensis*; ○, *Candida, Trichosporon* and *Pichia* spp.; _____, sample 1; -------, sample 2.

Steinbuch 1965; Walker & Ayres 1970). Fred & Peterson (1922) reported pink discolorations to be due to growth of non-ascospore forming yeasts or torulae. The extent of discoloration depended on the type and amount of sugar, salt concentration, pH, temperature and oxygen supply. High temperature (20°C), high salt concentration and high acid content enhanced the development of pink sauerkraut.

Steinbuch (1965) provided evidence that growth of *Sacch. exiguus* resulted in pink discoloration but also suggested that *C. krusei* and *T. holmii* may also be involved. Sauerkraut inoculated with *Sacch. exiguus* resulted in a higher pH and lower lactic acid level. The discoloration can range from pink to grey and vary between varieties of cabbage (Steinbuch 1966). The occurrence of yeast growth and subsequent discoloration results from retarded development of the lactic acid bacteria which is in turn influenced by trace metals, particularly manganese (Steinbuch 1967, 1969). In agreement with Fred & Peterson (1922) high fermentation temperatures (20°C) and high salt content were reported to stimulate growth of yeasts and thus encourage discoloration (Steinbuch 1970).

Although Mundt (1978) indicates that pectinolytic yeasts as well as moulds are involved in softening of brined cucumbers, Etchells *et al.* (1958) considered pectolytic enzymes from moulds to be largely responsible for such softening.

Numerous yeasts have however been isolated from surface films and from the interior of pickle brines (Etchells *et al.* 1953, 1958). The fermentative yeasts (especially *T. holmii*) are numerous during the first few days of brining when the brines have a relatively high content of sugars which have been leached from the cucumber, and when the competing lactic acid bacteria have not become predominant. Etchells *et al.* (1952) showed that *T. holmii* predominated during the early stages of fermentation (2–30 days) and that *T. versatilis* was most prevalent during later stages (70–100 days). *Hansenula subpelliculosa, Sacch. rosei* and *Sacch. rouxii* were active between the extremes. The carbon dioxide produced during this fermentation can become entrapped in the cucumber to form 'bloaters'. Recent work has shown that inoculation with the homofermentative bacteria *Pediococcus cerevisiae* and *Lactobacillus plantarum* and purging the brine tanks with gaseous nitrogen suppresses yeast growth and rapidly removes the carbon dioxide produced (Fleming *et al.* 1975; Costilow *et al.* 1977).

Vaughn *et al.* (1972) reported yeasts to be the cause of a massive commercial outbreak of softening and gas pocket formation in olives that had been stored in acidified low salt brines (3–4% as opposed to 6–7%) in an attempt to reduce the problem of brine disposal. *Saccharomyces oleaginosus* and *Sacch. kluyveri* were pectolytic and caused severe softening, while these species together with the strongly fermentative *Hansenula anomala* var. *anomala* caused gas-pocket formation. Earlier, Vaughn *et al.* (1969) reported the pink yeasts *Rhodotorula glutinis* var. *glutinis, Rh. minuta* var. *minuta* and *Rh. rubra* to produce polygalacturonases which caused slow softening of olives in strong brines (8%).

C. Acetic acid preserves

Saccharomyces acidifaciens (*Sacch. bailii*, van der Walt 1970) and to a much lesser extent *P. membranaefaciens* were reported to cause spoilage of British acetic acid preserves, including onions, gherkins, mixed clear pickles, red beetroot, red cabbage and piccalilli (Dakin & Day 1958). These authors described the yeast spoilage as usually characterized by gas formation, with rusty, sulphurous or yeast-like off-odours. In clear pickles they observed a creamy or brown sediment and on vegetables cream-coloured colonies produced above the surface of the liquid. In more viscous products the surface was partially or completely covered by a layer of cream-coloured to brown-coloured yeasts.

D. Sulphited fruits

A wide variety of fruits are preserved in sulphite liquor for use in the manufacture of jams, marmalades, candied and glacé fruits and maraschino cherries (Dennis & Harris 1979). The sulphite liquor consists of sulphurous acid and calcium bisulphite solutions, each containing equal concentrations of sulphur dioxide although the exact composition of the liquor can be varied according to the fruits concerned and their condition at harvest (Doesburg 1965; Watters 1975).

There have been reports of softening or even complete disintegration of certain fruits (e.g. cherries, strawberries) during storage in sulphite liquor prior to their use for manufactured products (Wiegand 1957; Dennis 1978) caused by the presence of microbial pectolytic enzymes. Steele & Yang (1960) implicated yeasts as being a source of polygalacturonase which caused softening of sulphited cherries. Later work by Lewis *et al.* (1963), however, showed that the softening of commercial samples of sulphited cherries was more likely caused by polygalacturonase produced during infection of the cherries by *Cytospora leucostoma*, *Aspergillus niger* and *Penicillium expansum*. Pectolytic enzymes of fungal origin have also been reported as the cause of breakdown of sulphited strawberries in the UK (Dennis & Harris 1979; Harris & Dennis 1979; Dennis *et al.* 1979; Archer 1979; Archer & Fielding 1979). Although the softening and breakdown of commercial samples of fruit is considered to be caused by polygalacturonases produced by *Mucor* and *Rhizopus* species (Dennis *et al.* 1979) pectolytic yeasts associated with strawberries have also been shown to produce sulphite-stable pectolytic enzymes (Dennis & Harris 1979). Culture filtrates of strains of both *Tr. pullulans* and *Cryptococcus albidus* var. *albidus* as well as the yeast-like mould *Aureobasidium pullulans* caused complete breakdown of sulphited fruit (Fig. 2). Where commercial samples of fruit are held overnight at ambient temperature between harvesting and sulphiting it is possible that the pectolytic yeasts may produce polygalacturonases which together with any produced by the fruit-rotting fungi will subsequently cause breakdown.

When used at the recommended concentrations, there is sufficient SO_2 present in the sulphite liquor to inhibit fungal growth completely. Dennis (1978), however, reported spoilage of sulphited strawberries as a result of fungal growth, primarily yeasts. This appears to be the only report of such spoilage and it is of interest that similar spoilage of other sulphited fruits does not appear to occur. Such yeast spoilage of sulphited strawberries results in the bleached fruit turning pink and having a fruity, winey smell which renders it unusable. Only the top few inches of fruit in the barrel is affected with the remainder being completely bleached and suitable for the manufacture of good quality jam. *Saccharomyces* species (*bailii, florentinus,*

Fig. 2. Breakdown of sulphited strawberries caused by culture filtrates of *Tr. pullulans*.

cerevisiae, kluyveri, uvarum) as well as the film yeasts *H. subpelliculosa* and *P. membranaefaciens*, have been isolated from spoiled samples of fruit. In addition, *Penicillium* species have also been observed to develop on the surface of severely spoiled fruit. In all of these cases the free SO_2 in the liquor, as determined by the method of Burroughs & Sparks (1964) ranged from 0.5 parts/10^6 to 26 parts/10^6 compared to > 750 parts/10^6 in liquor from uninfected barrels.

Saccharomyces bailii has been widely reported as a preservative-tolerant yeast (Pitt & Richardson 1973; Pitt 1974) growing in the presence of 500 parts/10^6 SO_2 (Reed & Peppler 1973). In contrast, Hammond & Carr (1976) reported that SO_2 rapidly kills certain yeasts. They showed that less than 25 parts/10^6 of free SO_2 (pH 3·4) markedly reduced the viable count of *Sacch. cerevisiae* after 8 h at 25°C. Similarly Reed & Peppler (1973) reported *P. membranaefaciens* to be sensitive to SO_2.

Since SO_2-sensitive yeasts have been isolated from the spoiled sulphited fruit, it would appear that spoilage results from inadequate mixing of liquor and fruit when a sufficient inoculum of the appropriate yeasts is present. More information is required on the incidence of the above yeasts in relation to fruit ripeness since Buhagiar & Barnett (1971) isolated the relevant yeasts from strawberries only very rarely. The sensitivity of these yeasts to SO_2 in relation to their growth on sulphited fruit also needs to be investigated.

4. References

ARCHER, S. A. 1979 Pectolytic enzymes and degradation of pectin associated with the breakdown of sulphited strawberries. *Journal of the Science of Food and Agriculture* **30**, 692-703.

ARCHER, S. A. & FIELDING, A. H. 1979 Polygalacturonase isoenzymes of fungi involved in the breakdown of sulphited strawberries. *Journal of the Science of Food and Agriculture* **30**, 711-723.

BEECH, F. W. & DAVENPORT, R. R. 1970 The role of yeast in cider making. In *The Yeasts* Vol. 3, ed. Rose, A. H. & Harrison, J. S. London: Academic Press.

BELL, T. A. & ETCHELLS, J. L. 1952 Sugar and acid tolerance of spoilage yeasts from sweet cucumber pickles. *Food Technology, Champaign* **6**, 468-472.

BUHAGIAR, R. W. M. & BARNETT, J. A. 1971 The yeasts of strawberries. *Journal of Applied Bacteriology* **34**, 727-739.

BURROUGHS, L. F. & SPARKS, A. H. 1964 The determination of the free sulphur dioxide content of ciders. *Analyst* **89**, 55-60.

COSTILOW, R. N., BEDFORD, C. L., MINGUS, D. & BLACK, D. 1977 Purging of natural salt-stock pickle fermentations to reduce bloater damage. *Journal of Food Science* **42**, 234-240.

DAKIN, V. C. & DAY, P. M. 1958 Yeasts causing spoilage in acetic acid preserves. *Journal of Applied Bacteriology* **21**, 94-96.

DE CAMARGO, R. & PHAFF, H. J. 1957 Yeasts occurring in *Drosophila* flies and in fermenting tomato fruits in Northern California. *Food Research* **22**, 367-372.

DENNIS, C. 1972 Breakdown of cellulose by yeast species. *Journal of General Microbiology* **71**, 409-411.

DENNIS, C. 1978 Post-harvest spoilage of strawberries. *Agricultural Research Council Research Review* **4**, 38-42.

DENNIS, C., DAVIS, R. P., HARRIS, J. E., CALCUTT, L. W. & CROSS, D. 1979 The relative importance of fungi in the breakdown of commercial samples of sulphited strawberries. *Journal of the Science of Food and Agriculture* **30**, 959-973.

DENNIS, C. & HARRIS, J. E. 1979 The involvement of fungi in the breakdown of sulphited strawberries. *Journal of the Science of Food and Agriculture* **30**, 687-691.

DOESBURG, J. J. 1965 *Pectic Substances in Fresh and Preserved Fruits and Vegetables* pp. 95-131. Wageningen, Netherlands: Institute for Research on Storage and Processing of Horticultural Produce.

ETCHELLS, J. L., COSTILOW, R. N. & BELL, T. A. 1952 Identification of yeasts from commercial cucumber fermentations in northern brining areas. *Farlowia* **4**, 249-264.

ETCHELLS, J. L., BELL, T. A. & JONES, I. D. 1953 Morphology and pigmentation of certain yeasts from brines and the cucumber plant. *Farlowia* **4**, 265-304.

ETCHELLS, J. L., BELL, T. A., MONROE, R. J., MASLEY, P. M. & DEMAIN, A. L. 1958 Populations and softening enzyme activity of filamentous fungi on flowers, ovaries and fruit of pickling cucumbers. *Applied Microbiology* **6**, 427-440.

FELLOWS, C. R. & CLAGUE, J. A. 1942 Souring of dried dates by sugar-tolerant yeasts. *Fruit Products Journal* **21**, 326, 327, 347.

FLEMING, H. P., ETCHELLS, J. L., THOMPSON, R. L. & BELL, T. A. 1975 Purging of CO_2 from cucumber brines to reduce bloater damage. *Journal of Food Science* **40**, 1304-1310.

FRED, E. B. & PETERSON, W. H. 1922 The production of pink sauerkraut by yeasts. *Journal of Bacteriology* **7**, 257-269.

HAMMOND, S. M. & CARR, J. G. 1976 The antimicrobial activity of SO_2 with particular reference to fermented and non-fermented fruit juices. In *Inhibition and Inactivation of Vegetative Microbes* ed. Skinner, F. A. & Hugo, W. B. London: Academic Press.

HARRIS, J. E. & DENNIS, C. 1979 The stability of pectolytic enzymes in sulphite liquor in relation to breakdown of sulphited strawberries *Journal of the Science of Food and Agriculture* **30**, 704–710.

INGRAM, M. 1958 Yeasts in food spoilage. In *The Chemistry and Biology of Yeasts* ed. Cook, A. H. New York: Academic Press.

KUNKEE, R. E., & AMERINE, M. A. 1970 Yeasts in winemaking. In *The Yeasts* Vol. 3, ed. Rose, A. H. & Harrison, J. S. London: Academic Press.

KURTZMAN, C. P., ROGERS, R. & HESSELTINE, C. W. 1971 Microbiological spoilage of mayonnaise and salad dressing. *Applied Microbiology* **21**, 870–874.

LAST, F. T. & PRICE, D. 1969 Yeasts associated with living plants and their environs. In *The Yeasts* Vol. 1, ed. Rose, A. H. & Harrison, J. S. London: Academic Press.

LEWIS, J. S., PIERSON, C. F. & POWERS, M. J. 1963 Fungi associated with softening of bisulphite-brined cherries. *Applied Microbiology* **11**, 93–99.

LODDER, J. & KREGER-VAN RIJ, N. J. W. 1952 *The Yeasts — A Taxonomic Study.* Amsterdam: North-Holland.

LOWINGS, P. H. 1956 The fungal contamination of Kentish strawberry fruits in 1955. *Applied Microbiology* **4**, 84–88.

LUH, B. S. & PHAFF, H. J. 1951 Studies on polygalacturonase of certain yeasts. *Archives of Biochemistry and Biophysics* **33**, 212–227.

LUH, B. S. & PHAFF, H. J. 1954*a* Properties of yeast polygalacturonase. *Archives of Biochemistry and Biophysics* **48**, 23–37.

LUH, B. S. & PHAFF, H. J. 1954*b* End product and mechanism of hydrolysis of pectin and pectic acid by yeast polygalacturonase. *Archives of Biochemistry and Biophysics* **51**, 102–113.

MILLER, M. W. & MRAK, E. M. 1953 Yeasts associated with dried-fruit beetles in figs. *Applied Microbiology* **1**, 174–178.

MILLER, M. W. & PHAFF, H. J. 1962 Successive microbial populations of calimyrna figs. *Applied Microbiology* **10**, 394–400.

MRAK, E. M., PHAFF, H. J., VAUGHN, R. H. & HANSEN, H. N. 1942 Yeasts occurring in souring figs. *Journal of Bacteriology* **44**, 441–450.

MUNDT, J. O. 1978 Fungi in the spoilage of vegetables. In *Food and Beverage Mycology* ed. Beuchat, L. R. Westport, Connecticut: AVI.

PITT, J. I. 1974 Resistance of some food-spoilage yeasts to preservatives. *Food Technology in Australia* **26**, 238–241.

PITT, J. I. & RICHARDSON, K. C. 1973 Spoilage by preservative-resistant yeasts. *CSIRO Food Research Quarterly* **33**, 80–85.

REED, G. & PEPPLER, H. J. 1973 *Yeast Technology.* Westport, Connecticut: AVI.

SALIK, H., ROSEN, B. & KOPELMAN, I. J. 1979 Microbial aspects and the deterioration process of soft dates. *Food Science and Technology* **12**, 85–87.

SPLITTSTOESSER, D. F. 1978 Fruits and fruit products. In *Food and Beverage Mycology* ed. Beuchat, L. R. Westport, Connecticut: AVI.

STEELE, W. F. & YANG, H. Y. 1960 The softening of brined cherries by polygalacturonase, and the inhibition of polygalacturonase in model systems by alkyl aryl sulphonates. *Food Technology, Champaign* **14**, 121–126.

STEINBUCH, E. 1965 Preparation of sauerkraut. *Sprenger Institute Annual Report* pp. 56–57. Wageningen, The Netherlands.

STEINBUCH, E. 1966 Manufacturing of sauerkraut. *Sprenger Institute Annual Report* pp. 47. Wageningen, The Netherlands.

STEINBUCH, E. 1967 Preparation of sauerkraut. *Sprenger Institute Annual Report* p. 47. Wageningen, The Netherlands.

STEINBUCH, E. 1969 Manufacturing of sauerkraut. *Sprenger Institute Annual Report* pp. 58–61. Wageningen, The Netherlands.

STEINBUCH, E. 1970 Preparation of sauerkraut. *Sprenger Institute Annual Report* pp. 43–46. Wageningen, The Netherlands.

STEVENS, B. J. H. & PAYNE, J. 1977 Cellulase and xylanase production by yeasts of the genus *Trichosporon. Journal of General Microbiology* **100**, 381–393.

VAN DER WALT, J. P. 1970 *Saccharomyces* Meyen emend. Reess. In *The Yeasts — A Taxonomic Study* 2nd edn, ed. Lodder, J. Amsterdam: North-Holland.

VAUGHN, R. H., JAKUBCZYK, T., MACMILLAN, J. D., HIGGINS, T. E., DAVE, B. A. & CRAMPTON, V. M. 1969 Some pink yeasts associated with softening of olives. *Applied Microbiology* **18**, 771–775.

VAUGHN, R. H., STEVENSON, K. E., DAVE, B. A. & PARK, H. C. 1972 Fermenting yeasts associated with softening and gas pocket formation in olives. *Applied Microbiology* **23**, 316–320.

WALKER, H. W. & AYRES, J. C. 1970 Yeasts as spoilage organisms. In *The Yeasts* Vol. 3, ed. Rose, A. H. & Harrison, J. S. London: Academic Press.

WATTERS, G. G. 1975 Brining cherries and other fruits. In *Commercial Fruit Processing* ed. Woodruff, J. G. & Luh, B. S. Westport, Connecticut: AVI.

WIEGAND, E. H. 1957 Brined cherry breakdown. In *Proceedings of the Oregon Horticultural Society, Corvallis, Oregon.*

ZEIN, G. N., SHEHATA, A. M. E. & SEDKY, A. 1956 Studies on Egyptian dates. 1. Yeasts isolated from souring soft varieties. *Food Technology, Champaign* **10**, 405–407.

Yeast Spoilage of Bakery Products

D. A. L. SEILER

*Flour Milling and Baking Research Association,
Chorleywood, Hertfordshire, UK*

Contents

1. Introduction

THE FUNDAMENTAL IMPORTANCE of yeasts in the manufacture of a wide range of foods, including fermented bakery products, is well recognized. What is perhaps less well appreciated is the role of yeasts in spoilage of bakery goods. The reason for this may be partly because the manifestations of yeast activity are often less objectionable than those of other micro-organisms and hence customer complaints due to this cause are less commonly encountered, and partly because there is no evidence that yeasts or the products of yeast action are in any way toxic. There seems little doubt, however, that yeasts can cause considerable wastage of both bakery ingredients and products and that the shelf life of certain goods can be limited by yeast spoilage rather than mould or bacterial spoilage.

This paper briefly discusses the beneficial uses of yeast in the baking industry and describes spoilage problems caused by non-osmophilic and osmophilic yeasts. Details of work carried out to determine the factors affecting the growth of yeasts in high sugar fillings and coatings are presented.

2. Beneficial Yeasts

A. Development of new strains of baker's yeast

Yeasts have been used since the earliest ages for leavening doughs. In the early days the expansion of the dough was caused by carbon dioxide produced by yeasts and bacteria naturally present in the flour. Even in those days some selection of strains unwittingly took place in that doughs which rose quickly and gave good flavoured bread would be retained and part used to seed future doughs. With the growth of the brewing and distilling industries during the Middle Ages the excess yeast was often used for bread-making purposes. However, as the baking industry also expanded, this source of yeast became limited and it became necessary to set up a separate industry for the production of yeast specifically for baking. At first home made barms were used for propagating the yeast but later it became more common to use a liquid medium similar to that used by the brewer, where the nutrients were derived from a mixture of malted barley and starch. More recently this 'grain liquor' has been replaced with a molasses-based medium. Allied with this development there was also a continuing improvement in the strain of *Saccharomyces cerevisiae*. Brewer's yeast, which was used initially, gave bread with a bitter taste and it was found that distiller's yeast was better. Although the quality of yeast continued to improve it was not until the end of the nineteenth century that the yeast manufacturer was able to overcome problems of flavour and contamination and give the baker the consistent product he receives today.

Until about 20 years ago, the strain of *Sacch. cerevisiae* used for bread making had an adaptive maltase system. In other words, it preferentially fermented the monosaccharides present in the dough and it took some time to adapt to fermentation of the disaccharide maltose. During this period of adaptation the rate of carbon dioxide production was reduced. This effect is shown in Fig. 1 from which it is apparent that there is a trough in the graph of gas production after the dough has been fermenting for 1–2 h. This trough becomes more pronounced with increasing yeast levels and dough temperatures. To overcome this deficiency a new strain of *Sacch. cerevisiae* was developed (Sykes 1971) which has a constitutive rather than an adaptive maltase system so that the sugars in the dough are fermented with equal facility and there is no hesitation in gassing rate. In addition, more gas is produced in the early stages of fermentation (Fig. 2).

Fortuitously, these developments took place a few years before the introduction of rapid doughmaking processes, such as the Chorleywood Bread Process (Chamberlain *et al.* 1962), which are now used by the majority of plant bakeries in the UK. With the bulk fermentation process,

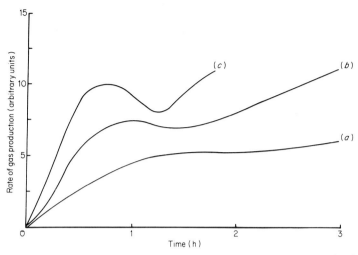

Fig. 1. Effect of yeast concentration and temperature on the rate of gas production in doughs containing the 'old' type yeast strain. *(a)* 1·1% yeast on flour, 26°C; *(b)* 1·8% yeast on flour, 27°C; *(c)* 2·1% yeast on flour, 32°C.

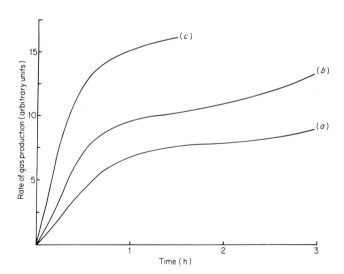

Fig. 2. Effect of yeast concentration and temperature on the rate of gas production in doughs containing the 'new' type yeast strain. *(a)* 1·1% yeast on flour, 26°C; *(b)* 1·8% yeast on flour, 27°C; *(c)* 2·1% yeast on flour, 32°C.

which was used up until that time, doughs were allowed to ripen by fermenting for several hours after mixing and before proving and baking. A brief reduction in gassing rate was of minor importance since it occurred during the long fermentation period. With rapid doughmaking processes, however, the use of yeast strains with an adaptive maltase system caused a reduction in gas production to occur during the critical stages of final proof and early baking. Moreover, because of the higher dough temperatures and higher levels of yeast used with these processes, the drop in gassing rate was more pronounced (Fig. 1). The availability of the new strain of yeast was therefore a major bonus in the development of these rapid processes.

B. Sour dough processes

With rye bread and other speciality breads, it is common practice to add a proportion of sour dough to the main dough mix to enhance flavour. To produce these sour doughs many bakers add a commercially prepared bacterial culture which usually consists of a mixture of *Lactobacillus brevis* and *L. plantarum*. Other bakers prefer to use a natural sour where the bacteria responsible for souring are derived from the ingredients in the dough. Even in these natural sours *L. brevis* and *L. plantarum* tend to dominate (Shultz 1948). Although these bacteria can produce sufficient carbon dioxide to aerate the dough, the leavening mainly results from the activity of yeasts. Kline *et al.* (1970) isolated several highly acid-resistant yeasts from sour dough used to make San Francisco French bread, including *Sacch. exiguus* and *Sacch. inusitatis* of which the former was considered to be principally responsible for leavening the dough. It is likely that these yeasts are also involved in aeration of sour doughs used for other types of bread.

3. Spoilage by Non-osmophilic Yeasts

In bakery coatings and fillings which have a very high a_w of over 0·97, such as cream, custard or meat, bacteria are generally at an advantage over yeasts and are the predominant cause of spoilage. With foods of rather lower a_w, i.e. 0·90–0·97, this situation is often reversed in that yeasts (or moulds) can develop more rapidly than bacteria. Products within this a_w range which are susceptible to yeast spoilage include bread and other fermented products, unbaked pastries and doughs, pizzas and some fresh fruit fillings. The problems associated with these products are discussed separately.

A. Bread

Three rather different forms of spoilage can result from contamination of bread with non-osmophilic yeasts.

(i) Contamination of dough

Bread of poor quality with an undesirable odour and flavour can result when yeasts other than those desired are present in the dough in large numbers. Such problems can arise if baker's yeast becomes contaminated with foreign yeasts during manufacture. In the past, contamination of this nature was commonplace but nowadays, with modern methods of yeast manufacture and control, the chance of such a problem arising is extremely small. The risk of foreign yeast contamination tends to be higher with long fermentation sponges or doughs where either faulty ingredients or build up of contamination on plant can be responsible.

(ii) Post-baking contamination

A more frequent problem is contamination of bread with yeasts after baking. The deterioration can take two forms depending on the yeast species present.

With active fermenting non-filamentous yeasts the first sign of spoilage is the development of an off-odour in the bread. In some cases the odour is readily identified as alcoholic but in others it is variously described as 'fruity', 'acid drops', 'acetone' etc., depending on the type of bread and species of yeast present. It is likely that many of the complaints concerning off-odours in bread are associated with yeast growths. With these non-filamentous yeasts, visible growth on the bread usually occurs somewhat belatedly and, except on dark breads, is difficult to detect. Many different yeasts are capable of growing on bread and causing such defects but not surprisingly baker's yeast, *Sacch. cerevisiae*, is most commonly encountered.

With certain filamentous yeasts the deterioration takes a different course in that visible growth rather than off-odour is the subject of complaint. With some filamentous yeasts, such as *Hansenula anomala* var. *anomala*, (Fig. 3c), the growth on brown breads is buff coloured and often speckled in appearance. With others a white spreading growth forms over the surfaces of the bread which is reminiscent of mould growth. The yeasts responsible for this type of white growth are commonly referred to as 'chalk moulds'. The most common and troublesome species is *Hyphopichia burtonii* (Fig. 3b)

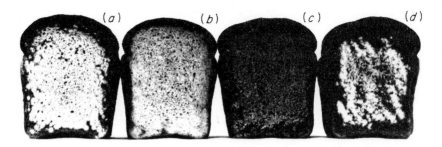

Fig. 3. Cut surfaces of brown bread artificially inoculated with different yeasts: *(a) Hyphopichia fibuligera; (b) Hyphopichia burtonii; (c) H. anomala* var. *anomala; (d) Tr. penicillatum.*

but *Hyphopichia fibuligera* (Fig. 3a) and *Trichosporon penicilatum* (Fig. 3d) are also encountered on occasion. Chalk mould problems tend to be more common with darker breads such as brown, wholemeal, rye, pumpernickel etc., on which the white growth is more obvious.

Endomycopsis burtonii in particular can multiply very rapidly on bread and become visible some time before mould growth occurs. Moreover, it tends to be more resistant to preservatives, such as propionates, than are moulds (Lamprecht 1955). Thus, the presence of this organism in the bakery can be a source of serious concern. Thankfully, chalk mould problems only occur in relatively few bakeries which suggests that the source of contamination arises within the bakery environment.

Gemeinhardt & Bergmann (1977) found that *Endomycopsis burtonii* was the most common organism present in the atmosphere and in settled dust in German ryebread bakeries. Our own experience from bakeries in the UK encountering chalk mould problems, suggests that contamination usually results from contact with surfaces containing these yeasts rather than from the atmosphere. Damp debris which may have built up on racks in coolers, conveyor belts or slicing machines is often to blame. Regular and thorough cleaning of these sites is often successful in eradicating the infection. Doughs, and particularly sour doughs, can often contain chalk moulds and other yeasts which will grow on bread. It is important, therefore, to ensure that workers preparing doughs are never allowed to handle the finished product or surfaces with which the product comes into contact.

B. Raw dough and pastry products

Raw short crust or puff pastry and chemically aerated dough products which have an a_w within the range 0·90–0·98 are sometimes subject to yeast spoilage. This usually takes the form of gas production which causes the pastry or dough to expand and results in an impairment in baking quality. If the initial inoculum of yeasts is high and the storage temperatures are low, yeast spoilage can often occur before bacterial spoilage.

The yeasts causing this type of spoilage are either naturally present in the dough ingredients or enter the pastry or dough during manufacture through contact with contaminated surfaces. With products such as pies which are sold with a cooked filling but raw pastry and baked by the customer, build up of contamination can result from the continuous incorporation of trimmings into the paste. A number of different yeasts are capable of fermenting such products but it is often found that baker's yeast is to blame.

When sold frozen, fermentation problems are unlikely since a proportion of the yeasts present will be destroyed during freezing and holding at deep freeze temperatures and, for spoilage to occur, the customer would have to hold the product for several days between defrosting and baking. Problems are much more likely when the products are sold refrigerated since many yeasts can grow under the conditions in commercial chill cabinets, i.e. 4–10°C, and there is more chance of the customer holding the product at chill or ambient temperature for some time before use. However, providing the pastry does not become excessively contaminated with yeasts, fermentation problems with such products are uncommon.

A recent development is the chemically aerated unbaked buns, rolls, doughnuts etc. which are sold refrigerated in special cannisters (Chen 1979). The long shelf life of 2–3 months is achieved because of the low a_w and pH of the products, the anaerobic conditions and high carbon dioxide concentration within the pack, and the use of ingredients with a low microbial count. Yeasts tolerant to the effects of anaerobiosis, carbon dioxide and low a_w and pH can occasionally cause fermentation spoilage.

C. Savoury products

The shelf life of unbaked pizzas which are sold refrigerated is often limited by the growth of yeasts. Sometimes the spoilage is due to visible yeast growth on toppings, such as olives, but more often contaminated cheese is to blame and fermentation during storage results in alcoholic off-odours and flavours, swelling of the pack and a general deterioration in the appearance of the product. To maximize shelf life it is important to purchase cheese of good quality, store at refrigerator temperatures, and use as quickly as possible. Stringent hygiene precautions must be taken during grating the cheese and applying the coatings and toppings to the base.

4. Spoilage by Osmophilic Yeasts

Fermentation problems caused by the growth of osmophilic yeasts in high sugar coatings and fillings used for flour confectionery products are fairly commonly encountered. The materials involved include jam, fruit fillings, fudge, fondant, marshmallow, marzipan and nut pastes. In most cases, the development of alcoholic odours and flavours is the main evidence of deterioration but with certain products the carbon dioxide gas produced as a result of fermentation can cause visible defects such as bubbling on fondant coatings, cracking of royal icing on birthday cakes etc.

A. Sources of contamination

It is of interest to examine how some of these coatings and fillings become contaminated with yeasts since this can have an important bearing on the fermentation-free shelf life of the product.

(i) *Jams*

Yeasts are usually absent from boxed or canned jams when delivered to the bakery because the temperatures achieved during manufacture are sufficient to destroy any yeasts present in the ingredients, and the jam is introduced into the containers while still hot enough to destroy any yeasts which may enter from the atmosphere or may be present on the inside surfaces of the box liner or can. Fermentation of jam in these boxes or cans is rare and usually results from faulty handling and storage of part-used containers; yeast contamination in such cases arises from the atmosphere or from utensils used to remove the jam.

For the large scale user, jam is often received in bulk tanks which can be fitted on to depositors for immediate use. If this jam is poured into these tanks directly after manufacture and allowed to set it is difficult to pump to the depositor. Thus, it is necessary to use 'broken set' jam which is achieved by allowing the jam to set partially in one tank and then breaking the set by pumping it into a second tank. With this process, yeast contamination can occur from the pump and pipeline used to transfer the jam from one tank to the other; from the inside surfaces of the second tank; from the valve and elbow joint at the base of this tank; and from exposure to the atmosphere. These sites are indicated by arrows in Fig. 4.

When these 'broken set' jams in bulk tanks were first made available problems of fermentation were not uncommon, but now the jam manufacturer takes elaborate precautions to clean and sterilize tanks so that the risk of yeast contamination is small.

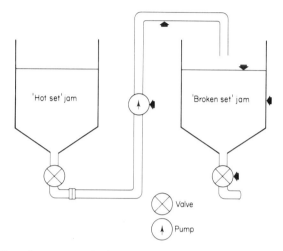

Fig. 4. Sites of yeast contamination during the preparation of 'broken set' jam.

Thus, jams as received by the baker, whether of the boxed, canned or 'broken set' variety, are likely to contain few if any yeasts. When fermentation problems occur, therefore, it is usually due to build up of contamination in the equipment used to transfer and dispense the jam on to the products. Like all high sugar coatings and fillings, jam is sticky and difficult to remove completely when washing equipment and it is all too easy to leave behind deposits where yeasts can multiply and contaminate further batches passing through the equipment. If such equipment is not dried carefully, yeasts will proliferate rapidly in the weak sugar solution which may remain behind.

(ii) Fondant and fudge type icings

Blocks of fondant as received by the bakery are rarely contaminated with yeasts and it is only when other ingredients, such as chocolate or sugar syrups, are added that yeasts may be present.

At the temperatures of 36–43°C which exist in holding tanks or enrobers containing fondant, yeasts will survive but growth, if any, will be slow. It is only when the fondant or fondant-coated goods are stored at ambient temperatures that multiplication will occur.

Most fermentation problems associated with fondant are caused by the continual feedback of material from the previous production. If such leftover fondant is stored for several days at ambient before re-use, any yeasts present will multiply so that large numbers will be added to the next mix. Infrequent and inadequate cleaning of equipment can also contribute to the problems.

(iii) *Marzipan and nut pastes*

Osmophilic yeasts can occur in the blanched and nibbed almonds used for manufacturing marzipan or almond paste. These yeasts are usually destroyed by the heat treatments during manufacture. Occasionally some yeasts survive but they do not multiply rapidly in the product if it is properly formulated.

Problems of fermentation often result from improper handling and storage of leftover paste. After coating and cutting products, such as battenburg, the trimmings of paste may be retained and fed back slowly into subsequent batches. Crumbs of cake attached to these trimmings have the effect of increasing the a_w so that it is more conducive to yeast multiplication. If successive batches of leftover paste are stored overnight or over weekends at ambient temperatures, a gradual build up of yeast contamination occurs so that eventually the product starts to show evidence of fermentation within a short period of manufacture. The situation is aggravated further if equipment used to handle the paste is not adequately cleaned after use.

B. Factors affecting the growth of yeasts in high sugar fillings and coatings

The time taken for a given high sugar coating or filling to show evidence of fermentation depends on a number of factors including the a_w of the substrate, the storage temperature, the number and type of yeasts initially present, the way in which the yeasts are introduced into the product, and the presence or absence of inhibiting substances.

(i) *Osmotic shock effects*

The rate at which yeasts will multiply in a high sugar filling or coating can be markedly affected by the way the yeasts are introduced. This is illustrated in Fig. 5 which gives the result from a test where two jams of different a_w were inoculated with *ca.* 800 cells/g of an osmophilic strain of *Sacch. bisporus* var. *bisporus*. Two methods of inoculation were used: with the first a portion of fermenting jam of known yeast count was thoroughly mixed into the jams; with the second a suspension of the yeast in 10% sucrose solution was mixed into the jams. The various inoculated jams were stored at 25°C and yeast counts made at regular intervals.

When the jams were inoculated with yeasts in a medium of similar a_w (i.e fermenting jam) the count increased regularly with storage time, the rate of multiplication, as expected, being more rapid in the higher a_w than in

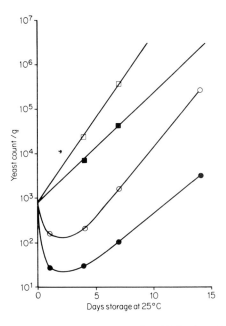

Fig. 5. Effect of the a_w of the inoculum on the rate of multiplication of yeast in jam. □, jam of a_w 0·78–0·79 inoculated with jam; ○, same jam with inoculum in sugar solution; ■, jam of a_w 0·75–0·76 inoculated with jam; ●, same jam with inoculum in sugar solution.

the lower a_w jam. On the other hand, when the yeasts were added in weak sugar solution, the count decreased during the early stages of storage before eventually increasing and multiplying at much the same rate as in the samples inoculated with fermenting jam.

The initial fall in the numbers of yeasts recovered from the jams inoculated with a weak sugar solution can be explained on the basis of 'osmotic shock'. In other words, when the yeasts in the sugar solution of high a_w and thus low osmotic pressure are mixed into jam with a lower a_w and higher osmotic pressure, a certain number of cells are killed because of the sudden change in pressure and those which survive take some time to adapt to the change in conditions. Similar effects with salt and sugars have been described by Onishi (1957).

From this test it is evident that the time taken for a high sugar filling or coating to show signs of fermentation will depend on the way it becomes contaminated with yeasts. No osmotic shock effect is likely to occur in cases where the yeasts survive heat treatment or where contamination arises through contact with deposits of fermenting material on equipment. On the other hand, an osmotic shock effect is likely to occur if residues of weak sugar solutions containing yeasts are left behind after equipment has been washed and these become mixed into the filling or coating. A third situation can occur if the weak sugar solution is not mixed into the filling or coating but remains in contact with it for some time, thus causing local dilution of the material. Exposure of hygroscopic high sugar materials to a humid atmosphere may also cause local dilution on the surfaces, as would storage under conditions which cause condensation. In these instances, there is less chance of osmotic shock and any yeasts present are likely to adapt quickly to growth in the filling or coating. Moreover, when large numbers of actively fermenting yeasts are present the sugars will be consumed, resulting in local conditions of higher a_w which will allow rapid yeast multiplication. While osmotic shock is unlikely to occur in these static situations, it is probable that shock, to a greater or lesser extent, will occur when the diluted sites become mixed into the bulk of the coating or filling.

(ii) a_w *and inoculum level*

Probably the most important factors governing the rate of fermentation are the a_w of the substrate and the number of yeast cells originally present. To determine the effect of these factors, a comprehensive series of tests has been carried out in which six jams of a_w varying from 0·73 to 0·83 were inoculated with different numbers of yeasts by mixing with suitable aliquots from a sample of freshly fermenting jam. The yeast present was *Sacch. baillii* var. *osmophilus*. After having carried out counts of the yeasts initially present, each sample of inoculated jam was stored in a plastic pot in an incubator at 25°C and examined at regular intervals for visible signs of fermentation, i.e. the appearance of small bubbles in the jam. The results from these tests are given in Fig. 6 where the fermentation-free shelf life in days at 25°C is plotted against the initial yeast count per gram for the six jams of different a_w.

As expected, the fermentation-free shelf life is increased with both reduction in a_w of the jam and reduction in the number of yeasts initially present. The effect of reduction in the initial inoculum level is particularly striking. In general, a 10-fold reduction in count is sufficient to increase the fermentation-free shelf life of the jam by about 50%. These results stress the importance of hygiene in dealing with fermentation problems.

Tests with other high sugar fillings and coatings such as fondant, marshmallow, sugar syrups etc. suggest that for a given a_w and initial inoculum level of osmophilic yeasts, the fermentation-free shelf life will be similar to that obtained in these tests with jams containing *Sacch. bailii* var. *osmophilus.*

Fermentation problems tend to be encountered more frequently with flour confectionery goods containing high sugar fillings and coatings than

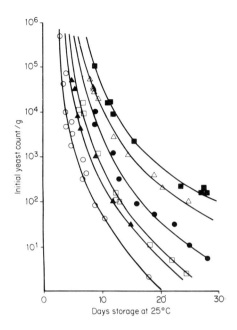

Fig. 6. Effect of a_w and initial yeast inoculum level on the rate of fermentation of jam. a_w Values of jams: ■, 0·73–0·74; △, 0·74–0·75; ●, 0·76–0·77; □, 0·77–0·78; ▲, 0·78–0·79; ○, 0·82–0·83.

with the fillings or coatings themselves. This is due partly to the additional yeast contamination which can occur during the processes of applying the fillings or coatings to the cake and partly to the fact that the a_w of these materials is often increased when they come in contact with the moister cake.

This latter effect is demonstrated in Table 1 which gives the results of one test where jams of a_w 0·78 and 0·73 were inoculated with 500–700 *Sacch. bailii* var. *osmophilus* cells/g and used to fill sponge cakes of a_w 0·84, 0·81 and 0·79. The jam filling from the cakes was examined at intervals during storage at 25°C for soluble solids concentration and evidence of a fermentation odour.

It is apparent that when the jam is introduced into the cakes the soluble solids level is reduced. The extent of this reduction depends on the difference between the a_w of the cake and the jam. As the soluble solids decrease, the a_w increases so that the fermentation-free shelf life of the jam in the cake is shorter than that of the jam on its own.

Similar effects can be expected with high sugar fillings other than jam. The a_w of coatings is less likely to be changed since only one surface comes in contact with the cake and they are subject to moisture loss during storage.

TABLE 1
Fermentation-free shelf life of cakes filled with jams

a_w		Refractometer soluble solids (%)		Approx. fermentation-free shelf life (days at 25°C)	
Jam	Cake	Jam	Jam from cake*	Jam	Cake
0·78	0·84	72·5	68·0	10	7
	0·81	—	69·8	—	9
	0·79	—	70·6	—	13
0·72	0·84	75·3	70·0	21	10
	0·81	—	70·6	—	14
	0·79	—	71·7	—	16

*after 24 h storage at 25°C.

(iii) *Storage temperature*

To determine the effect of storage temperature on the rate at which fermentation occurs in high sugar fillings and coatings, tests have been carried out where a sample of jam of a_w 0·78 was inoculated with approximately 10^3, 10^4 and 10^5 cells/g of *Sacch. rouxii* and separate samples stored at 27, 21, 16 and 10°C until the first signs of fermentation were noted. The fermentation-free shelf life of these jams at the various temperatures is given in Table 2.

TABLE 2
Effect of storage temperature on the fermentation-free shelf life of jam

Initial inoculum level/g	Fermentation-free shelf life (days) at storage temperatures of (°C)			
	27	21	16	10
10^3	10	14	26	>60
10^4	7	10	18	56
10^5	4	6	12	36

The fermentation-free shelf life is increased on average by 44%, 88% and 225% by reducing the storage temperature from 27°C to 21°C, 21°C to 16°C and 16°C to 10°C, respectively.

It is interesting to note that the effect of temperature on the rate of fermentation is similar to that found for the rate of mould growth on bread and cake (Seiler 1976) except at 10°C where fermentation is retarded to a greater extent than mould growth.

These findings emphasize the importance of storage at low temperatures in preventing fermentation spoilage problems associated with high sugar coatings and fillings.

(iv) *pH*

Tests in which jams were buffered to pH 3·5, 3·2 and 2·9 and inoculated with *Sacch. baillii* var. *osmophilus* showed that there was no significant difference in the fermentation-free shelf life at 27°C. English (1954) and Onishi (1963) have shown that the growth of the osmophilic yeast *Sacch. rouxii* in glucose solutions of 46–50% is only marginally affected by pH within the range 3·0–7·0. Thus, in respect of pH it would appear that rate of fermentation closely parallels the rate of growth in the substrate. On this assumption, it is unlikely that pH is a factor of importance in governing the rate of fermentation of most high sugar coatings and fillings.

(v) *Species of osmophilic yeast*

There is a large number of different species of osmophilic and osmo-tolerant yeasts which are capable of growing and causing fermentation of high sugar fillings and coatings. Tilbury (1976) lists 38 of the more commonly encountered which have been involved in spoilage of high sugar products. In our experience the species most frequently isolated from fermented bakery fillings and coatings are *Sacch. rouxii, Sacch. baillii* var. *osmophilus, Sacch. bisporus* var. *mellis* and *Sacch. bisporus* var. *bisporus*.

In the tests to determine the effect of a_w, inoculum level, temperature and pH described previously, the yeast species present in the jams used was *Sacch. baillii* var. *osmophilus*. Work is at present in hand to determine to what extent the results obtained with this organism are typical. Preliminary findings indicate that in a high sugar syrup of a_w 0·78, strains of *Sacch. rouxii, Sacch. bisporus* var. *mellis, Schizosaccharomyces pombe* and *Schiz. octosporus* produce visible evidence of fermentation rather earlier than *Sacch. baillii* var. *osmophilus*. The difference in fermentation-free shelf life between the various osmophilic species was small but sufficient to suggest that, for a given a_w and initial inoculum level, the shelf life may sometimes be slightly shorter than that indicated in Fig. 6.

(vi) *Preservatives*

Under the Preservatives in Food Regulations 1975, (Anon. 1975), flour confectionery products are allowed to contain up to 1000 parts/10^6 of sorbic acid or propionic acid or equivalent amounts of their calcium, potassium or sodium salts. Ingredients intended for use in the preparation of flour confectionery, such as high sugar coatings and fillings, are also permitted to contain these preservatives.

The results from one test carried out to determine the effectiveness of potassium sorbate in delaying fermentation of jam are given in Table 3.

A sample of jam with a_w of 0·82 was inoculated with *ca.* 10^3 cells of *Sacch. baillii* var. *osmophilus*/g. This was subdivided into three portions to which were added 0, 100 and 250 parts/10^6 of potassium sorbate. At each level of sorbate, samples were prepared with pH adjusted to 3·6, 3·2 and 3·0 and stored at 25°C until visible signs of fermentation were evident.

TABLE 3

Effect of pH and potassium sorbate concentration on the
fermentation-free shelf life of jam

Conc. of potassium sorbate in jam (parts/10^6)	Fermentation-free shelf life (days at 25°C) at pH		
	3·6	3·2	3·0
0	4·5	4·5	4·5
100	7	8·5	10·5
250	24	26	29

As expected, the effectiveness of sorbate is increased as the pH decreases. With 100 parts/10^6 sorbate the increase in shelf life is small and unlikely to be of commercial significance but with 250 parts/10^6 sorbate, the fermentation-free shelf life is increased by 430–540%. At the 250 parts/10^6 level a slight off-odour was noticeable in the jam.

Further tests using jam of rather lower a_w than that used in the above test, indicate that sufficiently long increases in fermentation-free shelf life can be obtained using 150–200 parts/10^6 potassium sorbate without any off-odour being noted.

It should be appreciated that when a high sugar coating or filling comes into contact with cake its pH may be changed. It has been shown for instance, that when jam of pH 3·6 is used to fill a cake with a pH of 6·0, the filling gradually becomes less acid during storage and eventually reaches a pH of 4·3. This change in pH affects the activity of a preservative such as potassium sorbate so that, whereas 150–200 might be required to prevent fermentation of the jam on its own, at least twice as much may be required to prevent fermentation in cake filled with this jam.

Little work appears to have been carried out to determine the effectiveness of sorbic acid or its salts in high sugar fillings and coatings other than jam. Since materials such as fudge, fondant, marshmallow, marzipan and nut pastes usually have a pH within the range 6·0–7·0 it is likely that concentrations approaching the maximum permitted level may be required to obtain a suitably long extension in the fermentation-free shelf life of these fillings and coatings.

C. Preventive measures

It is usually found that excessive contamination with osmophilic yeasts is the major cause of fermentation problems. This contamination frequently results from inadequate cleaning and disinfection of plant and equipment used to handle high sugar fillings and coatings. To ensure thorough cleaning of machinery such as depositors and enrobers they need to be dismantled completely at regular intervals and each part brushed to remove food deposits. All hoppers, holding tanks, kettles, conveyor belts, small utensils, savoy bags etc. should be thoroughly cleaned after use.

Where fermentation problems are being encountered it is desirable to disinfect after cleaning. Disinfectants such as hypochlorite or quaternary ammonium compounds can be used for this purpose but, since osmophilic yeasts tend to be more resistant than bacteria, it is necessary to use a concentration some 3–4 times higher than that recommended for destroying bacteria. After cleaning and disinfection, it is important to ensure that all plant and equipment is properly dried.

All stocks of high sugar fillings and coatings should be stored in a cool place and used in strict rotation. Any part-used containers must be placed in a refrigerator, kept covered and used as soon as possible. Material which is to be fed back into subsequent mixes, such as marzipan or nut pastes and pie pastry should be stored in a refrigerator and used within 24 h. At least once a week a virgin batch should be prepared which contains no 'leftovers'.

I would like to thank all my colleagues of the Research Association who assisted in the work to determine the factors affecting the fermentation-free shelf life of high sugar coatings and fillings described in this paper.

5. References

ANON. 1975 Preservatives in Food Regulations 1975. Statutory Instruments 1975 No. 1487. London: HMSO.

CHAMBERLAIN, N., COLLINS, T. H. & ELTON, G. A. H. 1962 The Chorleywood bread process. *Bakers' Digest* **36**, 52–53.

CHEN, R. W. 1979 Refrigerated doughs. *Cereal Foods World* **24**, 46–47.

ENGLISH, M. P. 1954 The physiology of *Saccharomyces rouxii*. *Journal of General Microbiology* **10**, 328–336.

GEMEINHARDT, H. & BERGMANN, I. 1977 Zum Vorkommen von Schimmelpilzen in Bäckereistäuben (Moulds in bakery dusts). *Zentralblatt für Bakteriologie, Parasitenkunde, Infektionskrankheiten und Hygiene* Abt. II **132**, 44–54.

KLINE, L., SUGIHARA, T. F. & McCREADY, L. B. 1970 Nature of San Francisco sour dough French bread process. *Bakers' Digest* **44**, 51–57.

LAMPRECHT, F. 1955 Zur Frage der Bekampfung des Brotschimmels insbesondere des Kreideschimmels. *Brot und Gëback* **9**, 26–29.

ONISHI, H. 1957 Studies on osmophilic yeasts, 1. Salt-tolerance and sugar-tolerance of osmophilic soy yeasts. *Bulletin of the Agricultural Chemical Society of Japan* **21**, 137–145.

ONISHI, H. 1963 Osmophilic yeasts. *Advances in Food Research* **12**, 53–94.

SEILER, D. A. L. 1976 The stability of intermediate moisture foods, with respect to mould growth. In *Intermediate Moisture Foods* ed. Davies, R., Birch, G. G. & Parker, K. J. pp. 166–181. London: Applied Science.

SCHULTZ, A. 1948 Uber neue Brotkrankheiten durch Kreide- und hoppenschimmel. *Getreide Mehl und Brot* **2**, 153–157.

SYKES, H. G. 1971 The role of yeasts in modern bakery practice. *British Baker* **163**, 28, 31–32, 48, 52, 55.

TILBURY, R. H. 1976 The microbial stability of intermediate moisture foods with respect to yeasts. In *Intermediate Moisture Foods*, ed. Davies, R., Birch, G. G. & Parker, K. J. London: Applied Science.

Xerotolerant (Osmophilic) Yeasts

R. H. TILBURY

*Tate & Lyle Limited, Group Research & Development,
Reading, Berkshire, UK*

Contents

1. Introduction

IN HIS REVIEW of eukaryotic water relations, Brown (1978) states that "water activity is a valuable and probably the most useful parameter currently available to describe microbial water relations in complex media". This largely reinforces the view of Mossel (1975) although the latter author expressed reservations about the use of this term and suggested that the relative water vapour pressure (p_w) might be more appropriate. A detailed discussion of water activity (a_w) is outside the scope of this chapter, but useful references are the classic paper of Scott (1957) and books edited by Duckworth (1975), Davies *et al.* (1976) and Troller & Christian (1978).

Yeasts capable of growth at low a_w values or high solute concentrations have been classified variously as osmophilic (Christian 1963), osmotophilic (van der Walt 1970), osmotolerant (Anand & Brown 1968), osmoduric (van der Walt 1970), osmotrophic (Sand 1973), xerophilic (Pitt 1975) and xerotolerant (Brown 1976). Strictly the term 'xerotolerant' should be used in preference to 'osmophilic' and other terms because these yeasts do not have a general requirement for decreased a_w or high osmotic pressure but merely tolerate drier conditions better than do non-osmotolerant yeasts (Anand & Brown 1968). However, the name osmophilic has been in common usage since it was first coined by von Richter (1912) and will be adopted here for convenience.

Various definitions of osmophilic yeasts have been proposed, usually based on an organism's ability to grow at a particular a_w value or sugar concentration. Christian (1963) defined osmophilic yeasts as those capable of multiplication in concentrated syrups of a_w below 0·85 whilst Scarr & Rose (1966) chose a limit of 65° Brix sucrose, equivalent to an a_w of 0·865 at 25°C. An a_w of 0·87 was regarded by Mossel (1971) as the lowest limit for growth of non-osmophilic yeasts. Van der Walt (1970) recommended the use of agar media containing 50% w/w and 60% w/w glucose to distinguish between 'osmotolerant' and 'osmotophilic' yeasts, equivalent to a_w values of *ca.* 0·90 and 0·85, respectively. In the sugar industry, osmophilic yeasts are conveniently defined in practical terms as those capable of growth in a saturated sucrose solution. This is approximately 67° Brix at 25°C, equivalent to an a_w of 0·85. Finally, Pitt (1975) defined xerophilic fungi, including osmophilic yeasts, as those "capable of growth under at least one set of environment conditions, at an a_w below 0·85". Choice of this limit was empirical, but in view of its practical value, it will be used here.

Since the paper by Scott (1957), several reviews have been published dealing with osmophilic yeasts and/or their role in food spoilage, notably those of Ingram (1957, 1958); Christian (1963); Onishi (1963); Walker & Ayres (1970); Kushner (1971) and Tilbury (1976). The Ph.D. theses of Anand (1969), Koh (1972) and Corry (1974) contain useful critical reviews, mainly concerned with physiological aspects of osmophilic yeasts. Spoilage of foods of plant origin by xerophilic fungi was reviewed by Pitt (1975), and Mossel (1975) discussed the relationships between water and micro-organisms in foods. Corry (1978) described the relationships of a_w to fungal growth in foods and beverages. Practical aspects of xerotolerant yeasts growing at high sugar concentrations were reviewed by the author (Tilbury 1980). Mechanisms of resistance of micro-organisms to water stress were fully discussed by Brown (1976, 1978).

The present chapter is a general review of osmophilic yeasts, mainly covering publications over the last decade. It attempts to highlight some unique aspects of this group, of interest to both academic and industrial microbiologists.

2. Principal Species

Table 1 lists those species of osmophilic yeast which are most commonly associated with spoilage of high solute content foods. The data are partly extracted from the review by Walker & Ayres (1970) and partly from additional or more recent publications (Tilbury 1976). The yeasts are named and classified according to Lodder (1970), but common synonyms and names of perfect/imperfect forms are included for ease of reference to earlier literature.

Some attempt has been made to indicate the sugar and salt tolerance of the organisms using maximum concentrations cited for the species in Lodder (1970). This information is incomplete as in some genera only salt tolerance is quoted, in others only sugar tolerance, and in several genera combined salt and sugar tolerance was tested. In Lodder (1970) sugar tolerance is given in terms of ability to grow in media containing 50% and 60% w/w glucose, equivalent to a_w values of *ca.* 0·90 and 0·845, respectively; these values are not necessarily the maximum sugar concentrations tolerated. For comparison, a_w values of salt concentrations are as follows: 10% w/v $= a_w$ 0·94; 15% $= 0·895$ and 20% $= 0·845$. Additional information on sugar tolerance is taken from three sources: Scarr & Rose (1966) identified yeasts capable of growth in 65% w/w sucrose syrups; Tilbury (1967) determined the minimum a_w for growth in sucrose/glycerol syrups; Windisch & Neumann (1965) tested isolates for ability to ferment fructose solutions containing 45, 60 or 75% sugar. According to the terminology of van der Walt (1970) yeasts that can grow at concentrations of 50% but not 60% w/w glucose are called osmoduric or osmotolerant, whereas those capable of growth at 60% w/w glucose are osmotophilic. This principle was adopted by Davenport (1975 and pers. comm.) but the names were shortened to 'osmotolerant' and 'osmophilic'. On this basis, only some of the species listed in Table 1 may be classified as truly osmophilic. According to Davenport (pers. comm.) the principal osmophilic yeasts are as follows: *Sacch. rouxii, Sacch. bailii* var. *osmophilus, Sacch. bisporus* var. *mellis, T. lactis-condensi, Schiz. pombe, Deb. hansenii, T. candida, P. ohmeri* and *H. anomala* var. *anomala.*

The most common and also the most xerotolerant osmophilic yeast is *Sacch. rouxii*, strains of which resist high concentrations of both sugar and

TABLE 1

Osmotolerant and osmophilic yeasts, according to the classification and description of Lodder (1970)

Organism	Synonyms: perfect (P) imperfect (I) forms	a_w Tolerance maximum solute concentration or minimum a_w for growth		Type of commodity commonly spoilt	
		Glucose (%w/w)	NaCl (%w/w)	High sugar	High salt
C. guilliermondii	(P) P. guilliermondii	57*	13	+	
C. krusei			10	+	+
C. parapsilosis		60	17	+	
C. tropicalis			13	+	
C. valida	(P) P. membranaefaciens		9	+	+
Cit. matritensis	(I) T. globosa	57*	(5 + 10)	+	
Deb. hansenii	(I) T. candida	50		+	+
E. burtonii	E. chodatii	50		+	
H. anomala var. anomala	(I) C. pelliculosa	0·75†	(5 + 10)	+	
H. polymorpha			(5 + 10)	+	
H. subpelliculosa			(5 + 10)	+	+
Kl. apiculata		50		+	
P. farinosa		50		+	
P. guilliermondii	(I) C. guilliermondii	50		+	
P. membranaefaciens	(I) C. valida	50		+	+
P. ohmeri	(I) E. ohmeri	50		+	+
Sacch. bailii var. bailii	Sacch. acidifaciens; Sacch. elegans	50		+	+
Sacch. bailii var. osmophilus		60		+	+
Sacch. bisporus var. bisporus		50		+	
Sacch. bisporus var. mellis	Sacch. mellis	0·70†, 60		+	
Sacch. microellipsoides var. osmophilus		60		+	
Sacch. rosei		60		+	+
Sacch. rouxii	many Zygosacch. spp.	0·65, 60		+	+
Schiz. octosporus		50		+	
Schiz. pombe		50		+	
T. apicola	T. bacillaris	57*	11	+	
T. candida	(P) Deb. hansenii, T. famata	0·65†	21	+	+
T. colliculosa		75	13	+	
T. dattila			15	+	
T. etchellsii		0·70†	21	+	+
T. glabrata			13	+	
T. kestoni		57*		+	
T. lactis-condensi		57*	11	+	
T. magnoliae			11	+	
T. versatilis	Br. versatilis	0·70†	13	+	+

*57% w/w glucose = 65° Brix sucrose = a_w 0·865. Data taken from Scarr & Rose (1966).
†Minimum a_w for growth in sucrose/glycerol syrups, determined by Tilbury (1967).

salt. Many strains formerly classified as *Zygosaccharomyces* sp. are now placed in this species. Less common but also very xerotolerant, is *Sacch. bisporus* var. *mellis*, formerly called *Sacch. mellis*. *Saccharomyces bailii* var. *osmophilus* is a xerotolerant variety which was previously known as *Sacch. acidifaciens*. *Hansenula anomala* is often found in materials of high solute concentration but it is not very xerotolerant. The perfect yeast *Deb. hansenii* is salt-tolerant and is commonly associated with fermentations of brined vegetables and occurs also in the marine environment (Norkrans 1966). Its imperfect form, *T. candida* (formerly *T. famata*) is encountered in high sugar products. Several other species of *Torulopsis* are xerotolerant, e.g. *T. etchellsii*, *T. versatilis*, but they are slower-growing than the *Saccharomyces* species.

Some important physiological properties of the main osmophilic yeasts are summarized in Table 2. This information is useful in distinguishing species and in assessing the spoilage potential of particular foods by osmophilic yeasts.

TABLE 2

Some physiological properties of common osmophilic yeasts (Lodder 1970)

Physiological characteristic		*Sacch. bailii* var. *osmophilus*	*Sacch. bisporus* var. *mellis*	*Sacch. rouxii*	*Deb. hansenii* and *T. candida*	*H. anomala* var. *anomala*
Sugar fermentation and assimilation	glucose	F	F	F	Fvw; A	F
	galactose	Av	Av	A	Fvw; A	F
	sucrose	Fv; Av	Fv; Avw	Fv; Av	Fvw; A	F
	maltose	NA	NA	F	Fvw; A	Fv; A
	lactose	NA	NA	NA	Av	NA
	raffinose	Fv; Av	NA	NA	A	F
Carbon assimilation	soluble starch	NA	NA	NA	Av	A
	ethanol	A	A	Av	A	A
	glycerol	A	A	Av	A	A
	sorbitol	A	A	A	A	A
Assimilation of KNO$_3$		NA	NA	NA	NA	A
Growth in vitamin-free medium		−	−	−	v	+
Growth at 37°C		−	−	−	v	v

F, fermented and assimilated; NA, not assimilated; A, assimilated but not fermented; w, weak; v, variable: some strains positive, others negative.

3. Occurrence

Where do osmophilic yeasts occur in nature? The obvious answer is in spoiling high-sugar or high-salt foods, but how do these become contaminated? Relatively little is known about the occurrence and survival of these organisms in air, soil or water. They have been found in the soil of apiaries (Lochhead & Farrell 1930) and vineyards (Davenport 1975), in the air of sugar factories (Tilbury 1967) and in mummified fruits (Davenport 1975). It is well known that insects such as bees and wasps carry an indigenous population of osmophilic yeasts, which will be transmitted to nectaries of fruits, honey etc. (Ingram 1958; Lund 1958; Walker & Ayres 1970). It is possible that piles of raw sugar may become infected in this way (Scarr & Rose 1966). Certain plant fluxes contain osmophilic yeasts such as *Citeromyces matritensis* (Ch. 5, this volume). However, the principal source of infection of foods appears to be by physical contact with previously contaminated residues on dirty processing plant (Ingram 1958; Tilbury 1967).

TABLE 3

Traditional intermediate moisture foods susceptible to spoilage by osmophilic yeasts

Product type	Examples	Water content (%w/w)	Solute content (%w/w)	Water activity (a_w)
Syrups, sugars, sweet spreads and preserves	Raw cane sugar	0·4–0·7	99·3–99·6	0·60–0·75
	Refined sucrose syrup	33·3	66·7	0·85
	Glucose or inverted syrup	20·0	80·0	0·72
	Barley syrup; malt extract	20–25	75–80	0·70–0·80
	Maple syrup	26–36	64–74	0·70–0·80
	Honey, jam, marmalade	20–35	65–80	0·75–0·80
Fruit juice, concentrates	Orange juice	35	65	0·80–0·84
	Raspberry	35	65	0·79–0·80
Confectionery products	Marzipan	15–17	83–85	0·75–0·80
	Glace cherries	30	70	0·75
	Toffees and caramels	8	92	0·60–0·65
Bakery products	Fruit cakes	20–28	72–80	0·73–0·83
	Christmas pudding	20–25	75–80	0·70–0·77
Dairy products	Sweetened condensed milk	30	70	0·83
Dried fruits	Prunes and figs	20	80	0·68
	Dates	12–25	75–88	0·60–0·65
Cereals	Flour, rice, pulses	16–19	81–84	0·80–0·87
	Rolled oats	10	90	0·65–0·75
Brined and pickled vegetables, sauces	Lactic acid fermentations: cucumbers, olives; acetic acid pickles	76–90	10–24	0·79–0·94
Meats	Fermented sausages, e.g. Hungarian salami, country-cured hams	74	26	0·83 0·87

Many types of food, raw material or food ingredient rely upon reduced a_w to inhibit bacterial spoilage. Traditional methods of food preservation often achieved a 'safe' a_w by addition of sugars, salt or by sun drying, whilst modern techniques such as vacuum evaporation and freeze-drying achieve the same effect by concentration. Such foods, with a_w values in the range of 0·60–0·85, are now grouped together as 'intermediate moisture foods' (Table 3). Although these products may be safe with respect to growth of food-borne pathogens (Leistner & Rödel 1976), they are susceptible to spoilage by xerotolerant yeasts and moulds unless pasteurized or chemically preserved (Seiler 1976; Tilbury 1976). The frequency and extent of these spoilage outbreaks is difficult to assess, due to commercial secrecy but undoubtedly they are of common occurrence. Literature up to 1969 was comprehensively reviewed by Walker & Ayres (1970) whilst information since that date was surveyed by the author (Tilbury 1976). Table 4 summarizes some of the data from the latter paper. Most of the unpublished results cited by the author arose from confidential work done on behalf of company customers. Similar data accumulated by Davenport (pers. comm.) are presented in Table 5.

TABLE 4

Osmophilic yeast spoilage of some high sugar foods

High sugar food	Predominant spoilage organisms	Reference
Raw cane sugar	*Sacch. rouxii; T. candida*	Tilbury (1967)
Refined sugars and syrups	*T. apicola; T. globosa* *T. lactis-condensi* *T. kestoni; C. guilliermondii* *Sacch. florentinus*	Scarr & Rose (1966)
Molasses	*Sacch. heterogenicus; T. holmii*	Scarr (unpublished)
Malt extract	*Sacch. rouxii*	
Fruit juice concentrates		Sand (1973)
low a_w, no preservative	*Sacch. rouxii; Sacch. bisporus*	
higher a_w, with preservative	*Sacch. bailii*	
Chocolate syrup	*T. etchellsii; T. versatilis* *C. pelliculosa*	Tilbury (unpublished)
Strawberry and apricot jams	*T. colliculosa; T. cantarellii*	Tilbury (unpublished)
Jams	*Sacch. bisporus*	Seiler (1976)
Strawberry concentrate	*T. versatilis; C. pelliculosa* *C. utilis*	Tilbury (unpublished)
Crystallized and syruped ginger	*Sacch. rouxii*	Lloyd (1975a)
Comminuted orange base and cordial	*Sacch. bailii* var. *bailii*	Lloyd (1975b)
Fruit concentrates	*Sacch. bailii* var. *bailii*	Pitt & Richardson (1973)
Glacé cherries	*Sacch. rouxii*	Tilbury (unpublished)
Marzipan and persipan	*Sacch. rouxii* *T. dattila*	Windisch & Neumann (1965)
Soft-centred Easter eggs	*Sacch. rouxii*	Tilbury (unpublished)

Data from Tilbury (1976).

TABLE 5

Occurrence of dominant spoilage yeasts in high sugar products,
Long Ashton Research Station, Bristol, 1969–1975

	Fruit concentrates (sp. gr. > 1305)				Sugar syrups	
Organism	Apple	Grape	Orange	Unknown	Brown	White
Osmophilic yeasts						
Sacch. rouxii	+	+	+	+	+	+
Sacch. bisporus var. *mellis*	−	+	+	−	−	−
Sacch. bailii var. *osmophilus*	+	+	+	+	+	+
Sacch. cerevisiae	+	+	+	−	−	−
Sacch. bayanus	−	−	−	+	−	−
Saccharomyces spp.	+	+	+	+	+	+
Schiz. pombe	−	−	+	−	−	−
T. versatilis	−	−	+	−	−	−
Osmotolerant yeasts						
C. valida	+	+	−	+	+	+
H. anomala var. *anomala*	+	+	−	+	+	+
Kl. apiculata	+	−	−	−	−	+
Candida spp.	+	+	−	+	−	+
Torulopsis spp.	+	+	+	+	−	+
Sacch. cerevisiae	+	+	+	+	−	+
Saccharomyces spp.	+	+	+	+	+	+

+, present; −, absent.
Data from Davenport (pers. comm.).

4. Economic Effects and Control of Spoilage

In foodstuffs with an a_w below 0·85, microbial spoilage is only possible by xerophilic fungi. In such a selective environment the relatively slow growth rate of osmophilic yeasts ceases to be a disadvantage. Yeast cells possess great metabolic activity so that fermentation and visible spoilage may be evident at relatively low population levels, e.g. 10^5–10^6 cells/ml (Ingram 1958). Whilst both xerotolerant yeasts and moulds are able to grow at low pH values and in high sugar concentrations, yeasts possess some competitive advantages over moulds. For example, their ability to grow anaerobically and to produce and tolerate ethanol, CO_2 and organic acids. Their disadvantage is the inability to penetrate and colonize solid substrates, which may explain why spoilage of liquid products tends to be by yeasts whilst solid products tend to be spoilt by moulds.

Typical symptoms of yeast spoilage are visible signs of fermentation such as turbidity, gas evolution with swelling and explosion of containers, alcoholic and fruity aromas, and discoloration (Fig. 1). Other effects include loss of solids and increase in moisture content.

Figures on the extent of financial losses caused by xerotolerant yeast

Fig. 1. Biodeterioration of packeted brown sugar by osmophilic yeasts: pale patches indicate areas of microbial growth

spoilage of foods are extremely hard to locate. No manufacturer likes to wash his dirty linen in public so there is nothing in the literature. Nevertheless it is known from personal experience dealing with customers in the sugar industry that problems are common, especially amongst the smaller producers, and the consequences can be severe. How then can this spoilage be prevented?

Methods such as heat sterilization, refrigeration or deep-freezing are usually unsuitable for high sugar content foods because of caramellization or high cost. More acceptable methods are elimination of infection by good manufacturing practice (GMP), pasteurization, filtration of liquids, u.v. irradiation, and use of chemical preservatives, humectants and acidulants. In many cases use is made of the 'hurdle' effect by combining several of these treatments (Leistner & Rödel 1976).

A. Good manufacturing practice

Advantage may be taken of the slow growth rate and increased lag phase of osmophilic yeasts at low a_w values, to increase the shelf life of products by minimizing the initial infection. This can be achieved by GMP, including the use of good quality raw materials, plant sanitation, hygienic handling, prevention of delays in processing, exclusion of insects, safe packaging etc. Details of specific applications may be found for fruit juice concentrates (Sand 1973), raw sugar production (Tilbury 1967), jams (Horner & Anagnostopoulos 1973; Seiler 1976), confectionery (Mossel & Sand 1968) and bakery products (Seiler 1966, 1975).

B. Pasteurization

It is well known through the work of Gibson (1973) and Corry (1974, 1976) that high solute concentrations enhance the heat resistance of yeasts, although at equal a_w values different solutes give differing degrees of protection. This has obvious implications for the design of pasteurization processes for high sugar foods. Values for the heat-resistance of yeasts such as those obtained by Put et $al.$ (1976) and Put (Ch. 10, this volume) at high a_w levels cannot be used safely to calculate pasteurization processes for low a_w materials. Furthermore, it has been shown (Wilson et $al.$ 1978) that response to heat treatment in the presence of solutes differs between yeast vegetative cells and ascopores.

In the light of these facts, it is essential that a manufacturer should do laboratory tests with his particular low a_w product to ensure that adequate pasteurization is achieved. For processes involving liquid products like syrups, malt extract and fruit juice, concentration by vacuum evaporation normally achieves an adequate pasteurization; likewise in manufacture of jams and confectionery products similar results are obtained by heating to dissolve the solids. For absolute safety, however, a flash pasteurization is advisable. Typical treatments are 20 s at 85°C for a 50° Brix concentrated orange juice and 30 s at 90°C for a 75° Brix chocolate-flavoured syrup. Lloyd (1975a) successfully 'pasteurized' syruped ginger by dipping it into 80° Brix sucrose syrup at 93 ± 1°C for a minimum of 2 min.

C. Antimicrobial preservatives

Many high sugar foods are susceptible to spoilage because their a_w is above the minimum for growth of xerotolerant yeasts and moulds. In products which are used periodically and which require a long shelf life at ambient temperature (e.g. jams, sweet pickles, fruit drink concentrates), initial pasteurization or GMP does not guarantee freedom from post-process contamination and subsequent spoilage. Here it is necessary to add anti-microbial preservatives, especially antimycotics.

Practical and legislative aspects of food preservatives were described by Jarvis & Burke (1976). Choice of antimycotics is restricted mainly to the following weak organic acids or their salts: benzoic acid; esters of *para*hydroxybenzoic acid; propionic acid; sorbic acid. Sulphurous acid in the form of SO_2, sulphite or bisulphite is widely permitted also. A problem in the application of these acid preservatives is that their antimicrobial activity is due to the undissociated acid and hence is strongly pH-dependent (Freese *et al.* 1973). Other weak organic acids such as acetic, citric and lactic acids are not classified as food preservatives but they are widely used in foods as acidulants. They exhibit some preservative activity which is primarily a function of their pH, although at the same pH value acetic acid is more inhibitory than lactic acid (Kimble 1977). There may be interactions between acidulants and other preservatives, e.g. lactic acid is synergistic with sorbate (Sahoo 1971).

In the UK, sulphite, benzoate and *para*hydroxybenzoate esters are the most widely permitted food preservatives. Within the EEC there is pressure to reduce permitted levels of SO_2 in foods, which has stimulated the search for effective alternatives, especially in fruit-based drinks (Ashworth & Jarvis 1975). It is recognized that SO_2 is not very effective against yeasts, especially in the presence of large amounts of reducing substances or where the pH exceeds 3·0. Mixtures of SO_2 and benzoate are thought to be synergistic (Rehm & Wittmann 1962), and are used successfully to preserve low acid foods and drinks against yeast spoilage. Sorbic acid is a better antimycotic than benzoic acid but at present its use in the UK is limited (Anon. 1977). Proposals to widen its application have been made and some have been approved (Anon. 1979).

Benzoate alone or sorbate alone were found to be successful alternatives to SO_2 alone or SO_2 plus benzoate in the preservation of orange concentrate and orange drinks, against the osmophilic yeasts *Sacch. bailii* and *Schizo. pombe* (Ashworth & Jarvis 1975). Further work confirmed that sorbate at 250–500 parts/10^6 gave a similar or better shelf life for fruit-based drinks than standard preservatives such as benzoic acid (550 parts/10^6) plus SO_2 (40 parts/10^6) (Patel 1977). In low solids jam, however, sorbate at 125–250 parts/10^6 inhibited xerophilic mould spoilage but did not prevent growth of *Sacch. bailii* and *Schiz. pombe* (Jarvis 1975).

It is now well known that some spoilage yeasts develop tolerance to organic acid preservatives (Pitt & Richardson 1973; Warth 1977). The most resistant organism is the acid-tolerant and sugar-tolerant yeast, *Sacch. bailii* (formerly *Sacch. acidifaciens*). Ingram (1960*a, b*) described strains of *Sacch. acidifaciens* resistant to 500 parts/10^6 of benzoic acid in fruit squashes at pH 3·5. In spoilt fruit juice concentrates, Sand (1973) reported that 500 strains of *Sacch. bailii* could tolerate 1000 parts/10^6 of benzoic acid at pH 3·0, a_w 0·94. Lloyd (1975*b*) obtained similar results in Australian comminuted orange products, but sorbic acid inhibited *Sacch. bailii* var. *bailii* at concentrations between 400 and 800 parts/10^6. Use of sorbic acid, however, was precluded as it interfered with product colour; SO_2 at the maximum permitted level of 230 parts/10^6 was found to be the only acceptable preservative, provided the pH was below 3·0. Pitt (1974) found that *Sacch. bailii* was the most preservative-resistant spoilage yeast, tolerating 600 parts/10^6 of both benzoic and sorbic acids at pH 2·5, in the presence of 10% glucose. Other yeasts, including *C. krusei* and *T. holmii* were inhibited by 300–400 parts/10^6 of these preservatives.

Sugar and salt are traditional food preservatives but their use concentration is limited by their solubility and organoleptic acceptability. At use concentrations they may inhibit bacterial growth but only retard yeast and mould spoilage. These disadvantages have stimulated the search for new humectants in intermediate moisture foods (Karel 1976). Ideally such compounds should have a bland flavour and possess antimicrobial properties in addition to their humectant role. The polyhydric alcohols appear promising in these respects and already glycerol and propylene glycol are used commercially as humectants in human and pet foods, respectively. A mixture of propylene glycol (5–15%) and potassium sorbate (0·1–0·2%) is reported to be a most effective mycostat in IMF petfoods (Burrows & Barker 1976). Sinskey (1976) showed that other aliphatic diols or their esters were more effective than glycerol against bacteria but did not test osmophilic yeasts. However, the approval of regulatory authorities remains to be granted for these and many other promising new preservatives.

D. 'Hurdle' effect — production of liquid sugars

In the production of liquid sugar, the 'hurdle' principle of preservation is used to good effect. The aim is to produce a product which conforms to the American Bottler's Association standard of less than 1 yeast cell/g of sugar (Scarr 1963). After vacuum evaporation to 67° Brix the liquid sugar is filtered through a diatomaceous earth pre-coat filter and cooled prior to filling into steam-sterilized, stainless steel, closed storage tanks. Sterile warm air may be blown over the headspace of the tanks to prevent condensation and dilution of the product, or u.v. lamps may be fitted in the headspace as an alternative. Microbiological air filters are fitted to the tanks, and in some cases both air and low coloured liquid products may also be passed through a u.v. sterilizer. All subsequent handling of product occurs in steam-sterilized pipes, tanks and road vehicles.

5. Methodology

A. Media and methods for isolation and enumeration

There are no commercially available dehydrated selective media for osmophilic yeasts, but a number of suitable media have been described in the literature. The important features of medium composition, as discussed by Sand (1973) are its a_w, pH, redox potential and nutrient availability. Obviously a_w is the key factor which enables selection of osmophilic yeasts and inhibition of bacteria, but a relatively low pH is also useful. Ingram (1959) described a 50% glucose–citric acid–tryptone agar for isolation of osmotolerant yeasts from concentrated orange juice, but it proved tedious to prepare. In our laboratory we routinely use Scarr's osmophilic agar for enumeration of osmophilic yeasts in sugar products (Scarr 1959). It is simple to prepare and easy to use, but its relatively high a_w of ca. 0·95 occasionally permits bacterial growth; certainly it will support growth of some non-osmophilic yeasts. Van der Walt (1970) recommended 50% and 60% w/w glucose agars, at a_w values of ca. 0·90 and 0·845 for characterization of osmotolerant and osmophilic yeasts, respectively. Mossel (1951) prefers a 60% w/w fructose–peptone–meat extract agar, at an a_w of 0·84 since fructose is more soluble than glucose. Most recently Pitt (1975) described a number of media suitable for cultivation of xerophilic fungi; one was a malt extract-yeast extract agar containing 60% w/w glucose, a_w 0·845.

Scarr's osmophilic agar is prepared as follows:

Make up a 45° Brix syrup containing sucrose, 35 parts; glucose, 10 parts; water, 55 parts. Dissolve Oxoid Wort Agar (CM 247) in this solution (50g/l syrup) and sterilize by autoclaving at 115°C for 15 min. Reinforcement of gel strength by addition of plain agar (10g/l) is desirable where plates are to be used for streaking or spreading. The pH is usually ca. 4·8.

Plates are incubated at 25–30°C and counted after 3 and 5 d. Colonies on this medium are usually well defined and opaque. Osmophilic *Saccharomyces* species form colonies 1–2 mm in diameter after 3 d, whilst the slower growing *Torulopsis* species require 5 d. Examples of colony morphology are shown in Fig. 2.

For routine enumeration of true osmotolerant yeasts in products, Scarr's agar has the convenience of giving a result within 3–5 d, but has the disadvantage of lack of specificity. On the other hand, media of a_w 0·85 possess the advantage of specificity but the disadvantage that incubation periods of up to 28 d are needed. In our laboratory we have attempted to formulate a medium which will select only true osmotolerant yeasts within

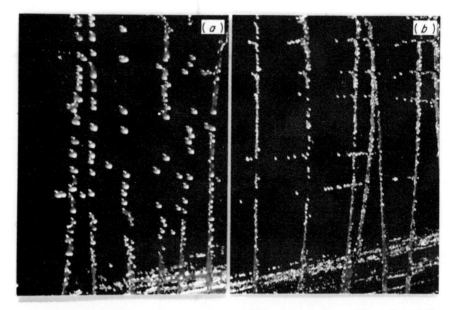

Fig. 2. Colony morphology of two common osmophilic yeasts growing on Scarr's Osmophilic Agar. Cultures were incubated for 3 d at 30°C *(a) Sacch. rouxii; (b) T. versatilis.* (× 2).

a period of 3 d at 30°C. Representative species of true osmotolerant yeasts (six isolates) and yeasts which could grow on Scarr's agar but not in 65° Brix sucrose syrup (12 isolates) were selected from the Tate & Lyle culture collection. Cultures were streaked on Scarr's agar plates modified by increasing the sucrose content in units of 5° from 45° to 65° Brix. The preliminary results (unpublished) showed that at 50° Brix all the true osmotolerant yeasts exhibited good growth within 3 d whilst none of the non-osmotolerant yeasts had grown. Confirmatory work needs to be done on this modification of Scarr's agar.

Viable counts of osmotolerant yeasts may be done by traditional techniques such as membrane filtration, direct plating or by serial dilution followed by pour plates or spread plates. Techniques suitable for use with fruit juice concentrates were described by Sand (1973) but they are of general applicability. Two special features should be noted. Firstly, in order to avoid loss of viability of cells due to osmotic shock in preparing serial dilutions of high sugar foods (Ingram 1957), the diluent should be adjusted to a low a_w value (Mossel & Sand 1968). Sterile glucose or sucrose (20%) solutions are suitable for this purpose, and should also be used for washing membranes. Secondly, where extended incubation periods of up to four weeks at 25–30°C are employed, drying out of the medium should be prevented by placing Petri dishes in closed boxes, polythene bags or desiccator jars.

Some osmophilic yeasts, particularly *Sacch. bisporus* var. *mellis* form clumps when growing in high sugar foods. It was reported by Eddy (1955) that use of mannose or maltose (20%) diluent is effective in breaking up clumps of *Sacch. cerevisiae*. Mild sonication may also be used (Seiler, Ch. 8, this volume).

Total counts of osmophilic yeasts may be made by the standard microscopic technique using a haemocytometer at a magnification of × 400. It was observed by Scarr (1968) that osmophilic yeasts growing in concentrated sugar solutions vary in cell density with age and activity. She found that phase contrast examination of 60° Brix syrups enabled distinctions to be made between live, actively growing cells (phase bright), dead cells (phase dark) and moribund cells (grey). Use can be made of this technique to estimate the viable count and activity of osmophilic yeasts in consignments of raw sugar on arrival; this is a rapid way of indicating whether the sugar is likely to deteriorate on storage.

B. Maintenance and identification

Cultures are best maintained on high sugar media, partly because some strains do not grow well on ordinary media like malt extract agar (Sand 1973), and also because some strains may lose their tolerance to high sugar concentrations if kept on ordinary media (Scarr 1954; Onishi 1963). A suitable maintenance medium is osmophilic MYGP broth (Wickerham 1951) which contains (g/l): malt extract, 3; yeast extract, 3; peptone, 5; glucose, 500; pH is not adjusted, sterilized at 115°C for 15 min. Cultures in this medium should be stored at 4°C and subcultured every six months. Some cultures may be freeze-dried successfully using the method of Rose (1970).

Identification of certain osmophilic yeasts, especially *Sacch. rouxii* and *Sacch. bisporus* var. *mellis*, is complicated by the fact that they usually exist in the haplophase and rarely produce ascospores (Davenport, pers. comm.). In recognition of this fact it is easier to omit some of the morphological tests described in Lodder (1970) and use instead the shorter identification schemes of Barnett & Pankhurst (1974) and Davenport (1974). Also, assimilation and fermentation patterns are best determined at sugar concentrations of 10% w/v rather than the usual 2% (Scarr & Rose 1965).

C. Stability and shelf life tests

The rationale of such tests for confectionery products was discussed by Mossel & Sand (1968). Similar principles apply to testing of other high sugar foods and will be mentioned here briefly in the light of experience gained in our laboratory. First, it is important to inoculate the test system with challenge organisms in an active physiological state. Ideally the inoculum should be prepared from cells actively growing in the same medium as the test system. Secondly, shelf life tests should be of long duration due to the slow growth rate and extended lag phase of organisms at low a_w. Thirdly, cycling of incubation temperatures is desirable to simulate day and night temperatures, as this may cause localized a_w changes in the food. Fourthly, adequate numbers of sub-samples should be examined, as yeast growth may be unevenly distributed in the food. Clumping of yeast cells may also lead to falsely low counts; vigorous agitation of dilutions is therefore recommended. Finally, resuscitation of damaged cells, following the principles outlined by Mossel (1975), is essential prior to their enumeration.

6. Physiology and Water Relations

A. Growth rate

Many factors influence the effects of a_w on growth of osmophilic yeasts in high solute foods, but in general, their rate of growth is directly proportional to a_w. Hence the lower the a_w of a food the longer its microbiological shelf life and the fewer the strains of yeasts able to grow in it.

Anand & Brown (1968) measured the growth rates of representative strains of osmophilic and non-osmophilic yeasts in a basal medium adjusted to various a_w levels with polyethylene glycol. None of the osmophilic yeasts showed a requirement for a reduced a_w, but they exhibited relatively broad a_w optima compared with the non-osmophilic yeasts and had growth rates only half as fast at their respective optimum a_w values.

The author (Tilbury 1967) measured the growth rate of *Sacch. rouxii* in sucrose broths whose a_w ranged from 0·997 to 0·935. Optimum growth occurred at an a_w of *ca.* 0·980 (25° Brix sucrose; mean generation time 1·30 h), whilst at a_w 0·935 (50° Brix sucrose) the mean generation time was 2·15 h. Growth rates and minimum a_w for growth were determined for 26 strains of osmophilic yeasts isolated from raw sugar, in sucrose glycerol syrups incubated at 27°C for 12 weeks. Representative results are summarized in Table 6.

Horner & Anagnostopoulos (1973) measured the radial growth rate and lag phase of colonies of *Sacch. rouxii* on sucrose agar over the a_w range 0·997–0·900. Lag of colony formation doubled to 40 h at a_w 0·900 compared with 20 h at a_w 0·960.

The implications of the low growth rates and extended lag phases of

TABLE 6

The effect of a_w *on growth of osmophilic yeasts in sucrose–glycerol syrups after 12 weeks incubation at 27°C*

Organism	Water activity (a_w)								Min a_w for growth
	0·800	0·750	0·725	0·700	0·675	0·650	0·625	0·600	
Sacch. rouxii (1)	NT	4	3	2	1	†	—	—	0·650
Sacch. rouxii (2)	4	3	NT	2	NT	1	NT	—	0·650
Sacch. bisporus var. *mellis*	4	3	NT	1	NT	—	NT	—	0·700
T. candida (1)	4	3	NT	1	NT	1	NT	—	0·650
T. candida (2)	NT	3	2	1	—	—	—	—	0·700
T. versatilis	NT	4	3	2	—	—	—	—	0·700
T. etchellsii	4	3	NT	†	NT	—	NT	—	0·700
H. anomala	2	†	NT	—	NT	—	NT	—	0·750

—, less than 1·5 × original count; †, 1·5–2·0 × original count; 1, 2–5 × original count; 2, 5–10 × original count; 3, 10–20 × original count; 4, more than 20 × original count; NT, not tested.

osmophilic yeasts at low a_w values are that high-solute foods may have a 'safe' shelf life of say, six months, but may spoil after 12 months. Once microbial growth begins, water is released, raising the a_w locally and accelerating the growth of the spoilage organisms (Mossel 1971, 1975).

B. Nature of the solute

Whilst a_w is a useful parameter for defining the effects of high solute concentrations on growth of osmophilic yeasts, it is now clear that the effect of 'total solids' in a food is better (Mossel 1975). At equal a_w values different solutes have different effects on growth. Anand & Brown (1968) showed that polyethylene glycol was more inhibitory than sucrose, whilst Horner & Anagnostopoulos (1973) found that glycerol was tolerated better than sucrose. Koppensteiner & Windisch (1971) determined the limiting values of osmotic pressure for growth of both osmophilic and non-osmophilic yeasts in solutions of a range of sugars and salts. The highest osmotic pressure tolerated (620 atm) occurred in fructose solutions; the salts were generally less well tolerated than the sugars. Salt tolerance increased in accordance with the Hofmeister series, i.e. NaCl was more inhibitory than KCl. Differences in the biological effects of humectants at equal a_w values were reported by Sinskey (1976). It may be concluded that the spoilage potential of a given food at a particular a_w value depends on the nature of the humectant and other food ingredients and should be determined by experiment.

C. Cell size and optical density

It was shown by Rose (1975) that both the cell size and optical density of osmophilic yeasts decreased with reduction in a_w of the suspending medium, using sucrose and polyethylene glycol 200 as solutes. Differences were observed between solutes and also between these yeasts and non-osmophilic yeasts. It was postulated that osmotic pressure and cell permeability of solutes may exert specific effects independently of a_w. Similar general observations were made by Corry (1976), who showed by electron microscopy of freeze-etched preparations that plasmolysis occurred in *Sacch. rouxii* cells in sucrose and sorbitol preparations, but less so in glucose, fructose and glycerol.

D. Temperature

Osmophilic yeasts are similar to mesophilic yeasts in that they grow over the range 0–40°C with an optimum temperature of about 27°C. In general, an increase in temperature increases the sugar concentration (and reduces the optimum a_w) and reduces the range of concentration tolerated (Ingram 1959). In the same way, both the optimum and maximum temperature increase with sugar concentration and reduction in a_w; this may explain why osmophilic spoilage of dried fruits etc. is usually associated with hot climatic conditions (Ingram 1958). For example, *Sacch. rouxii* and *T. kestoni* grow at 37°C on osmophilic agar (45% w/w sugar) but not on media containing 2% sugar (Scarr & Rose 1966). The phenomenon may explain the existence of apparently obligately osmophilic yeasts (Ingram 1957; Onishi 1963). Osmophilic yeasts probably behave like moulds in that they are most tolerant of low a_w near their optimum growth temperature (Pitt 1975).

An important phenomenon occurs when a packaged food is stored at fluctuating temperatures. This results in moisture migration so that certain parts of the food increase in a_w at the expense of others leading to considerable local shifts in both growth rate and selection of microbial types (Mossel 1975).

E. pH

At high levels of a_w, the effects of pH, acidity and buffering capacity of food on osmophilic yeasts are the same as for non-osmophilic yeasts, i.e. growth usually occurs over the range pH 2·0–7·0 with an optimum pH of 4·0–4·5. Hence the spoilage of medium and high acid foods, such as fruit juice concentrates and pickles, is usually caused by yeasts rather than by less acid-tolerant bacteria. It is thought that a reduction in pH below the optimum reduces the maximum concentration of sugar tolerated; similarly, it is likely that a reduction in a_w of the food lessens the tolerance of osmophilic yeasts to high acidity (Ingram 1959). Nevertheless, a few strains are able to tolerate both low a_w and high acidity, e.g. fermentation of 65° Brix concentrated lemon juice at pH 2·0 by certain *Saccharomyces* strains.

F. Redox potential (E_h) and gaseous environment

The majority of osmophilic yeasts are facultative anaerobes that grow most rapidly under aerobic conditions, but they may also produce vigorous fermentation under anaerobic conditions (Pitt 1975). An exception is *Debaryomyces hansenii* which grows poorly in the absence of oxygen. The ecological significance of this is that liquid foods are more susceptible to spoilage by osmophilic yeasts than by xerophilic moulds, which, being obligately aerobic, grow best near or on the surface of solid foods where the E_h is higher. Osmophilic yeasts grow faster on the surface of solid media than in liquid media; in fermenting liquids, growth is often confined to the superficial layers (Ingram 1959). Foods which possess both low E_h and high acidity can only be spoilt by osmophilic yeasts. In sealed containers fermentative spoilage by yeast may yield CO_2 which is inhibitory to many aerobic organisms (Mossel 1971).

G. Chemical composition of the substrate

The chemical composition of a food affects its spoilage potential by osmophilic yeasts primarily with regard to the influence of 'total solids' on its a_w. A secondary factor is nutrient availability. In general, osmophilic yeasts are not fastidious in their nutritional requirements and they are able to grow on a wide range of foods. All osmophilic yeasts can utilize some simple sugars as a carbon source, e.g. glucose, fructose, maltose or sucrose, whilst some can also assimilate organic acids such as lactic and acetic, and a few can hydrolyse starch. In contrast to many bacteria, yeasts are not proteolytic but can feebly attack a wide range of organic nitrogen compounds. Yeasts generally need only small amounts of nitrogenous compounds for growth, and are definitely fermentative rather than putrefactive. Consequently, foods with a high C:N ratio tend to be spoilt by yeasts, whereas foods with a high N:C ratio tend to be spoilt by bacteria. Similarly, yeasts need only small amounts of minerals for growth. As a group, yeasts possess good vitamin-synthesizing ability and hence are largely independent of an external supply (Ingram 1958); many osmophilic yeasts can grow in vitamin-free media. Natural antimicrobial substances may occur in foods and help to inhibit growth of spoilage yeasts, e.g. furfural and its derivatives produced in concentrated sugar syrups which have browned during storage (Ingram *et al.* 1955).

H. *Physical state of the substrate*

There is evidence that the structure and physical form of a food can influence the spoilage flora. Christian (1963) noted that osmophilic yeasts, in contrast to xerophilic moulds, may grow in liquid media at a_w values below 0·65, but not below a_w 0·75 on solid media. Possibly yeasts are less able than moulds to colonize solid substrates due to their lack of motility and non-mycelial mode of growth. An alternative explanation may be that total water content is greater in liquids than in solids at the same a_w. It has been shown that total water content as well as a_w influences microbial growth response and that the minimal a_w for growth is higher in foods whose water content is adjusted by adsorption rather than by desorption (Acott & Labuza 1975). Additionally, oxygen tension varies with the substrate. In nature, the combination of these factors results in the fact that spoilage of 'solid' foods like cereals, nuts and dried fruits tends to be caused by xerophilic moulds, whereas 'liquid' foods like fruit juice concentrates, syrups, jams and brines tend to be spoiled by osmophilic yeasts.

I. *Fructophily*

An unusual physiological property of many osmophilic yeasts is that when they are grown on invert sugar as carbon source, containing equimolecular amounts of glucose and fructose, the fructose is preferentially assimilated; this is termed fructophily. The symptoms of this phenomenon had long been observed in the cane sugar industry but the first taxonomic studies by Peynaud & Domercq (1955) arose from similar observations in the wine industry. They screened 700 strains of yeast isolated from Sauternes grapes and found that *Sacch. bailii* (*Sacch. acidifaciens*), *Sacch. elegans*, *Sacch. rouxii* and *T. bacillaris* were responsible. Later work (Tilbury 1967) on yeasts isolated from raw sugar showed that nearly all the xerotolerant species were fructophilic whilst none of the non-xerotolerant species were (Table 7). A possible explanation is that the cell-membrane sugar permeabilities of xerotolerant species differ from those of non-tolerant species, perhaps due to differences in transport mechanism.

The practical significance of this in the sugar industry is that in deteriorated raw sugar the selective removal of the laevorotatory fructose by osmophilic yeasts increases the dextrorotation of the sugar. This means that a sugar refiner, who buys his raw sugar on the basis of a polarimeter reading, pays for more sucrose than he gets.

TABLE 7

Selective utilization of fructose in invert sugar by yeasts isolated from raw sugar

Group	Organism	No. of strains tested	Mean sugar consumption g/100 ml		
			fructose (F)	glucose (G)	ratio F/G
Osmophilic yeasts	Sacch. rouxii	39	6·6	0·9	7·3
	Sacch. bisporus var. mellis	4	6·5	0·9	7·2
	T. candida	21	2·0	1·3	1·5
	T. etchellsii	3	2·0	1·8	1·1
	T. gropengiesseri	1	1·3	0·8	1·6
	T. versatilis	9	2·1	1·1	1·9
Non-osmophilic yeasts or 'semi-osmophilic yeasts'	H. anomala	1	0	6·0	0
	P. farinosa	1	0·6	1·4	0·43
	Sacch. cerevisiae	3	3·4	10·4	0·33

Yeasts were grown in 35° Brix invert syrup for 9 days at 30°C.
Data from Tilbury (1967).

7. Mechanisms of Xerotolerance

Our knowledge on the mechanisms of xerotolerance is largely due to the work of Brown and his colleagues in Australia, who published a series of papers over the last decade. This work was comprehensively reviewed by Brown (1976, 1978) and a brief summary of his conclusions follows.

Micro-organisms growing in an environment of low a_w have an interior of comparably low a_w. Obviously for growth to occur the enzyme complement must also be functional at this low a_w. This can be achieved either by producing enzymes which are inherently resistant to inhibition or by producing an intracellular environment which is not inhibitory. The halophilic bacteria employ both mechanisms, but the eukaryotes depend solely on modification of their interior by accumulating a compatible solute. It was shown that enzymes in tolerant and non-tolerant eukaryotes are essentially similar in their water relations.

Xerotolerant eukaryotes accumulate high concentrations of polyols in response to a water stress; when the stress is extreme, the polyol is usually glycerol. It is also known that polyols, especially glycerol, confer a remarkable degree of protection on enzymes at low a_w; hence they function as the essential compatible solutes. They also function as osmoregulators under water stress. The problem is that non-tolerant species of the same eukaryotic genera can also accumulate glycerol in response to water stress.

Xerotolerant yeasts accumulate at least one acyclic polyol, commonly arabitol, when grown in a dilute medium; non-tolerant yeasts do not. When grown at low a_w tolerant species such as *Sacch. rouxii* respond by accumulating glycerol whilst its arabitol content remains constant. *Saccharomyces cerevisiae* which is not xerotolerant, also responds to water stress by accumulating glycerol. Over the range of solute concentrations which it can tolerate, *Sacch. cerevisiae* accumulates as much glycerol as does *Sacch. rouxii* under the same conditions.

There is, however, a major difference in the mechanisms by which glycerol accumulation is regulated in the two yeasts. *Saccharomyces rouxii* synthesizes relatively small amounts of additional glycerol in response to diminishing a_w; its major response is to retain within the cell a progressively greater proportion of the glycerol which it synthesizes. On the other hand, *Sacch. cerevisiae* leaks to the growth medium a very high proportion of the glycerol which it produces. It responds to diminishing a_w value by synthesizing much more glycerol and retaining an approximately constant proportion within the cell.

A major cause of the different tolerance ranges of the two yeasts is probably that *Sacch. cerevisiae* eventually directs an unacceptably high proportion of its biosynthetic capacity to glycerol production. Another important difference between the yeasts is that *Sacch. cerevisiae* seems to be generally more permeable than *Sacch. rouxii* to a variety of solutes, a factor which could itself be important in initiating a major metabolic response to water stress in *Sacch. cerevisiae*.

8. A Case History

The following case history based on the author's experience (Tilbury 1967) concerns the biodeterioration of Guyanese raw cane sugar, a serious problem at the time. It is used to illustrate many of the features of osmophilic yeasts mentioned in the preceding pages.

Guyanese sugar was heavily infected with osmophilic yeasts, causing losses of up to 0·5% of sugar in bulk storage, and problems to the refiners due to the fructophily phenomenon previously mentioned. Counts of 10^7 yeasts/g sugar were typical, accompanied by pleasant alcoholic aromas. The principal species were identified as *Sacch. rouxii* and *T. candida* which were shown to be capable of growth at a_w values down to 0·65. A survey of the Guyanese sugars from eleven factories revealed that most of them had an a_w value in excess of 0·65, typically in the 0·70–0·75 region. Not surprisingly a good correlation was found between moisture content and a_w of the sugar. Sugar leaving the crystallizers was free of osmophilic yeasts, but post-process contamination occurred readily in the factory by contact with infected residues of sugar on conveyor belts, bins etc. Growth then occurred quite rapidly in bulk storage.

Why was it that Guyanese raw sugar had this problem of high a_w when sugar from most other countries did not? It transpired that during the separation of crystals from molasses by centrifugation, the sugar was washed with water, which diluted the molasses film surrounding the crystals. This was done to increase the purity of the product to 98% as specified by the refiners. Normally this procedure is unnecessary with modern high speed centrifuges, but the then existing centrifuges were old, low speed models.

Attempts to solve the problem by improvements in factory hygiene were unsuccessful, as was the use of bactericides such as SO_2 in the wash water. Irradiation treatment was discounted on economic grounds. The best solution appeared to be either to install sugar driers or to replace the centrifuges with faster machines. Finally the latter policy was adopted and the biodeterioration problem disappeared.

9. Commercial Applications

Although most of our knowledge concerning osmophilic yeasts originated from their harmful role in causing biodeterioration of high solute foods, their unique properties may also be put to beneficial use. Two such applications are briefly mentioned.

The first concerns the potential commercial production of polyhydric alcohols by fermentation. Early processes for the production of glycerol were developed in Germany during the First World War, based on modification of the ethanolic fermentation by *Sacch. cerevisiae*, but various technical problems prevented them from being economic. Interest was reawakened at the end of the Second World War when groups in the USA and Canada developed improved processes using xerotolerant yeasts such as *Sacch. bailii, Sacch. rouxii* and *T. magnoliae*. This work was reviewed by Spencer (1968) and more recent developments again by Spencer & Spencer (1978). With the increasing hydrocarbon shortage the time may now be ripe to look again at the economics of producing glycerol by fermentation. In addition, other polyols such as arabitol, erythritol and xylitol could also be made by fermentation with xerotolerant yeasts.

Finally, it is relevant to look at fermentation processes which are not new but perhaps a 1000 years old. These are the production of traditional food fermentations of the Orient, such as soy sauce and miso paste. Both these products are staple food-stuffs in Japan, and both involve a primary fermentation by the mould *Aspergillus oryzae* to hydrolyse protein and carbohydrates, followed by a secondary fermentation with halophilic strains of *Sacch. rouxii*. Detailed studies on this fermentation were reviewed by Onishi (1963). Interest in these products is now awakening in the West, and commercial plants are already operating in the USA. It will not be surprising if European food manufacturers soon follow suit. It is to be hoped that such developments will occur, because they should stimulate further research on a remarkable group of micro-organisms.

The author thanks the Directors of Tate & Lyle Ltd for their kind permission to publish this paper.

10. References

ACOTT, K. M. & LABUZA, T. P. 1975 Microbial growth response to water sorption preparation. *Journal of Food Technology* **10**, 603–611.

ANAND, J. C. 1969 *The physiology and biochemistry of sugar-tolerant yeasts*. Ph.D. Thesis, University of New South Wales.

ANAND, J. C. & BROWN, A. D. 1968 Growth rate patterns of the so-called osmophilic and non-osmophilic yeasts in solutions of polyethylene glycol. *Journal of General Microbiology* **52**, 205–212.

ANON. 1977 FAC/Rep/24 1977 Food Additives and Contaminants Committee Report on the Review of the use of Sorbic Acid in Food. London: HMSO.

ANON. 1979 Preservatives in Foods Regulations 1979. Statutory Instruments No. 752/79. London: HMSO.

ASHWORTH, J. & JARVIS, B. 1975 The replacement of sulphur dioxide as the microbial preservative in orange concentrate and orange drinks. *British Food Manufacturing Industries Research Association Research Report No. 227.*

BARNETT, J. A. & PANKHURST, R. J. 1974 *A New Key to the Yeasts.* Amsterdam: North-Holland.

BROWN, A. D. 1976 Microbial water stress. *Bacteriological Reviews* **40**, 803–846.

BROWN, A. D. 1978 Compatible solutes and extreme water stress in eukaryotic micro-organisms. *Advances in Microbial Physiology* **17**, 181–242.

BURROWS, I. E. & BARKER, D. 1976 Intermediate moisture petfoods. In *Intermediate Moisture Foods* ed. Davies, R., Birch, G. G. & Parker, K. J. pp. 43–53. London: Applied Science.

CHRISTIAN, J. H. B. 1963 Water activity and the growth of micro-organisms. In *Recent Advances in Food Science* Vol. 3, ed. Leitch, J. M. & Rhodes, D. N. pp. 248–255. London: Butterworths.

CORRY, J. E. L. 1974 *The effect of sugars and polyols on the heat resistance of salmonellae and osmophilic yeasts.* Ph.D. Thesis, University of Surrey.

CORRY, J. E. L. 1976 The effect of sugars and polyols on the heat resistance and morphology of osmophilic yeasts. *Journal of Applied Bacteriology* **40**, 269–276.

CORRY, J. E. L. 1978 Relationships of water activity to fungal growth. In *Food and Beverage Mycology* ed. Beuchat, L. R. pp. 45–82. Westport, Connecticut: AVI.

DAVENPORT, R. R. 1974 A simple method, using Stripdex equipment, for the assessment of yeast taxonomic data and identification keys. *Journal of Applied Bacteriology* **37**, 269–271.

DAVENPORT, R. R. 1975 *The distribution of yeasts and yeastlike organisms in an English vineyard.* Ph.D. Thesis, University of Bristol.

DAVIES, R., BIRCH, G. G. & PARKER, K. J. ed. 1976 *Intermediate Moisture Foods.* London: Applied Science.

DUCKWORTH, R. B. ed. 1975 *Water Relations of Foods.* London: Academic Press.

EDDY, A. A. 1955 Flocculation characteristics of yeasts. II. Sugars as dispersing agents. *Journal of the Institute of Brewing* **61**, 313–317.

FREESE, E., SHEU, C. W. & GALLIERS, E. 1973 Function of lipophilic acids as antimicrobial food additives. *Nature, London* **241**, 321–325.

GIBSON, B. 1973 The effect of high sugar concentrations on the heat resistance of vegetative micro-organisms. *Journal of Applied Bacteriology* **36**, 365–376.

HORNER, K. J. & ANAGNOSTOPOULOS, G. D. 1973 Combined effects of water activity, pH and temperature on the growth and spoilage potential of fungi. *Journal of Applied Bacteriology* **36**, 427–436.

INGRAM, M. 1957 Micro-organisms resisting high concentration of sugars or salts. In *Microbial Ecology* ed. Pollock, M. R. & Richmond, M. H. pp. 90–133. Symposium of the Society for General Microbiology No. 7. Cambridge: Cambridge University Press.

INGRAM, M. 1958 Yeasts in food spoilage. In *The Chemistry and Biology of Yeasts* ed. Cook, A. H. pp. 603–633. London: Academic Press.

INGRAM, M. 1959 Physiological properties of osmophilic yeasts. *Revue des fermentations et des industries alimentaires* **14**, 23–33.

INGRAM, M. 1960*a* Studies on benzoate-resistant yeasts. *Acta Microbiologica* **7**, 95–105.

INGRAM, M. 1960*b* An influence of carbon source on the resistance of a yeast to benzoic acid. *Annales de l'Institut Pasteur, Paris* **11**, 167–178.

INGRAM, M., MOSSEL, D. A. A. & DE LANGE, P. 1955 Factors produced in sugar-acid browning reactions, which inhibit fermentation. *Chemistry and Industry* **63**.

JARVIS, B. 1975 Sorbic acid as a preservative for 'low solids' jam. *British Food Manufacturing Industries Research Association Technical Circular No. 594.*

JARVIS, B. & BURKE, C. S. 1976 Practical and legislative aspects of the chemical preservation of foods. In *Inhibition and Inactivation of Vegetative Microbes* ed. Skinner, F. A. & Hugo, W. B. pp. 345–367. London: Academic Press.

KAREL, M. 1976 Technology and application of new intermediate moisture foods. In *Intermediate Moisture Foods* ed. Davies, R., Birch, G. G. & Parker, K. J. pp. 4–31. London: Applied Science.

KIMBLE, C. E. 1977 Chemical food preservatives. In *Disinfection, Sterilization and Preservation* 2nd edn, ed. Block, S. S. pp. 834–858. Philadelphia: Lea & Febiger.

KOH, T. Y. 1972 *Studies on the osmophilic yeast.* Ph.D. Thesis, University of Cambridge.

KOPPENSTEINER, G. & WINDISCH, S. 1971 Osmotic pressure as a limiting factor for growth and fermentation of yeasts. *Archiv für Mikrobiologie* **80**, 300–314.

KUSHNER, D. J. 1971 Influence of solutes and ions on micro-organisms. In *Inhibition and Destruction of the Microbial Cell* ed. Hugo, W. B. pp. 259–283. London: Academic Press.

LEISTNER, L. & RÖDEL, W. 1976 The stability of intermediate moisture foods with respect to micro-organisms. In *Intermediate Moisture Foods* ed. Davies, R., Birch, G. G. & Parker, K. J. pp. 120–137. London: Applied Science.

LLOYD, A. C. 1975*a* Osmophilic yeasts in preserved ginger products. *Journal of Food Technology* **10**, 575–581.

LLOYD, A. C. 1975*b* Preservation of comminuted orange products. *Journal of Food Technology* **10**, 565–574.

LOCHHEAD, A. G. & FARRELL, L. 1930 Soil as a source of infection of honey by sugar-tolerant yeasts. *Canadian Journal of Research* **3**, 51–64.

LODDER, J. ed. 1970 *The Yeasts — A Taxonomic Study* 2nd edn. Amsterdam: North-Holland.

LUND, A. 1958 Ecology of yeasts. In *The Chemistry and Biology of Yeasts* ed. Cook, A. H. pp. 63–91. London: Academic Press.

MOSSEL, D. A. A. 1951 Investigation of a case of fermentation in fruit products rich in sugar. *Antonie van Leeuwenhoek* **17**, 146–152.

MOSSEL, D. A. A. 1971 Physiological and metabolic attributes of microbial groups associated with foods. *Journal of Applied Bacteriology* **34**, 95–118.

MOSSEL, D. A. A. 1975 Water and micro-organisms in foods — a synthesis. In *Water Relations of Foods* ed. Duckworth, R. pp. 347–361. London: Academic Press.

MOSSEL, D. A. A. & SAND, F. E. M. J. 1968 Occurrence and prevention of microbial deterioration of confectionery products. *Conserva* **17**, 23–32.

NORKRANS, B. 1966 Studies on marine occurring yeasts: growth related to pH, NaCl concentration and temperature. *Archiv für Mikrobiologie* **54**, 374–392.

ONISHI, H. 1963 Osmophilic yeasts. *Advances in Food Research* **12**, 53–94.

PATEL, M. 1977 Use of sorbic acid as a preservative in fruit-based drinks. *British Food Manufacturing Industries Research Association Technical Circular No. 645.*

PEYNAUD, E. & DOMERCQ, S. 1955 Sur les espéces de levures fermentant sélectivement le fructose. *Annales de l'Institut Pasteur, Paris* **89**, 346–351.

PITT, J. I. 1974 Resistance of some food spoilage yeasts to preservatives. *Food Technology, Australia* **26**, 238–241.

PITT, J. I. 1975 Xerophilic fungi and the spoilage of foods of plant origin. In *Water Relations of Foods* ed. Duckworth, R. pp. 273–307. London: Academic Press.

PITT, J. I. & RICHARDSON, K. C. 1973 Spoilage by preservative-resistant yeasts. *CSIRO Food Research Quarterly* **33**, 80–85.

PUT, H. M. C., DE JONG, J., SAND, F. E. M. J. & VAN GRINSVEN, A. M. 1976 Heat resistance studies on yeast spp. causing spoilage in soft drinks. *Journal of Applied Bacteriology* **40**, 135–152.

REHM, H. J. & WITTMANN, H. 1962 Beitrag zur Kenntnis der antimicrobielle Wirkung der schwefligen Säure. I. Ubersicht uber einflussnehmende Faktoren auf die antimikrobielle Wirkung der schwefligen Säure. *Zeitschrift für Lebensmitteluntersuchung und-Forschung* **118**, 413–429.

ROSE, D. 1970 Some factors influencing the survival of freeze-dried yeast cultures. *Journal of Applied Bacteriology* **33**, 228–232.

ROSE, D. 1975 Physical responses of yeast cells to osmotic shock. *Journal of Applied Bacteriology* **38**, 169–175.

SAHOO, B. N. 1971 Ph.D. Thesis, University of Missouri. Cited by SINSKEY, A. J. 1976 in *Intermediate Moisture Foods* ed. Davies, R. R., Birch, G. G. & Parker, K. J. p. 262. London: Applied Science.

SAND, F. E. M. J. 1973 Recent investigations on the microbiology of fruit juice concentrates. In *Technology of Fruit Juice Concentrates — Chemical Composition of Fruit Juices* Vol. 13, pp. 185–216. Vienna: International Federation of Fruit Juice Producers, Scientific-Technical Commission.

SCARR, M. P. 1954 *Studies on the taxonomy and physiology of osmophilic yeasts isolated from the sugar cane.* Ph.D. Thesis, University of London.

SCARR, M. P. 1959 Selective media used in the microbiological examination of sugar products. *Journal of the Science of Food and Agriculture* **10**, 678–681.

SCARR, M. P. 1963 Microbiological standards for sugar. *Journal of the Science of Food and Agriculture* **14**, 220–223.

SCARR, M. P. 1968 Studies arising from observations of osmophilic yeasts by phase contrast microscopy. *Journal of Applied Bacteriology* **31**, 525–529.

SCARR, M. P. & ROSE, D. 1965 Assimilation and fermentation patterns of osmophilic yeasts in sugar broths at two concentrations. *Nature, London* **207**, 887.

SCARR, M. P. & ROSE, D. 1966 Study of osmophilic yeasts producing invertase. *Journal of General Microbiology* **45**, 9–16.

SCOTT, W. J. 1957 Water relations of food spoilage micro-organisms. *Advances in Food Research* **7**, 83–127.

SEILER, D. A. L. 1966 Fermentation problems in high sugar coatings and fillings. *British Baking Industries Research Association Bulletin No. 2 April*, pp. 49–53.

SEILER, D. A. L. 1975 Fermentation of fudge and fondant. *British Baking Industries Research Association Bulletin No. 2 April* pp. 58–60.

SEILER, D. A. L. 1976 The stability of intermediate moisture foods with respect to mould growth. In *Intermediate Moisture Foods* ed. Davies, R., Birch, G. G. & Parker, K. J. pp. 166–181. London: Applied Science.

SINSKEY, A. J. 1976 New developments in intermediate moisture foods: humectants. In *Intermediate Moisture Foods* ed. Davies, R., Birch, G. G. & Parker, K. J. pp. 260–280. London: Applied Science.

SPENCER, J. F. T. 1968 Production of polyhydric alcohols by yeasts. *Progress in Industrial Microbiology* **7**, 1–42.

SPENCER, J. F. T. & SPENCER, D. M. 1978 Production of polyhydroxy alcohols by osmo-tolerant yeasts. In *Economic Microbiology* Vol. 2, *Primary Products of Metabolism* ed. Rose, A. H. pp. 393–425. London: Academic Press.

TILBURY, R. H. 1967 *Studies on the microbiological deterioration of raw cane sugar.* M.Sc. Thesis, University of Bristol.

TILBURY, R. H. 1976 The microbial stability of intermediate moisture foods with respect to yeasts. In *Intermediate Moisture Foods* ed. Davies, R., Birch, G. G. & Parker, K. J. pp. 138–165. London: Applied Science.

TILBURY, R. H. 1980 Xerotolerant yeasts at high sugar concentrations. In *Microbial Growth and Survival in Extremes of Environment* ed. Corry, J. E. L. & Gould, G. pp. 103–128. Society for Applied Bacteriology Technical Series No. 15. London: Academic Press.

TROLLER, J. A. & CHRISTIAN, J. H. B. 1978 *Water Activity and Food.* London: Academic Press.

VAN DER WALT, J. P. 1970 Criteria and methods used in classification. In *The Yeasts — A Taxonomic Study* 2nd edn, ed. Lodder, J. pp. 34–113. Amsterdam: North-Holland.

VON RICHTER, A. A. 1912 Uber einem osmophilen Organisms den Hefepilz *Zygosaccharomyces mellis acidi* sp. n. *Mykologisches Zentralblatt,* **1,** 67.

WALKER, H. W. & AYRES, J. C. 1970 Yeasts as spoilage organisms. In *The Yeasts* Vol. 3, ed. Rose, A. H. & Harrison, J. S. pp. 463–527. London: Academic Press.

WARTH, A. D. 1977 Mechanism of resistance of *Saccharomyces bailii* to benzoic, sorbic and other weak acids used as food preservatives. *Journal of Applied Bacteriology* **43,** 215–230.

WICKERHAM, L. J. 1951 The taxonomy of yeasts. *Technical Bulletin No. 1029, U.S. Department of Agriculture, Washington, D.C.*

WILSON, J. M., WOOD, J. M. & JARVIS, B. 1978 The heat resistance of vegetative and ascospore forms of yeasts in relation to their environmental conditions. *British Food Manufacturing Industries Research Association Research Report No. 275.*

WINDISCH, S. & NEUMANN, I. 1965 Zur mikrobiologischen Untersuchung von Marzipan. *Susswaren* **7,** 355–358, 484–490, 540–546.

The Heat Resistance of Selected Yeasts Causing Spoilage of Canned Soft Drinks and Fruit Products

HENRIËTTE M. C. PUT AND J. DE JONG

Carl C. Conway Laboratories, Thomassen & Drijver-Verblifa NV, Deventer, The Netherlands

Contents

1. Introduction

ALTHOUGH THE BACTERIOLOGICAL quality of foods has been assessed for many years only recently has the occurrence of yeasts and moulds in foods caused concern. Whereas many fungi (moulds) are able to grow, form mycotoxins and cause diseases under diverse conditions (Graham 1968), yeasts are rarely pathogenic when taken by the oral route.

Both yeasts and moulds are able to spoil food under conditions that are unfavourable for bacterial growth, that is: low pH (< 3.5); low a_w (< 0.90); low temperature ($< 5°C$), (Pederson *et al.* 1961; Senser *et al.* 1967).

The minimal growth conditions for a few yeasts are much lower; i.e. pH 1.5, a_w 0.60, $-34°C$ (Walker 1977). Standards for yeasts in foods have only been established for some non-heat-preserved food products such as dried and frozen eggs, apple juice, margarine (Walker 1977; Roberts 1978).

Yeasts are the most likely cause of spoilage in soft drinks, fruit juices and fruit products with high sugar content (Sand & Kolfschoten 1969; Walker & Ayres 1970; Sand 1973, 1976; Anderson 1977). In addition, yeasts can be important spoilage agents under conditions where bacteria and moulds are not in competition, for example in frozen meat and fish (McCormack 1950); refrigerated foods (Walker & Ayres 1970), dried and crystallized fruits (Ingram 1957, 1958), food brines, fat and food oils (Phaff *et al.* 1966). Spoilage of these products can be controlled by good sanitary practice and storage conditions in combination with appropriate use of heat treatment or preservative agents.

For some time it was thought that both yeast vegetative cells and ascospores showed little difference in resistance to heat, disinfectants and preservatives. Hence, yeast sporulation investigations, unlike bacterial spore studies, have been confined to taxonomy, genetics and hybridization (Fowell 1969; Kreger-van Rij 1969; Lodder 1970; Anderson & Martin 1975; Haber *et al.* 1975; Novak & Zsolt 1977).

This paper deals with the heat resistance of yeasts and yeast ascospores causing spoilage of canned soft drinks and fruit products, and the practical consequences for the heat preservation of these products.

2. Heat Resistance of Sporing and Non-sporing Yeasts Isolated from Canned Soft Drinks and Fruit Products

Some data on the heat resistance of micro-organisms causing spoilage of soft drinks and fruit products are available (Jensen 1960; Dakin 1962; Lott & Carr 1964; Peter 1965; Kozaki *et al.* 1968; Mrozek & Roth 1968; Dittrich 1972; Kopelman & Schyer 1976), although little has been published concerning the efficiency of heat pasteurization or the relationship between initial microflora and its heat resistance and the spoilage of soft drinks and fruit products (Del Vecchio *et al.* 1951; Dayharsh & Del Vecchio 1952; Patashnik 1953; Meyrath 1962; Shapton 1966; Shapton *et al.* 1971; Savel 1974). Preliminary investigations reported here deal with the heat resistance of 120 yeast strains (Table 1).

A standardized method was used to examine isolates from canned spoiled soft drinks, related fruit products and plant equipment. Most attention was focused on the predominant spoilage-causing yeast species (*Saccharomyces cerevisiae, Sacch. bailii, Sacch. uvarum, Sacch. chevalieri*).

TABLE 1

Types of yeast genera and number of species tested

Genus	No. species	No. strains
Asporogenous yeast genera		
Brettanomyces	4	7
Candida	10	18
Kloeckera	2	2
Rhodotorula	2	4
Torulopsis	3	4
Ascomycetous yeast genera		
Debaryomyces	1	1
Hansenula	1	5
Kluyveromyces	2	3
Lodderomyces	1	1
Pichia	2	2
Saccharomyces	10	67
Saccharomycopsis	1	6

The results of these heat resistance studies showed that amongst the ascomycetous yeast strains tested a higher heat resistance was observed than in asporogenous strains. From 35 asporogenous yeast strains tested only *ca.* 25% survived 10 min at 62·5°C and only one strain survived 20 min at 62·5°C. Amongst the 85 ascomycetous yeast strains tested, however, *ca.* 60% survived 10 min at 62·5°C, *ca.* 30% 20 min at 62·5°C, while *ca.* 16% survived 10 min at 65°C. It therefore was suggested that ascospores might determine heat resistance to some extent (Roth 1969). Besides, it could be concluded that the most dominant spoilage yeasts (*Sacch. cerevisiae* and *Sacch. chevalieri*) are also the most heat-resistant species both in heat processing practice (appertization) and in laboratory investigations. Other yeasts showing high heat resistance included strains of *Sacch. bailii, Sacch. uvarum* and *Kluyveromyces bulgaricus* (Put *et al.* 1976).

However, the practical consequences of laboratory studies can only be assessed more clearly after additional investigations of factors that can influence heat resistance of sporing yeasts. Thus, the following factors are of importance when considering heat resistance: the extent and synchronization of sporulation (Roth 1969; Haber & Halvorson 1972); the type and origin of the strain, its laboratory handling and possibly acquired adaptation to heat (Christophersen & Precht 1952); composition of the heating and recovery medium (pH, a_w, CO_2 & O_2 partial pressure) (Perigo *et al.* 1964; Kopelman & Schyer 1976; Stevenson & Richards 1976; Graumlich & Stevenson 1978; Juven *et al.* 1978); and presence of cell clusters in the heating medium (Fisher 1975; Cerf 1977).

3. Heat Resistance of *Saccharomyces* spp.

A. Ascospore density

A standardized Thermal Death Time (TDT) tube test method was used to study the influence of the ascospore density on the heat resistance of equal total cell concentrations of *Sacch. cerevisiae* strain 195 and *Sacch. chevalieri* strain 215 (Put *et al.* 1977).

The results of these heat resistance studies clearly indicate that the thermal death rate (*D* value) of both these strains depends upon the initial ascospore number when equal total cell concentrations and four different proportions of ascospores are heated, for temperatures between 57·5°C and 62·5°C (Table 2). This increase in *D* value is probably more apparent than real because there are two populations of cells present, heat-resistant spores and the much less heat-resistant vegetative cells.

In addition, these data show that *D* values of ascospores are about 10 times higher than those of vegetative cells.

TABLE 2

	Temp. °C		D-values min. (End Point Method)*			
Asci: veg. cells		1:1	$1:5 \times 10^2$	$1:5 \times 10^4$	$<1:5 \times 10^4$	
Sacch. cerevisiae (195)						
	58	23·2	18·1	9·25	3·1	
	60	16·8	11·4	2·75	1·9	
	62·5	7·5	5·0	2·0	0·7	
Asci: veg. cells		1:1	$1:10^2$	$1:10^3$	$1:10^5$	$<1:10^5$
Sacch. chevalieri (215)						
	57·5	25·7	23·0	15·0	10·1	2·6
	59	15·5	13·0	8·7	6·75	1·7
	60	9·0	8·0	5·0	4·0	0·8
	62·5	3·4	2·8	2·2	1·3	0·5

$$*D = \frac{U}{\log a - \log b} \text{ (Schmidt 1957)}$$

when D = decimal reduction time or death rate constant at T °C. This means: time in min to obtain 90% reduction in surviving cell number or time for the thermal death rate curve to transfer one log, and U = heating time in min; a = initial number of vegetative cells, asci or ascospores and b = most probable number of surviving vegetative cells, asci or ascospores.

B. Decimal dilutions of ascospores

Further investigations into the heat resistance of *Saccharomyces* ascospores (strains 195 & 215), using decimal dilutions prepared from the same ascospore stock suspension, showed that the shape and the slope of the

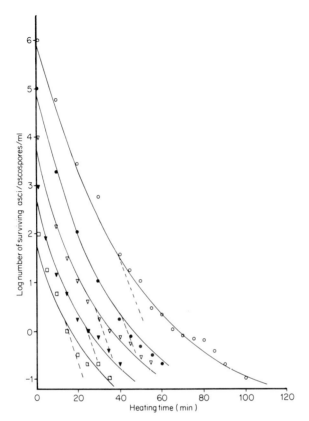

Fig. 1. Thermal Death Rate curves of *Sacch. cerevisiae* 195 at 60°C in 1/20 M phosphate-citrate buffer of pH 4·5. Initial number of asci/ascospores per ml: ○, 10^6; ●, 10^5; ▽, 10^4; ▼, 10^3; □, 10^2. Asci: veg. cells = 2 : 1 (Put *et al.* 1977).

thermal death rate (TDR) curves (*D* values and tail) remain equal and parallel (Fig. 1) showing that the heat resistance of ascospores does not depend on the density of the spore population.

These observations lead to the conclusion that heat resistance is homogeneously distributed among the *Saccharomyces* ascospores tested. (Han 1975; Put *et al.* 1977).

C. D *and* z *values of ascospores*

Using the same standardized TDT tube test method as mentioned above, heat survival and heat destruction times of *Saccharomyces* ascospores of strains 195 & 215 were determined (Fig. 2).

End point *D* values were calculated and plotted on the same graph.

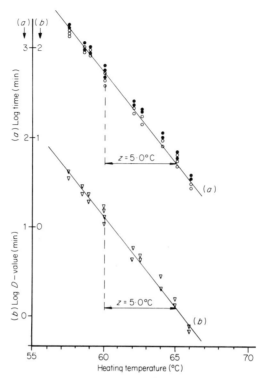

Fig. 2. Thermal Death Time curves of *Sacch. cerevisiae* 195 in 1/20 M phosphate-citrate buffer of pH 4·5. Initial number of asci/ascospores 10^6 per ml. Asci: veg. cells = 2 : 1. ○, survival time; ●, destruction time; ▽, D values (min) (Put *et al.* 1977).

By this calculation method, z values of 5·0° and 5·4° (for strain 195 & 215), respectively, were observed (z = number of °C for a TDT curve to shift by one $\log_{10}$ cycle). Most of the D values of yeast ascospores observed by other authors are much lower, while published z values vary between 3·5° and 10° (Beamer & Tanner 1939*b*; Dakin 1962; Roth 1969).

D. $D_{60°}$ *values of ascospores and vegetative cells*

A 10-fold greater heat resistance of the yeast ascospores than of the corresponding vegetative cells might not only be a phenomenon of strains 195 & 215 but a property common to *Saccharomyces* spp. in general. To test this hypothesis $D_{60°}$ values of 18 *Saccharomyces* strains were determined (Table 3). All except strains 175, 180 and 182 were isolated from heat-processed, bottled or canned soft drinks and fruit products or their raw materials.

TABLE 3

Heat resistance of Saccharomyces *ascospores versus their vegetative cells*

Saccharomyces species	Source	Strain code	$D_{60°}$ value (min) Ascospores	$D_{60°}$ value (min) Veg. cells
S. bailii	I	242	8·1	0·3
S. bailii var. *bailii*	I	164	14·2	0·10
S. bailii var. *osmophilus*	I	144	11·0	0·14
	II	175	19·2	0·11–<0·35
	III	177	8·1	<0·40
	II	180	8·2	<0·35
	III	194	17·0	<0·35
S. cerevisiae	III	195	10·8	<0·35
	III	198	5·1	<0·30
	III	213	12·0	<0·40
	III	219	7·5	<0·35
	III	220	8·5	<0·35
	III	271	17·5	0·32–<0·35
	III	65	16·4	0·10–<0·35
S. chevalieri	III	215	14·3	<0·35
	III	216		<0·30
			$D_{55°}$ *value (min)*	
S. uvarum	II	182	15	0·30
	III	189	8	<0·35

*I, raw material; II, cold-sterilized carbonated soft drink; III, heat-processed soft drink or fruit product (Put & de Jong, unpublished).

Ascospore formation was obtained by the method of Put & de Jong (unpublished). The sporing medium was 2% Difco Malt Extract Agar (*Sacch. bailii*) and Kleyn's acetate agar (remaining *Saccharomyces* species). Ascospore-free cells were grown in BBL Mycophil Broth to which 1000 μg/ml of erythromycin ethylsuccinate (Eli Lilly Benelux, Belgium) were added to inhibit sporulation (Puglisi & Zennaro 1971; Kuenzi *et al.* 1974).

Asco-sporulation inhibition was thought to be necessary since it had been observed that curves of vegetative yeast cells often ended in a short tail, the curve being parallel with that of the corresponding ascospore TDR curve at the same temperature. This phenomenon led us to believe that a low number of ascospores might have been formed even in a typically vegetative growth medium (White 1951; Tingle *et al.* 1973; Corry 1976a).

From the results as given in Table 3, it was calculated that $D_{60°}$ values of the *Saccharomyces* ascospores tested are 25-fold (min.) and 175-fold (max.) higher than $D_{60°}$ values of their corresponding vegetative cells (Put & de Jong, unpublished). Simultaneously z values of some *Saccharomyces* strains have been determined: viz. *Sacch. cerevisiae* and *Sacch. chevalieri* ascospores $z = 4\cdot5°–5\cdot5°$, vegetative cells $z = 5\cdot0°–5\cdot5°$; *Sacch. bailii* ascospores $z = 4\cdot5°–5\cdot0°$, vegetative cells $z = ca.$ $4\cdot0°$ (Put & de Jong, unpublished).

E. Enzymatic digestion of the ascus wall and of vegetative cells

Washed suspensions of *Saccharomyces* ascospores can only be stored at 5°C for 14–21 days. If stored longer they lose their heat resistance. Mature ascospores of *Saccharomyces* spp. remain enclosed in their ascus. Since loss of heat resistance was not due to ascospore germination — as far as could be observed by phase contrast microscopic examination — it was thought that biochemical properties of the ascus epiplasm and of resting vegetative cells might have some influence on the ascospore heat resistance.

We therefore traced the enzymatic activity of glusulase (β-glucuronidase) and mercapto-ethanol for the release of ascospores of *Sacch. bailii* strain 164 and *Sacch. cerevisiae* strain 271 and the digestion of their vegetative cells (Bachmann & Bonner 1959; Johnson & Mortimer 1959; Rousseau & Halvorson 1969; Savarese 1974, 1975).

Simultaneously *K. bulgaricus* strain 120 has been studied, a yeast species showing spontaneous release of mature ascospores which, unlike the *Saccharomyces* ascospores, did not show any loss of heat resistance after *ca.* 2 months storage at 5°C (Lodder 1970; see Section 3).

TABLE 4

Enzymatic release of yeast ascospores and digestion of vegetative cells

	Treatment with mercapto-ethanol and Glusulase for							
Mercapto-ethanol 0·07 mol/l, pH 4·5, 25°C	Nil‡	30 min	Nil			30 min		
Glusulase* 50,000 UF†/ml	Nil	Nil	2 h	4 h	6 h	2 h	4 h	6 h
Release of ascospores (%)								
K. bulgaricus	100	100§	100			100§		
Sacch. bailii	0	0	60	100	—	100	100	—
Sacch. cerevisiae	0	0	30	60	75	60§	90§	100§
Digestion of veg. cells (%)								
K. bulgaricus	0	0	25	35	50	50	60	90
Sacch. bailii	0	0	50	70	—	60	80	—
Sacch. cerevisiae	0	0	0	20	25	0	20	50
D values (min)								
K. bulgaricus $D_{64°}$	9	8	1·5	<1	<1	6	5	4·5
Sacch. bailii $D_{57·5°}$	12·5	11	9·5	9	—	10	9	—
Sacch. cerevisiae $D_{60°}$	15	15	10	10	10	12	10·5	10

*Lyophilized juice of *Helix pomatia* (Industrie Biologique Française, 92 Gennivilliers, France).
†UF, units Fishman.
‡No treatment with mercapto-ethanol (upper row) or Glusulase (lower row).
§Clumping of ascospores.
—, not tested.
Data from Put & de Jong (unpublished).

From the results (Table 4), it can be concluded that 100% release of ascospores can be obtained within 2–4 h by enzymatic activity of Glusulase (*Sacch. bailii*, strain 164) or by Glusulase and mercapto-ethanol (*Sacch. cerevisiae*, strain 271). However, at the same time the ascospore heat resistance will be decreased to a certain extent and this method is therefore not suitable in heat resistance studies.

Other methods, such as French pressure cell, membrane filtration, ultrasonic disruption, glass beads and glass powder and lipolytic agents, are also less suitable on account of their small effect on the release of ascospores, or the loss of ascospore heat resistance resulting from damage of the ascospore wall (Emeis 1958*a*, *b*; Emeis & Gutz 1958; Rousseau & Halvorson 1969; Michener & King 1974; Dawes 1976).

Separation of asci and vegetative cells by centrifugation in a discontinuous sucrose density gradient and electrophoresis is another possible method but it has not been tested by us (Resnick *et al.* 1967; Lewis & Patel 1978).

4. Heat Resistance of *Kluyveromyces* spp.

A. Introduction

Heat resistance studies of *Kluyveromyces* ascospores are reported separately from those on *Saccharomyces* since both genera belong to different taxonomic groups (genera 8 & 16 respectively, Lodder 1970); and they show some typical differences in heat resistance properties (see below). *Kluyveromyces* spp. form one to many ascospores in fragile asci which rupture readily at maturity. Ascospores tend to agglutinate, they are not spherical and are weakly acid-fast. *Saccharomyces* spp. on the contrary, form 1–4 spherical, acid-fast spores in asci that are not ruptured easily at maturity. *Kluyveromyces* ascospore heat resistance studies have not been reported previously.

Kluyveromyces strains tested were: *K. bulgaricus* strain 120 and *K. cicerisporus* strain 277, isolated from canned apple-sauce and strawberries in sugar syrup.

B. Spore production

Spore formation was obtained on Difco Potato Dextrose Agar (PDA, pH 5·5), Difco 2% Malt Extract Agar (MA, pH 4·7) and vegetable juice agar (V8, pH 5·7), incubated for 2–15 days at 25°C (Lodder 1970).

Maximal spore production was observed after 1–2 d at 25°C. Maximal heat resistance was obtained after 2–3 d at 25°C. Ascospores (1–4) in asci were visible only after 1 d at 25°C. Spore suspensions were prepared as previously described for *Saccharomyces* (Put *et al.* 1977). They can be stored at 5°C for 6 weeks without loss of heat resistance.

Free ascospores can be distinguished from vegetative cells by phase-contrast microscopy. Acid-fast staining by Fleming's or Ziehl Neelsen's methods is not suitable, because some spores, being only weakly acid-fast, lose their stain (Miller & Kingsley 1961; Miller & Hoffmann-Ostenhof 1964). Measurement of turbidity and enzyme activity are unsuitable methods (von Bärwald 1972; Bestic & Arnold 1976; see Section 3.*E*).

C. Decimal dilutions of ascospores

While testing the TDR at 60° in decimal dilutions from the same ascospore stock suspension of *K. bulgaricus*, we found that the shape and the slope of the survivor curves did not remain equal and parallel over the whole range (Fig. 3).

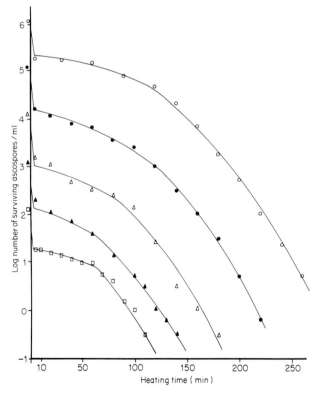

Fig. 3. Thermal Death Rate curves of *K. bulgaricus* 120 at 60°C in 1/20 M phosphate-citrate buffer of pH 4·5. Initial number of ascospores/ml: ○, 10^6; ●, 10^5; △, 10^4; ▲, 10^3; □, 10^2. Ascospores: veg. cells = 1 : 2 (Put & de Jong, unpublished).

Moreover, it was observed that the curves are non-linear, showing two break points: first, a rapid heat inactivation rate of *ca.* 90% of the ascospores during the first few minutes of the heating process ($D_{60°}$ of *ca.* 5 min); secondly, a relatively large shoulder over *ca.* 30% of the total thermal destruction time, shifting the curve by about a half log cycle, and slowly inactivating another 50% of the ascospores present; and thirdly, a linear curve, the slope of which is almost parallel with the death rate curves of the other decimal dilutions of ascospores tested. This part of the curve shows $D_{60°}$ values of *ca.* 40 min, being the heat resistance of the remaining *ca.* 5% of ascospores initially present. By the end point method calculated $D_{60°}$ values vary between 36 and 50 min.

D. The influence of sporing medium on ascospore heat resistance

The highest ascospore heat resistance of *Kluyveromyces* strains 120 and 277 was observed when spores were produced on PDA and incubated for 3–14 d at 25°C (see Figs 4, 5, Table 5). Also it was noted that the shape of the

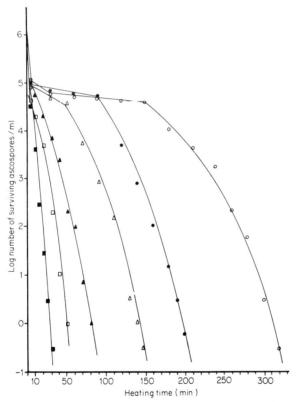

Fig. 4. Thermal Death Rate curves of *K. bulgaricus* 120 in 1/20 M phosphate-citrate buffer of pH 4·5. Initial number of ascospores 5 × 10^5/ml. Heating temp. (°C): ○, 60°; ●, 61°; △, 62°; ▲, 63°; □, 64°; ■, 65°. Ascospores: veg. cells = 1 : 1. (Put & de Jong, unpublished).

TABLE 5
Kluyveromyces ascospore heat resistance

Sporing medium	Incubation at 25°C (d)	$D_{60°}$ values (min) min.	max.	mean	No. tests	z-values (°C)	No. tests
K. bulgaricus 120							
PDA	3–14	32	50	40	20	5–6*	4
MA	3–14	15	30	20	5	—	1
MA †	3–4		30		2	—	—
	12		16		2	—	—
PDA †	3–4		44		2	—	—
	12		36		2	—	—
V8	7		19		1	—	—
	1		3·3		2	—	—
	1·5		20		2	—	—
	2	35	40		2	—	—
PDA ‡	3		48		2	—	—
	5		42		2	—	—
	7		42		2	—	—
	10		42		2	—	—
	14	32	37		2	—	—
K. cicerisporus 277							
PDA	3–7	24	45	35	4	ca. 7	2
PDA	4		45		1	—	—
MA	4		35		1	—	—

*See Fig. 5.
†See Fig. 6.
‡See Fig. 4.
F, end point method (Schmidt 1957); PDA, Potato Dextrose Agar (Difco); MA, Malt Extract Agar (Difco); V8, Vegetable Juice Agar (Lodder 1970); —, not tested.
Data from Put & de Jong (unpublished).

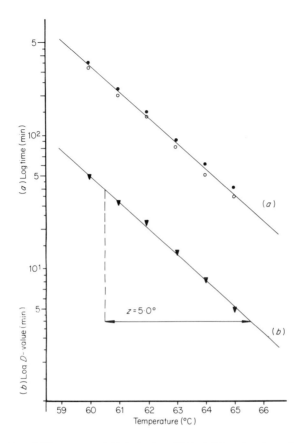

Fig. 5. Thermal Death Time curves of *K. bulgaricus* 120 in 1/20 M phosphate-citrate buffer of pH 4·5. Initial number of ascospores 5 × 10⁵/ml. ○, survival time; ●, destruction time; ▼, *D* values (min) (Put & de Jong, unpublished).

survivor curve is related to the sporing medium used (Fig. 6), PDA showing a typical shoulder, MA and V8 showing an "ever"-decreasing rate of inactivation (Roberts & Hitchins 1969; Toda 1970). On the other hand, time of incubation in the sporing medium, and thus the sporulation time, is another important factor in the mechanism of *Kluyveromyces* ascospore heat resistance (Fig. 7).

E. D *and* z *values of ascospores and vegetative cells*

For these studies the same method as described for *Saccharomyces* spp. was used (see Section 3.*D*).

The results for vegetative cells of *Kluyveromyces* strains 120 and 277 show $D_{60°}$ values of *ca.* 0·20 and 0·1 min and z values of 4·5 and 4·8°, respectively.

The average $D_{60°}$ values of the corresponding ascospores observed are 40 and 35 min, respectively (Fig. 5); *ca.* 200–350-fold higher than the $D_{60°}$ values of vegetative cells (Table 5).

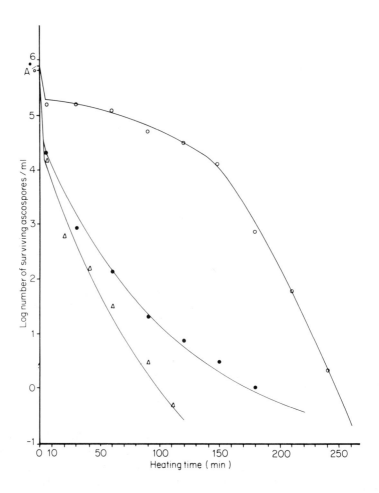

Fig. 6. Thermal Death Rate curves of *K. bulgaricus* 120 at 60°C in 1/20 M phosphate-citrate buffer of pH 4·5. Initial number of ascospores, 10^6 per ml. Sporulation medium: ○, PDA (Potato Dextrose Agar); ●, MA (Malt Juice Extract Agar); △, V8 (Vegetable Agar). Ascospores: veg. cells = 1 : 2; 1 : 3; 1 : 4 respectively. (Put & de Jong, unpublished).

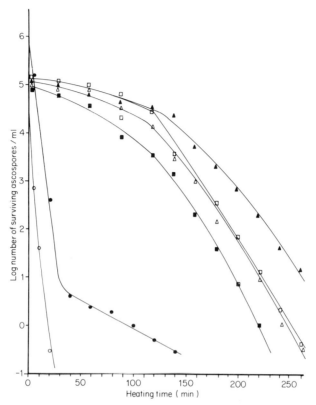

Fig. 7. Thermal Death Rate curves of *K. bulgaricus* 120 at 60°C in 1/20 M phosphate-citrate buffer of pH 4·5. Initial number of ascospores, 10^6 per ml. Incubation time of sporulation medium (PDA) at 25° (d): ○, 1; ●, 1·5; △, 2; ▲, 3; □, 7; ■, 14 (Put & de Jong, unpublished).

5. The Influence of Sugar and Salt (a_w) on the Heat Resistance of Yeast Ascospores

A. Introduction

It is known that the heat resistance of vegetative cells of osmophilic yeasts (*Sacch. rouxii*) is increased in solutions of sucrose (Gibson 1973). Corry (1976a) established that the heat resistance of vegetative yeast cells does not depend directly on the a_w of the solution, but on the type of solute used.

When testing the effect of sugars, maximal heat resistance was found in solutions of sucrose. In addition, microscopic studies indicated that the sugar solutions cause shrinkage of the yeast cells as a result of plasmolysis (cell dehydration). At the same time there is a weak penetration of the sugar solute into the cell (Rose 1975; Corry 1976b). The yeast cell shrinkage was greater in non-osmophilic, than in osmophilic yeast strains (Brown 1974), while the greatest increase in heat resistance corresponded with the greatest cell shrinkage or degree of plasmolysis (Corry 1976b).

Resistance of bacterial endospores to heat may also result from dehydration, in this case, however, of the central protoplast. The heat resistance is increased when the a_w of the heating menstruum is decreased, although this depends on the type of solute and not on the a_w. Comparing solutions at the same a_w, sucrose, a large molecule, shows a low bacterial spore penetration ability, corresponding with a greater increase in heat resistance than is obtained by small NaCl molecules which penetrate much more easily into the spore (Gould & Dring 1975; Gould 1977).

As it was thought that certain similarities might exist between the mechanisms of heat resistance of vegetative yeast cells, yeast ascospores and bacterial spores, investigations into the influence of a_w and a_w-decreasing solutes on the heat resistance of yeast ascospores were thought necessary (Put *et al.* 1977).

Additionally, natural fruit juices, fruit juice concentrates, soft drink concentrates, fruits in sugar syrup, soft drinks and jellies and jams that are heat preserved, vary in a_w (0·75–0·995) and in solute, e.g. sugar, concentrations (1–65%). Thus data on the heat inactivation of yeasts and yeast ascospores in relation to a_w, solute type and solute concentration should be available for calculating the effect of the application of an industrial heat treatment (Windish & Neumann 1969, 1970).

B. Test method

Yeast strains tested were: *K. bulgaricus* 120; *K. cicerisporus* 277; *Sacch. bailii* 144, 164; *Sacch. cerevisiae* 175, 271; *Sacch. chevalieri* 65.

Spore suspensions were prepared as described in Sections 3 and 4. Spore equilibration was obtained by centrifuging the ascospore suspension at 5°C and resuspending it in a buffered (pH 4·5) solution of sucrose or NaCl (a_w 0·995, 0·975 or 0·950), followed by storage at 5°C for 16 h, after which the initial spore concentration was assessed by plate count on BBL Mycophil Agar incubated for 10 d at 25°C. At the same time the heat resistance was tested by the TDT tube method (Put *et al.* 1977).

Surviving ascospores were counted on BBL Mycophil Agar. Decimal dilutions were prepared in tap water. A Sina equihygroscope (Sina AG, Zürich) was used for a_w measurements (Leistner & Rödel 1974; Gál 1975).

C. $D_{60°}$ values

$D_{60°}$ values, calculated by the end point method (Schmidt 1957) are tabulated in Table 6. Some z values are given in the same table.

TABLE 6
D60° values* of yeast ascospores in different solutes

Species		Buffer 1/20 mol/l	NaCl solutions			Sucrose solutions		
			0·995 / 1	0·975 / 4	0·950 / 8	0·995 / 10	0·975 / 30	0·950 / 45
K. bulgaricus	120	44 8–9† (z = 5·6°)	45–50 8–9† (z = 5·0°)	50–55 9–10‡ (z = 5·0°)	10–12‡	8‡	10‡	16‡
K. cicerisporus	277	24 (z = 7·3°)	24 (z = 7·3°)	30 (z = 8·2°)	36 (z = 8·4°)	30	40 (z = 4·0°) 70† (z = 4·0°)	54 (z = 7·2°)
Sacch. bailii	144	7–8·5 (z = 4·8°)	8·2	9·4	—	8	11	21
	164	10 (z = 4·5°)	8	20	—	8	16	20
Sacch. cerevisiae	175	15·5 (z = 5·5°)	20	20	—	26	30	—
	271	12·5 (z = 4·5°)	16	16	—	42	38	—
Sacch. chevalieri	65	10·5 (z = 5·0°)	15	16·5	—	36	41	—
	215	10 (z = 5·4°)	9	10	12	11	15	—

Values of a_w (upper row) and % w/v of solute (lower row) of

*End Point Method (Schmidt 1957).
†In strawberry juice, see Fig. 6 (Put & de Jong, unpublished).
‡D_{64°
—, not tested.

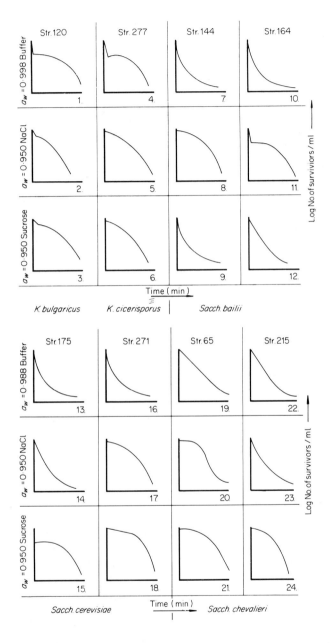

Fig. 8. Schematic shapes and slopes of Thermal Death Rate curves of *Kluyveromyces* and *Saccharomyces* ascospores at 60°C (pH 4·5) as influenced by solutes of $a_w = 0.950$ (Put & de Jong, unpublished).

The TDR curves are not similar in shape, (Fig. 8): they are influenced by the a_w, the type of solute and the yeast species.

D. Scanning electron microscope (SEM) studies

Yeast ascospores, equilibrated in 1/40 mol/l buffer, pH 4·5 (a_w of 0·995) in polyethyleneglycol, sucrose or NaCl (a_w of 0·950) were prepared for SEM studies by a membrane method as described by Davenport (1976) and Kurtzman et al. (1975). SEM preparations and photomicrographs were made by Dr R. R. Davenport, University of Bristol Research Station, Long Ashton, Bristol, UK. The preparations were examined in a Cambridge Type MK IIA SEM at an accelerating voltage of 10 kV; the final aperture was 142·5 μm, and the specimen stage was inclined 45° to the incident beam (Kurtzman et al. 1975).

Measurements of the size of the long axis showed enlargement of the ascospores at lower a_w in solutes of sucrose and NaCl, while at the same time cracks and gaps between cracks in the ascospore walls were observed (Figs 9, 10 & 11).

E. Transmission electron microscope (TEM) studies

In addition some TEM examinations of K. bulgaricus 120 ascospores before and after heat inactivation in a buffered suspension of pH 4·5 and $a_w = 0·998$ were made by Dr N. J. W. Kreger-van Rij (Fig. 12).

Heat-inactivated ascospores show a greyish layer around the protoplasm membrane, perhaps indicating leakage of the membrane as a result of multiple membrane injuries (Kreger-van Rij, pers. comm.).

6. Conclusions

Spoilage of heat preserved soft drinks, fruit juices, fruits in sugar syrup, fruit concentrates, jellies and jams is predominantly caused by ascosporogenous yeasts. The thermal death rate of ascomycetous yeasts depends upon the initial ascospore number; the higher the initial ascospore density of a cell suspension, the higher the D value (Put et al. 1977; Table 2).

The heat resistance ($D_{60°}$ values) of ascospores of 20 yeast strains: Sacch. bailii (3); Sacch. cerevisiae (10); Sacch. chevalieri (3); Sacch. uvarum (2); K. bulgaricus (1); K. cicerisporus (1); proved to be 25-fold (minimum) to 350-fold (maximum) higher than the $D_{60°}$ values of the corresponding vegetative cells (Put & de Jong, unpublished; Table 3). Such a pronounced difference between the heat resistance of vegetative cells and that of the corresponding ascospores has only exceptionally been observed by other authors (Kozaki et al. 1968; Mori et al. 1971). This phenomenon can probably be related to the diploid character of the yeast cell cycle, giving rise to spontaneous sporulation of yeasts growing in a complete medium (Tingle et al. 1973; Hartwell 1974). Consequently, small numbers of yeast ascospores that can be formed among a vegetative population, exhibiting a considerably higher heat resistance, may have led several authors to report only slight differences between the heat resistance of vegetative cells and the corresponding ascospores (Beamer & Tanner 1939a, b; Lund 1951; Roth 1969; Mori et al. 1971; Kuenzi et al. 1974; Fowell 1975).

Heat resistance of yeast ascospores is less dependent on the species than on the strain. It should therefore be taken into account that industrial applied (sublethal) heat processes may show a typical selection of yeasts which have a high heat resistance, either innate or acquired by adaptation.

Some other factors that may influence the yeast ascospore heat resistance are: temperature adaptation of the species or strain during the heat treatment, resulting in a so-called continuously decreasing death rate or tailing of the survivor curve (Komemushi et al. 1968; Han et al. 1976; Tomlins & Ordal 1976; Sharpe & Bektash 1977); a pre-selection of a heat resistant ascospore followed by crossing or hybridization (Emeis & Gutz 1958; Fowell 1969); sporing medium and incubation time/temperature — the higher the temperature for growth and sporulation, the higher the ascospore heat resistance (Fowell 1967; Fowell & Moorse 1960; Warth 1978); sporulation asynchrony (Haber & Halvorson 1972; Fast 1973); clumping of spores during the heat treatment (Fisher 1975; Cerf 1977; Cerf & Métro 1977); the enumeration method for the recovery of heat-surviving ascospores (Stumbo 1965; Roberts & Hitchins 1969; Graumlich & Stevenson 1978). Also, a decrease in the a_w of the heating menstruum ($a_w = 0.950$) increases the heat resistance, which is less dependent on the a_w than on the a_w-decreasing solute (von Schelhorn 1956; Corry 1976a; Smittle 1977).

Most of the heat survival curves of the yeast ascospores tested are not straight logarithmic, but tend to tail off (*Saccharomyces* spp.) or to form a shoulder (*Kluyveromyces* spp.; see Fig. 4), which is probably due to heterogeneity of the ascospore population. Conversely, it was observed that the shape and the slope of survival curves of *Sacch. cerevisiae*, *Sacch. chevalieri* and *K. bulgaricus* remain almost equal and parallel when decimal dilutions out of the same stock suspension are tested (Figs 1, 3), which indicates homogeneity of the ascospore population (Han 1975; Han et al. 1976). Account must be taken of the possibility that there might exist some genetically-induced innate heterogeneity in the heat resistance of yeast

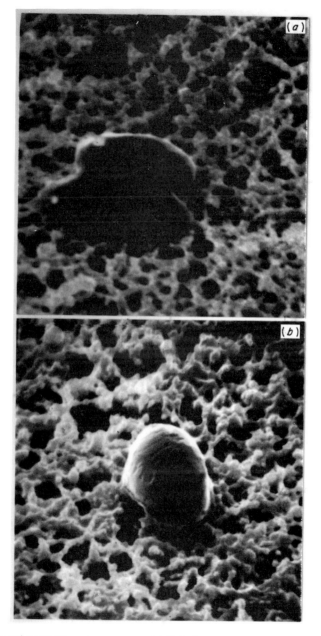

Fig. 9. Heat resistance of yeast ascospores. *(a) K. bulgaricus* 120 ascospores SEM in 0·9%
NaCl a_w 0·998, × 8000. *(b) K. bulgaricus* ascospores SEM in 45% sucrose a_w 0·995,
× 7900. (Davenport, pers. comm.).

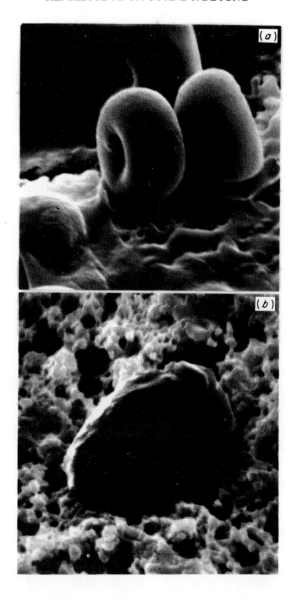

Fig. 10. Heat resistance of yeast ascospores. *(a) Sacch. bailii* 144 ascospores SEM in 0·93 NaCl a_w 0·998, × 8000. *(b) Sacch. bailii* 144 SEM in 45% sucrose a_w 0·995 × 8000. (Davenport, pers. comm.).

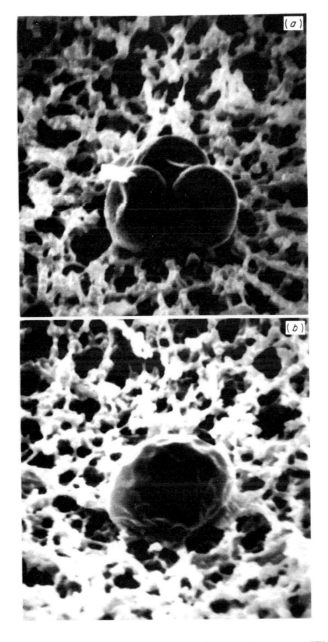

Fig. 11. Heat resistance of yeast ascospores. *(a) Sacch. chevalieri* 65 ascospores SEM in 0·9%
NaCl a_w 0·998, × 8000. *(b) Sacch. chevalieri* ascospores SEM in 45% sucrose a_w 0·995,
× 8000. (Davenport, pers. comm.).

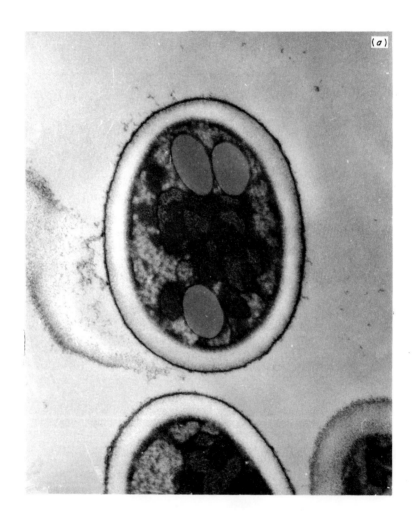

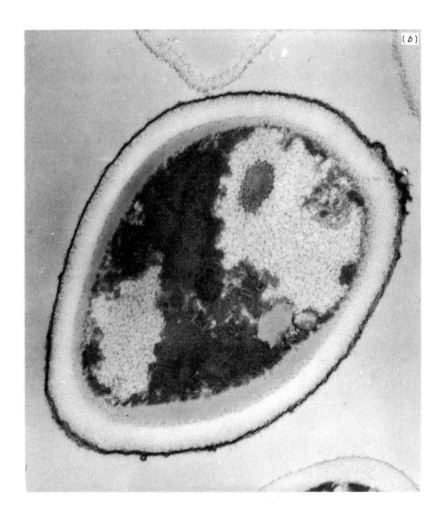

Fig. 12. Heat resistance of yeast ascospores. *(a) K. bulgaricus* 120 ascospores TEM, × 28,000, before heat inactivation. *(b) K. bulgaricus* 120 ascospores TEM, × 42,000, after heat inactivation. (Kreger-van Rij, pers. comm.).

ascospores, due to their diploid character (Fast 1973). It is difficult, however, to distinguish whether the heterogeneity in ascospore populations is caused by the innate characteristics of ascospores existing before heating or by acquisition of heat resistance (*Kluyveromyces*) as a result of heating (Moats *et al.* 1971).

Other possible factors affecting survival curves are: biochemical factors of vegetative cells in the heating menstruum; physiological factors (Moats *et al.* 1971); clumping of ascospores of *Kluyveromyces* spp. (although *K. bulgaricus* showed *ca.* 50% free ascospores and ≤ 50% clusters of maximum 2–4 ascospores, while at the same time heated suspensions of the *K. cicerisporus* contained a maximum of 25% free ascospores and *ca.* 75% clusters of 4–20 ascospores); heat protection of *Saccharomyces* ascospores by the epiplasm of the intact ascus; artefacts in establishing the survival curve (Cerf & Métro 1977); artefacts in establishing the survival of ascospores at a_w < 0·998 (Fig. 8). Heat survival curves of *Sacch. bailii* for instance, change from concave (a_w = 0·998) into convex at a_w = 0·950, in an NaCl solution, but they do *not* change at a_w = 0·950 in a solution of sucrose, which can probably be related to the osmophilic character (a relatively high sucrose and a low NaCl tolerance) of the species as compared to osmotolerant *Sacch. cerevisiae* and *Sacch. chevalieri* (Murdock *et al.* 1953; von Schelhorn 1956; Witter *et al.* 1958; Brown 1974; Smittle 1977; Juven *et al.* 1978). On the other hand, a reduced number of survivors in a spore clump and spores enclosed by an intact ascus may not have been exactly assessed by the MPN or plate count method applied. At the same time gradual accumulated effects of the heating may have resulted in injury of the spores, but not in loss of viability.

Free ascospores of *Saccharomyces* spp. can be acquired by enzymatic digestion of the ascus wall and the vegetative cells by Glusulase (β-glucuronidase, Table 4). However, at the same time we found that the heat resistance of the ascospores was decreased by the hydrolytic process, although the ascospore wall remained undamaged and the dormancy unbroken as far as could be observed by light microscope examinations (Johnston & Mortimer 1959; Put & de Jong, unpublished). Consequently enzymatic hydrolysis of vegetative cells and the ascus wall is not suitable for obtaining clean, free ascospores for heat resistance studies and for quantitative examination of their possible plasmolysis or shrinkage rate in various solutes (a_w) by turbidity measurements as described by Corry (1976*b*) for vegetative yeast cells. Other physical, chemical and mechanical methods show disadvantages of the same order — they affect the heat resistance of the spore (Dawes 1976). Separation of vegetative cells and ascospores by centrifugation easily leads to spore clumping (Emeis & Gutz 1958).

In order to establish whether heat resistance of yeast ascospores is associated with dehydration of the cell rather than with replacement of cell water by solutes, as has been observed in vegetative cells of osmophilic yeasts (Brown 1974; Corry 1976b), further research is essential.

From the results of the reported studies it can be concluded that, in spite of the eukaryotic nature of yeasts accounting for some fundamental differences in the structure and the chemical composition of yeast ascospores (e.g. a true nucleus, meiotic division, no cortex, no dipicolinic acid), some similarities exist in the physiology of sporulation as well as in the outer structure and biochemical function of the yeast ascospore wall and the bacterial spore cortex (Snider & Miller 1966; Fowell 1975; Gould 1977). However, much more fundamental research is required to understand the mechanism of the heat resistance of yeast ascospores.

Since statistical data and laboratory models for studying yeast ascospore heat resistance are not yet available, the practical implications of yeast ascospore heat resistance are at present neither predictable nor calculable. Many more investigations should be done to assess: the initial yeast ascospore load on freshly harvested fruit (e.g. berries, stone fruit, citrus fruit etc.); the inactivation rate of yeast ascospores during the various industrially applied heat processes; the chance of re-infection of fruit products with yeasts or yeast ascospores after an industrially applied heat treatment or other processes; the sporulation ability of the yeasts initially present as well as of the yeasts (re)infecting the product during and after industrially applied processes; z values of yeast ascospores assessed in relation to chemical composition, a_w and pH of fruit substrates (Wucherpfennig & Franke 1968; Kopelman & Schyer 1976; Juven et al. 1978).

To illustrate the above, classic beer PU-values are related to $z = 7°$ (Del Vecchio et al. 1951), but on the other hand P values for pasteurization processes as described by Shapton et al. (1971) are related to $z = 10°$. So considerable confusion will arise unless the reference temperatures and z values are specified and corrections are made before plotting the slope of a reference lethal curve (Patashnik 1953; Shapton 1966). In a normal regulated beer pasteurization process, a heat treatment of ca. 60 PU is given. However, by applying a more accurate temperature measurement and control, an even better product with respect to flavour stability and sterility, can be achieved with a heat treatment of only 8 PU (Harrison 1970).

Industrially applied heat processes for fruit products and soft drinks show a still wider range of heat process values which seldom derive from model studies. Calculations of the minimum F values for inactivation of the initial microflora in relation to their heat resistance are therefore not possible.

Some other examples of inconsistencies in the application of heat pasteurization are:

(a) For the preparation of (citrus) fruit juices and their concentrates, inactivation of pectin esterase enzyme is obtained by means of pasteurization at 85–95°C in vessels (10^4–10^5 PU) or by flash-heating at 80–85°C ($ca.$ 10^3 PU) in a plate or tubular heat exchanger followed by the concentration process. This is effected by heating in an open kettle or by evaporation under vacuum at temperatures between 45–70°C resulting in $a_w = 0.73$–0.94 and pH $= 1.6$–3.8 (Tressler & Joslyn 1971; Sand 1973; Rothschild et $al.$ 1975). It is surprising that after this heat treatment concentrated orange juice still contains a rich variety of yeast species, among which $Saccharomyces$ is predominant (Sand 1977).

(b) Carbonated cassis drinks (blackcurrant) are pasteurized at 75°C (min. 10^3 PU) whereas other carbonated fruit drinks are heated at $ca.$ 65°C (max. 10^2 PU). However, it is not known if the differences in heat pasteurization are related to the initial yeast ascospore load or to the heat inactivation rate, which in its turn may be affected by the heating menstruum and recovery substrate (Witter et $al.$ 1958; Senser et $al.$ 1967; Sand 1971; Dawes 1976; Stevenson & Richards 1976).

(c) Pasteurization temperatures of $ca.$ 65°C are used for soft drinks although pre-pasteurization in a thin layer during 3–5 s at 95°C ($ca.$ 7000 PU) may have been applied, in which case a grossly excessive heat treatment is given. Nevertheless by applying well regulated hygienic conditions and an intensive process control the PU values required may be significantly lower and, moreover, a more stable flavour will be achieved.

Although some factors relating to yeast ascospore heat resistance are now better understood, the mechanism of the thermomicrobiology of yeast ascospores is still shrouded in mystery.

7. References

ANDERSON, A. W. 1977 The significance of yeasts and molds in foods. $Food$ $Technology,$ $Champaign$ **31**, 47–51.

ANDERSON, E. & MARTIN, P. A. 1975 The sporulation and mating of brewing yeasts. $Journal$ of the $Institute$ of $Brewing$ **81**, 242–247.

BACHMANN, B. J. & BONNER, D. M. 1959 Protoplasts from $Neurospora$ $crassa.$ $Journal$ of $Bacteriology$ **78**, 550–556.

BEAMER, P. R. & TANNER, F. W. 1939a Resistance of non-spore-forming bacteria to heat. $Zentralblatt$ $für$ $Bakteriologie,$ $Parasitenkunde,$ $Infektionskrankheiten$ und $Hygiene$ Abt. II, **100**, 81–89.

BEAMER, P. R. & TANNER, F. W. 1939b Heat resistance studies on selected yeasts. $Zentralblatt$ $für$ $Bakteriologie,$ $Parasitenkunde,$ $Infektionskrankheiten$ und $Hygiene$ Abt. II, **100**, 202–211.

BESTIC, P. B. & ARNOLD, W. N. 1976 Linear transformation of standard curves for yeast turbidity. $Applied$ and $Environmental$ $Microbiology$ **32**, 640–641.

BROWN, A. D. 1974 Microbial water relations: Features of the intracellular composition of sugar-tolerant yeasts. *Journal of Bacteriology* **118**, 769–777.

CERF, O. 1977 Tailing of survival curves of bacterial spores. A review. *Journal of Applied Bacteriology* **42**, 1–19.

CERF, O. & MÉTRO, F. 1977 Tailing of survival curves of *Bacillus licheniformis* spores treated with hydrogen peroxide. *Journal of Applied Bacteriology* **42**, 405–415.

CHRISTOPHERSEN, J. & PRECHT, H. 1952 Untersuchungen zum Problem der Hitzeresistenz. II. Untersuchungen an Hefezellen. *Biologisches Zentralblatt* **71**, 585–601.

CORRY, J. E. L. 1976*a* The effect of sugars and polyols on the heat resistance and morphology of osmophilic yeasts. *Journal of Applied Bacteriology* **40**, 269–276.

CORRY, J. E. L. 1976*b* Sugar and polyol permeability of *Salmonella* and osmophilic yeast cell membranes measured by turbidimetry and its relation to heat resistance. *Journal of Applied Bacteriology* **40**, 277–284.

DAKIN, J. C. 1962 Pasteurization of acetic acid preserves. In *Recent Advances in Food Science* Vol. 2, ed. Hawthorn, J. & Leitch, J. M. pp. 128–141. London: Butterworths.

DAVENPORT, R. R. 1976 Experimental ecology and identifications of micro-organisms. In *Microbial Ultrastructure* ed. Fuller, R. & Lovelock, D. W. London: Academic Press.

DAWES, W. 1976 Inactivation of yeasts. In *Inhibition and Inactivation of Vegetative Microbes* ed. Skinner, F. A. & Hugo, W. B. pp. 279–304. London: Academic Press.

DAYHARSH, C. A. & DEL VECCHIO, H. W. 1952 Thermal death time studies on beer spoilage organisms II. *Proceedings of American Society of Brewing Chemistry* **1952**, 48–52.

DEL VECCHIO, H. W., DAYHARSH, C. A. & BASELT, F. C. 1951 Thermal death time studies on beer spoilage organisms I. *Proceedings of American Society of Brewing Chemistry* **1951**, 45–50.

DITTRICH, H. H. 1972 Mikroorganismen als Schädlinge in Fruchtsäften und Fruchtsaftgetränken. *Flüssiges Obst* **39**, 518–522.

EMEIS, C. C. 1958*a* Untersuchungen an durch Massenisolation gewonnenen Sporen von *Saccharomyces carlsbergensis* (*uvarum*). *Die Brauerei* **11**, 160–163.

EMEIS, C. C. 1958*b* Die Gewinnung von Askosporenmassen von *Saccharomyces*-Arten auf Grund besonderer Oberflächen-eigenschaften. *Naturwissenschaft* **45**, 441.

EMEIS, C. C. & GUTZ, H. 1958 Eine einfache Technik zur Massenisolation von Hefesporen. *Zeitschrift für Naturforschung* **13**, 647–650.

FAST, D. 1973 Sporulation synchrony of *Saccharomyces cerevisiae* grown in various carbon sources. *Journal of Bacteriology* **116**, 925–930.

FISHER, D. J. 1975 Flocculation — Some observations on the surface charges of yeast cells. *Journal of the Institute of Brewing* **81**, 107–110.

FOWELL, R. R. 1967 Factors controlling the sporulation of yeasts. II. The sporulation phase. *Journal of Applied Bacteriology* **30**, 450–474.

FOWELL, R. R. 1969 Sporulation and hybridization of yeasts. In *The Yeasts* Vol. 1, ed. Rose, A. H. & Harrison, J. S. pp. 303–383. London: Academic Press.

FOWELL, R. R. 1975 Ascospores of yeasts. In *Spores VI* ed. Gerhardt, P., Costilow, R. N. & Sadoff, H. L. pp. 124–131. Washington: American Society for Microbiology.

FOWELL, R. R. & MOORSE, M. E. 1960 Factors controlling the sporulation of yeasts. I. The presporulation phase. *Journal of Applied Bacteriology* **23**, 53–68.

GÁL, S. 1975 Recent advances in techniques for the determination of sorption isotherms. In *Water Relations of Foods,* ed. Duckworth, R. B. pp. 139–154. London: Academic Press.

GIBSON, B. 1973 The effect of high sugar concentrations on the heat resistance of vegetative micro-organisms. *Journal of Applied Bacteriology* **36**, 365–376.

GOULD, G. W. 1977 Recent advances in the understanding of resistance and dormancy in bacterial spores. *Journal of Applied Bacteriology* **42**, 297–309.

GOULD, G. W. & DRING, G. J. 1975 Heat resistance of bacterial endospores and concept of an expanded osmoregulatory cortex. *Nature, London* **258**, 402–405.

GRAHAM, H. D. 1968 *The Safety of Foods.* Westport, Connecticut: AVI.

GRAUMLICH, T. R. & STEVENSON, K. E. 1978 Recovery of thermally injured *Saccharomyces cerevisiae*. Effects of media and storage conditions. *Journal of Food Science* **43**, 1865–1870.

HABER, J. E. & HALVORSON, H. O. 1972 Cell cycle dependency of sporulation in *Saccharomyces cerevisiae*. *Journal of Bacteriology* **109**, 1027–1033.

HABER, J. E., ESPOSITO, M. S., MAGEE, P. T. & ESPOSITO, R. E. 1975 Current trends in genetic and biochemical study of yeast sporulation. In *Spores VI* ed. Gerardt, P., Costilow, R. N. & Sadoff, H. L. pp. 132–137. Washington: American Society for Microbiology.

HAN, Y. W. 1975 Death rates of bacterial spores: nonlinear survivor curves. *Canadian Journal of Microbiology* **21**, 1464–1467.

HAN, Y. W., ZHANG, H. I. & KROCHTA, J. M. 1976 Death rate of bacterial spores: mathematical models. *Canadian Journal of Microbiology* **22**, 295–300.

HARRISON, J. G. 1970 Pasteurization and flavour. *Brewer's Guardian* (supplement) *October* pp. 16–17, 35.

HARTWELL, L. H. 1974 *Saccharomyces cerevisiae* cell cycle. *Bacteriological Reviews* **38**, 164–198.

INGRAM, M. 1957 Microorganisms resisting high concentrations of sugar or salt. In *Microbial Ecology* pp. 90–133. 6th Symposium of the Society for General Microbiology. Cambridge: Cambridge University Press.

INGRAM, M. 1958 Yeasts in food spoilage. In *Chemistry and Biology of Yeasts* ed. Cook, A. H. pp. 603–633. London: Academic Press.

JENSEN, M. 1960 Experiments on the inhibition of some thermoresistant moulds in fruit juices. *Annales de l'Institut Pasteur, Lille* **11**, 179–182.

JOHNSTON, J. R. & MORTIMER, R. K. 1959 Use of snail digestive juice in isolation of yeast spore tetrads. *Journal of Bacteriology* **78**, 292.

JUVEN, B. J., KANNER, J. & WEISSLOWICZ, H. 1978 Influence of orange juice composition on the thermal resistance of spoilage yeasts. *Journal of Food Science* **43**, 1074–1076, 1080.

KOMEMUSHI, S., OKUBU, K. & TERUI, G. 1968 Kinetic studies on the thermal death of microorganisms (IV).(III) On the change of death rate constant of bacterial spores in the course of heat sterilization. *Journal of Fermentation Technology* **46**, 249–256.

KOPELMAN, I. J. & SCHYER, M. 1976 Thermal resistance of endogenous flora in reconstituted orange juice. *Lebensmittel-Wissenschaft und -Technologie* **9**, 91–92.

KOZAKI, M., OHARA, N. & KITHARA, K. 1968 Yeasts isolated from swelled canned apple. *Abstracts of the Annual Meeting of the Agriculture Chemical Society of Japan, Tokyo* pp. 90–95.

KREGER-VAN RIJ, N. J. W. 1969 Taxonomy and systematics of yeasts. In *The Yeasts* Vol. 1, ed. Rose, A. H. & Harrison, J. S. pp. 5–78. London: Academic Press.

KUENZI, M. T., TINGLE, M. A. & HALVORSON, H. O. 1974 Sporulation of *Saccharomyces cerevisiae* in the absence of a functional mitochondrial genome. *Journal of Bacteriology* **117**, 80–88.

KURTZMAN, C. P., SMILEY, M. J. & BAKER, F. L. 1975 Scanning electron microscopy of ascospores of *Debaryomyces* and *Saccharomyces*. *Mycopathologia* **55**, 29–34.

LEISTNER, L. & RÖDEL, W. 1974 The significance of water activity for microorganisms in meats. In *Water Relations of Foods* ed. Duckworth, R. B. pp. 309–324. London: Academic Press.

LEWIS, M. J. & PATEL, P. C. 1978 Isolation and identification of the cytoplasmic membrane from *Saccharomyces carlsbergensis* by radioactive labelling. *Applied and Environmental Microbiology* **36**, 851–856.

LODDER, J. ed. 1970 *The Yeasts — A Taxonomic Study* 1st edn. Amsterdam: North-Holland.

LOTT, A. F. & CARR, J. G. 1964 Characteristics of an organism causing spoilage in an orange juice beverage. *Journal of Applied Bacteriology* **27**, 379–384.

LUND, A. 1951 Some beer spoilage yeasts and their heat resistance. *Journal of the Institute of Brewing* **52**, 36–41.

McCORMACK, G. 1950 Technical note No. 5 — "Pink Yeast" isolated from oysters grown at temperatures below freezing. *Commercial Fisheries Review* **12**, 11a, 28.

MEYRATH, J. 1962 Problems in fruit juice pasteurization. *Recent Advances in Food Science* **2**, 117–127.

MICHENER, D. H. & KING, D. A. 1974 Preparation of free heat resistant ascospores from *Byssochlamys* asci. *Applied Microbiology* **27**, 671–673.

MILLER, J. J. & HOFFMANN-OSTENHOF, O. 1964 Spore formation and germination in *Saccharomyces*. *Zeitschrift für allgemeine Mikrobiologie* **4**, 273–294.

MILLER, J. J. & KINGSLEY, V. V. 1961 A membrane filter mounting method and spore stain for *Saccharomyces*. *Stain Technology* **36**, 1–4.

MOATS, W. A., DABBAH, R. & EDWARDS, V. M. 1971 Interpretation of non-logarithmic survivor curves of heated bacteria. *Journal of Food Science* **36**, 523–526.

MORI, H., NASUNO, S. & IGUCHI, N. 1971 A yeast isolated from tomato catchup. *Journal of Fermentation Technology* **49**, 180–187.

MROZEK, H. & ROTH, K. 1968 Untersuchungen zur Haltbarkeitsgefährdung von Limonaden durch Hefen. *Zeitschrift für Lebensmitteluntersuchung und -Forschung* **138**, 343–351.

MURDOCK, D. I., TROY, V. S. & FOLINAZZO, J. F. 1953 Thermal resistance of lactic acid bacteria and yeast in orange juice and concentrate. *Food Research* **18**, 85–89.

NOVAK, E. K. & ZSOLT, J. 1977 Some remarks in the sexual process of yeasts. *Proceedings of the 5th International Symposium on Yeasts, Keszthely, Hungary* pp. 31–32.

PATASHNIK, M. 1953 A simplified procedure for thermal process evaluation. *Food Technology* **7**, 1–6.

PEDERSON, C. S., ALBURG, M. N. & CHRISTENSEN, M. D. 1961 The growth of yeasts in grape fruit juice stored at low temperature. IV Fungistatic effects of organic acids. *Applied Microbiology* **9**, 162–167.

PERIGO, J. A., GIMBERT, B. L. & BASHFORD, T. E. 1964 The effect of carbonation, benzoic acid and pH on the growth rate of a soft drink spoilage yeast as determined by a turbidostatic continuous culture apparatus. *Journal of Applied Bacteriology* **27**, 315–332.

PETER, A. 1965 Fruchtsaft Infektionen durch Schimmelpilze und ihre Bekämpfung. *Industrielle Obst und Gemüseverwertung* **50**, 841–845.

PHAFF, H. J., MILLER, M. W. & MRAK, E. M. 1966 *The Life of Yeasts: Their Nature, Activity, Ecology and Relation to Mankind.* Cambridge, Massachusetts: Harvard University Press.

PUGLISI, P. P. & ZENNARO, E. 1971 Erythromycin inhibition of sporulation in *Saccharomyces cerevisiae*. *Experientia* **27**, 963–964.

PUT, H. M. C., DE JONG, J. & SAND, F. E. M. J. 1977 The heat resistance of ascospores of *Saccharomyces cerevisiae* strain 195 and *Saccharomyces chevalieri* strain 215, isolated from heat preserved fruit juice. In *Spore Research 1976 II* ed. Barker, A. N., Wolf, L. J., Ellar, D. J., Dring, G. J. & Gould, G. W. pp. 545–563. London: Academic Press.

PUT, H. M. C., DE JONG, J., SAND, F. E. M. J. & VAN GRINSVEN, A. M. 1976 Heat resistance studies on yeasts causing spoilage in soft drinks. *Journal of Applied Bacteriology* **40**, 135–152.

RESNICK, M. A., TIPPETTS, R. D. & MORTIMER, R. K. 1967 Separation of spores from diploid cells of yeast by stable flow free-boundary electrophoresis. *Science* **158**, 803–804.

ROBERTS, H. R. 1978 Canned fruit juices. Lemon juice; Standards of identity and fill of container. *Federal Register* **43**, 14678–14682.

ROBERTS, T. A. & HITCHINS, A. D. 1969 Resistance of spores. In *The Bacterial Spore* ed. Gould, G. W. & Hurst, A. pp. 611–670. London: Academic Press.

ROSE, D. 1975 Physical response of yeast cells to osmotic shock. *Journal of Applied Bacteriology* **38**, 169–175.

ROTH, K. 1969 Zur Hitzeresistenz von vegetativen und sporulierten Hefezellen. *Das Erfrischungsgetränk* **22**, 1079–1090.

ROTHSCHILD, G., VAN VLIET, C. & KARSENTY, A. 1975 Pasteurization conditions for juices and comminuted products of Israeli citrus fruits. *Journal of Food Technology* **10**, 29–38.

ROUSSEAU, P. & HALVORSON, H. O. 1969 Preparation and storage of single spores of *Saccharomyces cerevisiae*. *Journal of Bacteriology* **100**, 1426–1427.

SAND, F. E. M. J. 1971 Zur biologischen Stabilität von Cassis-Getränken. *Brauwelt* **111**, 1235–1240.

SAND, F. E. M. J. 1973 Recent investigations on the microbiology of fruit juice concentrates. In *Technology of Fruit Juice Concentrates Chemical Composition of Fruit Juices XIII* pp. 185–216. Vienna: International Federation of Fruit Juice Producers.

SAND, F. E. M. J. 1976 Zum gegenwärtigen Stand der Getränke-Mikrobiologie. *Brauwelt* **116**, 220–230.

SAND, F. E. M. J. 1977 Yeasts isolated from concentrated orange juice. *Proceedings of the 5th International Symposium on Yeasts, Keszthely, Hungary* pp. 121–122.

SAND, F. E. M. J. & KOLFSCHOTEN, G. A. 1969 Investigations on the yeast flora of soft drinks. *Antonie van Leeuwenhoek* **35** (Supplement), D11–D12.

SAVARESE, J. J. 1974 Germination studies on pure yeast ascospores. *Canadian Journal of Microbiology* **20**, 1517–1521.

SAVARESE, J. J. 1975 Evidence for change in the ascospore wall during yeast germination. *Journal of General and Applied Microbiology* **21**, 123–126.

SAVEL, J. 1974 Grundlagen der Pasteurization. Verallgemeinerung der Pasteurisiereinheit und Berechnung des Abtötungseffekts im Gemisch von zwei verschiedenen Mikroorganismen. *Brauwissenschaft* **27**, 43–48.

SCHMIDT, C. F. 1950 A method for the determination of the thermal resistance of bacterial spores. *Journal of Bacteriology* **59**, 433–437.

SCHMIDT, C. F. 1957 Thermal resistance of microorganisms. In *Antiseptics, Disinfectants, Fungicides and Chemical and Physical Sterilization* ed. Reddish, C. F., pp. 831–900. Philadelphia: Lea and Febiger.

SENSER, F., REHM, H. J. & WITTMANN, H. 1967 Zur Kenntnis Fruchtsaftverderbender Mikroorganismen. 1. Vorkommen von Mikroorganism (besonders Schimmelpilzen) in verschieden Fruchtsaftarten. *Industrielle Obst- und Gemüseverwertung* **52**, 175–178.

SHAPTON, D. A. 1966 Evaluating pasteurization processes. *Process Biochemistry* **1**, 121–124.

SHAPTON, D. A., LOVELOCK, D. W. & LAURITA-LONGO, R. 1971 The evaluation of sterilization and pasteurization processes from temperature measurements in degrees Celsius (°C). *Journal of Applied Bacteriology* **34**, 491–500.

SHARPE, K. & BEKTASH, R. M. 1977 Heterogeneity and the modelling of bacterial spore death: the case of continuously decreasing death rate. *Canadian Journal of Microbiology* **23**, 1501–1507.

SMITTLE, R. B. 1977 Influence of pH and NaCl on the growth of yeasts isolated from high acid food products. *Journal of Food Science* **42**, 1552–1553.

SNIDER, I. J. & MILLER, J. J. 1966 A serological comparison of vegetative cell and ascus walls and the spore coat of *Saccharomyces cerevisiae*. *Canadian Journal of Microbiology* **12**, 485–488.

STEVENSON, K. E. & RICHARDS, L. J. 1976 Thermal injury and recovery of *Saccharomyces cerevisiae*. *Journal of Food Science* **41**, 136–137.

STUMBO, C. R. 1965 *Thermobacteriology in Food Processing*. London: Academic Press.

TINGLE, M., SINGH KLAR, A. J., HENRY, S. A. & HALVORSON, H. O. 1973 Ascospore formation in yeasts. In *23rd Symposium of the Society for General Microbiology (London)* ed. Ashworth, J. M. & Smith, J. E. pp. 209–243. Cambridge: Cambridge University Press.

TODA, K. 1970 Studies on heat sterilization (VIII) Complex reaction model for heat inactivation of bacterial spores. *Journal of Fermentation Technology* **48**, 811–818.

TOMLINS, R. I. & ORDAL, Z. J. 1976 Thermal injury and inactivation in vegetative bacteria. In *Inhibition and Inactivation of Vegetative Microbes* ed. Skinner, F. A. & Hugo, W. B. pp. 153–190. London: Academic Press.

TRESSLER, D. K & JOSLYN, M. A. 1971 *Fruit and Vegetable Juice Processing Technology*. Westport, Connecticut: AVI.

VON BARWALD, G. 1972 Bestimmung lebendiger Hefezellen durch biochemischen Test. *Brauwissenschaft* **25**, 192–195.

VON SCHELHORN, M. 1956 Untersuchungen über das Asterben von Hefen und anderen Mikroorganismen in Substanzen von hohem NaCl bzw. Zuckergehalt. *Archiv für Mikrobiologie* **25**, 39–57.

WALKER, H. W. 1977 Spoilage of food by yeasts. *Food Technology, Champaign* **31**, 57–61.

WALKER, H. W. & AYRES, J. C. 1970 Yeasts as spoilage organisms. In *The Yeasts* Vol. 3, ed. Rose, A. H. & Harrison, J. S. pp. 464–527. London: Academic Press.

WARTH, A. D. 1978 Relationship between the heat resistance of spores and the optimum and maximum growth temperature of *Bacillus* spp. *Journal of Bacteriology* **134**, 699–705.

WHITE, J. 1951 Effects of common bactericidal substances and of heat on yeast growth and viability. *Journal of the Institute of Brewing* **57**, 470–479.

WINDISCH, S. & NEUMANN, I. 1969 Hefen in Mandelmassen und Süsswaren. *Süsswaren* **13**, 1406–1410.

WINDISCH, S. & NEUMANN, I. 1970 Über Resistenz, Bedeutung und Bekämpfung von

Saccharomyces carlsbergensis (uvarum) und *Saccharomyces rouxii* in Marzipan. *Bericht Symposium für Technische Mikrobiologie, Berlin* pp. 415–422.

WITTER, L. D., BERRY, J. M. & FOLINAZZO, J. F. 1958 The viability of *Escherichia coli* and a spoilage yeast in carbonated beverages. *Food Research* **23**, 133–142.

WUCHERPFENNIG, K. & FRANKE, I. 1968 Ergebnisse von Untersuchungen über die Hitzeresistenz eines Stammes von *Saccharomyces cerevisiae* in Abhängigkeit von den Aciditätsverhältnissen. *Die Fruchtsaft-Industrie* **13**, 194–203.

Cold-tolerant Yeasts and Yeast-like Organisms

R. R. DAVENPORT

*University of Bristol, Long Ashton Research Station,
Long Ashton, Bristol, UK*

Contents

1. Introduction

MANY YEASTS and yeast-like organisms have been isolated from a variety of cold natural, cultivated crop and food environments. Some species are restricted to cold habitats while other species include strains which have been isolated from cold environments but are able to grow at higher temperatures.

Low temperature incubation (other than 25°C, the standard yeast incubation temperature) is essential for these cold-tolerant organisms.

2. Terminology

The terms 'psychrotrophic' and 'psychrophilic' have been avoided for two reasons. First, both words are liable to misinterpretation; second, the boundaries are not always clear between certain mesophiles and their psychrotrophic variants (Schmidt-Lorenz 1967). Schmidt-Lorenz (1967) discussed various terms and he welcomed the use of 'psychrotrophic' for strains of typical mesophilic bacteria that can multiply easily below 5°C (Eddy 1960; Mossel & Zwart 1960). On the other hand 'psychrophilic' micro-organisms are essentially those able to multiply at temperatures near 0°C (Schmidt-Lorenz 1967): other organisms with multiplication optima and maxima below 20°C were designated as facultative psychrophilic types. Yet another definition was given by Morita (1975) who proposed that psychrophiles be re-defined as "organisms having an optimal temperature for growth at about 15°C or lower, a maximal temperature for growth at about 20°C, and a minimal temperature for growth at 0°C or below".

He used the term 'psychrotrophic' for organisms that do not meet the above definition except that those organisms formerly referred to as obligate psychrophiles should be termed psychrophiles. The situation with yeasts is more difficult than with bacteria since some species or strains can fit either definition while other yeasts cannot be defined by these terms at present. Therefore in this account the term 'cold-tolerant' is used to include all yeasts and yeast-like organisms capable of growth at 5°C and below, irrespective of their upper temperature limits. It is always necessary however to indicate whether an organism can grow at low temperatures or just tolerate them. As with bacteria, transitional forms can be found.

The maximum and minimum temperatures for yeast growth are within the general range 0–37°C depending on the characteristics of yeast species and strains and various environmental parameters (Ch. 1, this volume). Table 1 lists the yeast and yeast-like genera reported in various cold environments and a check list of species (with their salient characteristics) is given in Table 5.

TABLE 1

Genera of yeasts and yeast-like organisms reported from various cold environments*

Basidiomycetous type organisms	Ascomycetous type organisms
Aessosporon	*Aureobasidium*‡ (Fig. 1*b*)†
Bullera‡	*Candida*‡ (Fig. 3*c*)
Candida‡ (Fig. 4*d*)	*Hansenula*
Cryptococcus‡	*Kloeckera/Hanseniaspora*
Leucosporidium	*Kluyveromyces*
Rhodospiridium (Fig. 4*b*)	*Lipomyces* (Fig. 4*c*)
Sporidiobolus	*Nadsonia* (Fig. 4*a*)
Sporobolomyces (Fig. 2*a, b*)	*Saccharomyces*
Sterigmatomyces	*Schwanniomyces*
Trichosporon‡ (Fig. 1*c, d*)	*Torulaspora (Debaryomyces)*
	Torulopsis‡
Yeast-like genus of unknown affinity	
Sympodiomyces	

*Data from various sources given in Table 2.
†Illustrations taken from Davenport (1975).
‡Imperfect genera.

The general features of cold-tolerant yeasts are:
(1) Most species are of basidiomycetous type (at least 75%); in general, these are more suited to colder habitats since their minimum temperatures for growth are lower than those for ascomycetous organisms. In addition the latter yeasts lack other characteristics such as absence of carotenoid pigmentation, and intracellular lipid storage (which prevents cells from

drying out), high osmotolerance, certain vegetative features (e.g. capsules) and a complex sexual slow-growing life-cycle (see Figs 2a, b, c; 4b).

(2) Many of the species are non-fermentative (74%) and nitrate positive (50%), irrespective of the group to which they belong.

(3) It is highly probable that reproduction of yeasts in cold environments is restricted to those strains which have a minimum growth temperature of 0–5°C (i.e. most basidiomycetous type yeasts). Higher growth temperature-requiring organisms must just survive rather than reproduce in cold habitats. These range from sources in Antarctica to a wide variety of frozen foods (Table 2) where yeasts can be among the dominant spoilage organisms, e.g. *Rhodotorula glutinis* in frozen peas (Mulcock 1955).

TABLE 2

Summary of cold environments where yeasts and yeast-like organisms can be detected

Antarctica	Soils, snow, sea water, marine animal and plant sources
Other cold regions	Sea, fresh water, marine animal and plant sources
Beverages	Fruit and vegetable products (juices, pulps, concentrates), beers and wines
Frozen foods	Meats and meat products (e.g. minced beef with vegetables), fish and fish products (e.g. cod, salmon and trout, fish fingers)
Dairy products	Cream, cheese, milk, yoghurt
Fruit and vegetables	Peas, strawberries, raspberries

Data from: Ayres (1960); Barnett & Pankhurst (1974); Barnett *et al.* (1979); Davenport (unpublished); Di Menna (1960, 1966a, b); Eklund *et al.* (1965, 1966); Fell & Phaff (1970); Fell *et al.* (1970); Fell & Hunter (1974); Fell & Statzell (1971); Fell *et al.* (1973); Komagata & Nakase (1965); Lodder (1970); Mulcock (1955); Nakase & Komagata (1969, 1970a, b 1971a, b, c, d, e, f, g); Newell & Hunter (1970); Phaff (1970a, b, c) ; Rodrigues de Miranda (1975); Schmidt-Lorenz (1970a, b); Schmidt-Lorenz & Gutschmidt (1968, 1969); Seeliger (1956); Sonck & Yarrow (1969); Watson *et al.* (1976).

(4) The ascomycetous yeasts which reproduce at 5°C and below, all have rough thick walled ascospores which contain lipid droplets (e.g. *Lipomyces* sp. Fig. 4c). One can infer that such structures would be more resistant to cold than the smooth ascospores of other genera (e.g. *Saccharomyces*).

(5) Some species are both consistently found in, and restricted to, cold environments (e.g. the majority of *Leucosporidium* spp.); these are true inhabitants. Other yeasts flourish in cooler environments but have a wider range of temperature tolerance for growth (e.g. *Rhodospiridium* spp.).

(6) Some yeasts which occur in a wide range of frozen food commodities can cause microbial changes (Schmidt-Lorenz & Gutschmidt 1968) (see also Table 3).

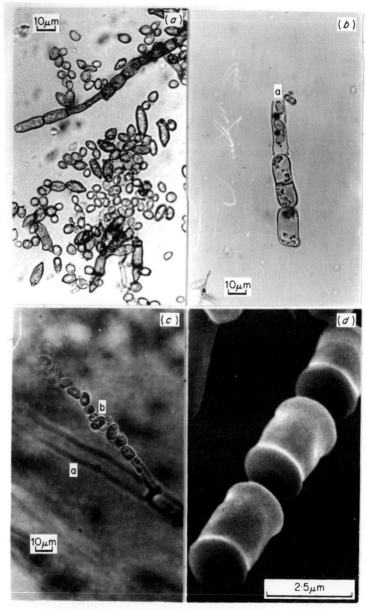

Fig. 1. Yeast-like organisms isolated from agar plates incubated at 5°C. (*a*) *Cladosporium herbarum* (from *Vitis* spp. leaves). Yeast cells and short chains of hyphal cells, i.e. 'Black yeast state'. This state is also formed with other micro-organisms, e.g. *Aureobasidium pullulans* and *Philaphora* spp. (*b*) *Aureobasidium pullulans* (from *Salix* spp. leaves). This organism has extremely variable morphological and physiological characteristics usually due to environmental conditions, e.g. inside healthy fruit buds (Davenport 1967, 1968) and (*b*) where these cells, containing endospores, a, were observed from growth in the agar beneath the main colony. (*c*) *Tr. pullulans* (from a collection of different invertebrates caught in a pitfall trap): a, mycelium; b, arthrospores. This strain produces mainly budding cells and very little mycelium which was only observed deep in the agar under and beside the main colony growth. (*d*) *Tr. pullulans*. SEM of arthrospores.

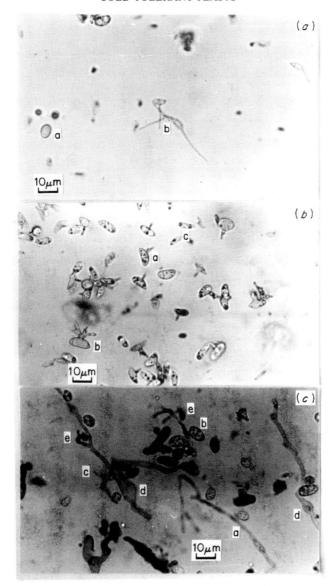

Fig. 2. Carotenoid ballistospore (from a collection of flying insects) forming yeasts which grow at 5°C but subsequent transfer fails to produce growth irrespective of either media or incubation temperature. (a) *Sporobolomyces* sp.: a, vegetative cell; b, mother cells with sterigmata. (b) *Sporobolomyces* sp.: a, mother cells with single sterigmata; b, mother cells with multiple sterigmata; c, ballistospores. (c) *Sporidiobolus* sp.: a, mycelium; b, vegetative cells; c, sterigmata; d, clamp connection; e, ballistospore on sterigmata.

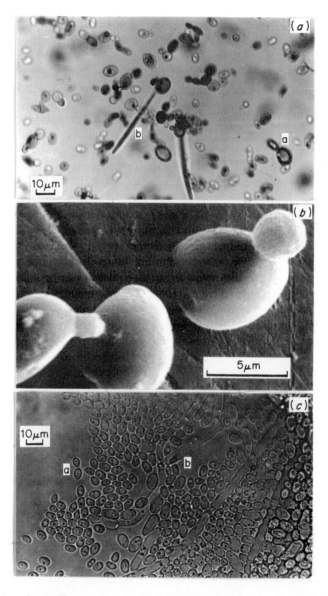

Fig. 3. Yeasts and yeast-like organisms, isolated from agar plates incubated at 5°C. (a) M. pulcherrima (from selected flying insects): a, budding yeasts; b, asci containing 1–2 needle
shaped ascospores. Ascospore formation was only observed from mounts of cells from
colonies grown on grape juice yeast extract agar at 2°C. (b) M. pulcherrima. SEM of
budding vegetative cells. (c) C. brumptii (from a collection of different invertebrates
caught in a pitfall trap). a, vegetative cells; b, pseudomycelium formation.

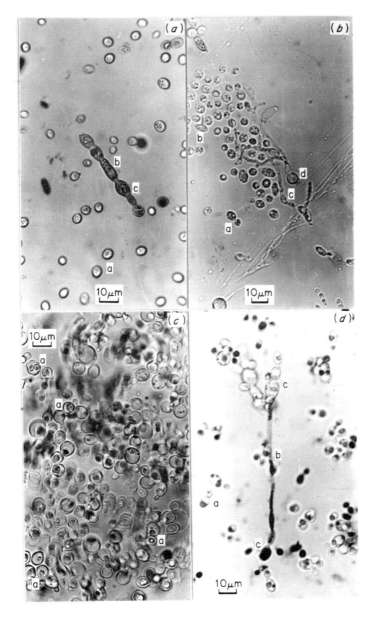

Fig. 4. Yeasts and yeast-like organisms which grew on isolation plates incubated at 5°C, but failed to grow on subsequent transfer irrespective of media or incubation temperatures used. (a) *Nadsonia* sp. (from leaves, *Salix* sp.): a, vegetative cells; b, heterogenic conjugation between mother cell and a bud to give an ascus containing a warty-walled ascus, c. (c) *Sporidiobolus* sp. (from a collection of flying insects): a, budding vegetative cells; b, mother cell with sterigmata (for ballistospore formation); c, mycelium; d, chlamydospores. (c) *L. starkeyii* (from acid soil): a, conjugating asci with 1-2 ascospores, each containing an oil droplet. (d) *Candida* sp. (from a collection of different invertebrates caught in a pitfall trap): a, budding vegetative cells; b, very unusual feature, true mycelium with budding cells (blastospores) either end, c.

TABLE 3

Examples of yeast numbers from cold environments

Environment/substrate	Yeasts/g × 10^3	Reference
Frozen foods		Schmidt-Lorenz &
Raw beef mince	9·4	Gutschmidt (1968)
Trout	9·4	
Mixed vegetables	18·8	
Peas*	TM	Mulcock (1955)
Polar regions		
Snow	0·01–0·2	
Marine sediment	0·3–5·9	
Soil (Antarctica)	0·0–5·0	Di Menna (1960)
	0·005–100	Di Menna (1966b)
(East Greenland)	0·2–56	

*Initial count was > 5 cells/g. Subsequent incubation at 4°C (12 d) or –6°C (12 d) plus 4 d at + 10°C gave too many colonies to count (TM).

Stokes & Remond (1966) concluded from their data that "psychrophiles are both ubiquitous and numerous in nature and probably play important roles in the cycles of matter". As far as it is known this subject has received little attention. However, some reports clearly show that certain yeasts exhibit biochemical activities at low temperatures: e.g. lipolysis below 0°C (*Endomyces candidum* and *Saccharomycopsis lipolytica*; Alford & Pierce 1961) and use of triglycerides, lecithin, casein, gelatin and crab meat protein at 0·5°C (*Trichosporon* sp. and *Aureobasidium pullulans*; Eklund *et al.* 1965, 1966). It is interesting to note that the latter investigations showed that yeasts increased rapidly in air-packed, irradiated (0·2 and 0·4 Mrad) crab meat, stored at 0·5 and 5·6°C. Similar samples which were vacuum packed contained no yeasts. Temperature optima were determined for bacteria and yeasts from cold mountain habitats (Mosser *et al.* 1976). They found that the lowest optimal temperatures were between 10 and 15°C even though most of the isolates were obtained from sites at or near 0°C. Evidence has been presented that the function and composition of the cellular membranes are important factors in temperature adaptation in psychrophilic yeasts (Watson *et al.* 1976). Their view is that these yeasts manipulate the fatty-acid unsaturation index, thus allowing the cells to alter their membrane fluidity and function with temperature changes. Davenport (1968, 1970, 1975) observed yeast counts at 5°C which were higher than the corresponding dilution plate counts at 15°C. This effect was for selected crop (apple, pear and grape) habitats in England. Table 4 gives a series of yeast counts from various plant organs which were processed, plated out and incubated.

TABLE 4

Comparison of yeast counts from shoot samples, surface plated and incubated at 5 and 15 °C.

Shoot sample/sample number	Yeasts/g at incubation temperature (°C)	
	5	15
Grape leaves		
1	161	87
2	212	96
Flowers		
1	1097	122
2	170	26
Pears (fruit)		
1	3500	2300
2	TM*	2700
3	TM	7700
4	7000	3800

*TM, too many to count.
Data from Davenport (1970, 1975).

The increase in numbers was two-fold. First, there was a general increase in species known to have optimal growth temperatures below 25°C (e.g. carotenoid yeasts, some *Candida, Cryptococcus* and *Torulopsis* spp.). Second, some yeasts were only present on 5°C plates (Figs 2a–c; 4a–d). Furthermore, some morphological structures were only observed (Figs 1b; 2a–c; 4a–d) at this temperature. Attempts to subculture many of these yeasts at 5°C, with the same media, failed and no satisfactory explanation could be found. It was concluded either that some strains became so adapted to the cold that incubation at higher temperatures was lethal or that there was a complex interaction between members of the mixed microflora present. In a review of some interaction studies between temperature and psychrophilic micro-organisms Inniss (1975) pointed out that most investigations had been concentrated on the effects of high rather than low temperatures, thereby giving various observations at superoptimal temperatures.

For physiological studies on yeasts however one should consult McMurrough & Rose (1973), Rose (1962, 1968) and Stokes (1963), but these do not clarify how the organisms function at low temperatures.

Finally, attention is drawn to the fact that both isolation and taxonomic methods are not standardized for cold-tolerant yeasts. This may account for some differences and/or discrepancies in various studies. Clearly more investigations on cold-tolerant yeasts are required.

TABLE 5

Check list of the principal yeasts and yeast-like species reported from cold environments*

Species	Min. °C	Max. °C	Ecology†	Pigmentation	Polysaccharide formed ± intracellular lipids	Urease activity	Fermentation	Nitrate	Nitrite	Salt tolerance (max. %)	Sugar tolerance (a_w)	Capsules &/or chlamydospores	Additional structures‡	G + C ratio
A. Optimum growth — temperature < 25°														
Basidiomycetous type organisms														
Leu. antarcticum	?	19	Ant	−	+	+	−	+	?	?	?	+	+	?
Leu. fridium	<−1	19	Ant, FF	−	+	+	+	+	?	?	?	+	+	50
Leu. gelidum	<−1	19	Ant, FF	−	+	+	−	+	?	?	?	+	+	52
Leu. nivalis	<−1	19	Ant, FF	−	+	+	−	+	?	?	?	+	+	?
Leu. scottii	−1	35	Ant, FF	−	−	+	−	+	?	?	0·91	+	+	59
Leu. stokesii	?	19	Ant, FF	C	+	+	−	+	?	?	?	+	+	?
Rhodosp. bisporidii	?	30	Ant	C	+	+	−	+	?	?	?	+	+	?
Rhodosp. capitatum	?	24	Ant	C	+	+	−	+	?	?	?	+	+	?
Rhodosp. dacryoidum	?	30	M	C	−	+	−	−	?	?	?	+	+	67
Rhodosp. diobovatum	?	35	M	C	−	+	−	+	?	?	?	+	+	68
Rhodosp. informo-miniatum	?	30	Ant	C	+	+	−	+	+	?	?	+	+	51
Rhodosp. malvinellum	?	30	M	C	−	+	−	+	?	?	?	+	+	65
Rhodosp. sphaerocarpum	?	30	M	C	−	+	−	+	?	?	?	+	+	?
Rhodosp. toruloides	?	40	M	C	−	+	1	+	+	?	?	+	−	?
Rh. aurantiaca	?	38	P	C	−	+	−	+	?	?	?	+	−	55
Rh. glutinis+	−2	39	P, F, FF	C	+	+	−	+	+	?	0·87	+	−	61–68
Rh. gramminis+	?	33	P	C	+	+	−	+	+	?	?	+	−	?
Rh. lactosa+	?	30	P	C	−	+	−	+	+	?	?	−	−	57

Species														
Rh. marina[+]	?	<30	M	C	−	−	−	+	−	?	?	−	−	51
Rh. minuta[+]	−2	38	P, FF	C	−	−	−	+	−	?	0·92	−	−	51
Rh. pallida[+]	?	34	P	C	+	+	−	+	−	?	?	−	−	51
Rh. rubra[+]	?	38	P, F, FF	C	+	+	−	+	+	?	0·94	+	−	60
Sp. albo-rubescens[+]	?	30	P, M	C	+	+	−	+	+	0·8	?	−	+	?
Sp. antarcticus[+]	?	30	Ant	C	−	+	−	+	−	?	?	−	−	?
Sp. gracilis[+]	?	<30	P	C	−	+	−	+	+	?	?	+	+	?
Sp. hispanicus[+]	?	30	P	C	+	+	−	+	+	?	?	−	−	?
Sp. holsaticus[+]	?	30	U	C	−	+	−	+	⧺	?	?	+	−	?
Sp. odorus[+]	?	30	P	C	+	+	−	+	+	?	?	−	−	?
Sp. pararoseus[+]	?	30	U	C	+	+	−	+	⧺	?	?	+	−	?
Sp. runicea[+]	?	<30	FF	C	+	+	−	+	+	?	?	−	−	54
Sp. roseus[+]	?	<30	U	C	+	+	−	+	⧺	?	?	+	−	?
Sp. singularis[+]	?	<30	P	C	−	−	−	+	+	?	?	+	−	?
Aessosporon salmonicolor	?	30	U	−	+	+	−	+	+	?	0·90	−	+	59
Cr. albidus[+]	−12	35	U	C	+	−	−	−	+	?	?	+	+	54
Cr. gastricus[+]	?	26	An, S	−	+	−	−	−	−	?	?	−	+	60
Cr. hungaricus[+]	?	24	Ant, S	C	+	+	−	+	⧺	?	0·90	+	−	?
Cr. laurentii[+]	?	37	U	−	−	?	−	+	⧺	?	0·91	−	+	51–59
Cr. luteolus[+]	?	34	U	C	+	+	−	+	⧺	?	?	+	−	51–59
Cr. macerans[+]	?	<30	P	−	+	−	−	+	+	?	?	+	+	?
Cr. skinneri[+]	?	30	An	−	+	+	−	+	+	?	?	−	+	?
Cr. terrus[+]	−10	35	S, FF	−	+	+	±	+	+	?	0·91	+	−	54
C. aquatica[+]	?	26	FW	−	+	+	−	?	?	8	?	+	?	?
C. curiosa[+]	?	23	Ant, FF	−	?	+	−	?	?	11	?	?	?	?
C. curvata[+]	?	38	An	−	+	+	−	+	?	10	?	−	+	61
C. diffluens[+]	−12	31	P, FF	−	−	−	−	−	+	6	?	+	−	61
C. glaebosa[+]	?	30	FF	−	−	−	−	+	−	11	?	−	+	43
C. humicola[+]	?	39	An, P, S, FF	−	+	+	−	+	?	8	?	+	−	63
C. rugosa[+]	?	43	An, F, FF	−	?	−	+	−	?	14	0·09	−	+	50
C. salmonicola[+]	?	32	P	−	+	+	−	+	−	11	?	+	−	40
B. alba[+]	−1	30	A, P	−	−	+	−	+	+	?	?	−	−	54
B. gradispora[+]	?	30	P	−	+	+	−	+	+	?	?	+	−	?
B. tsgae[+]	?	30	P	C	+	+	−	+	+	?	?	−	−	?
Sporid. johnsonii	?	37	P	C	+	+	−	+	+	?	?	−	−	!
Sporid. ruinenii	?	30	P	C	−	+	+	+	+	?	0·35	+	+	?

TABLE 5 (*cont'd*)

	Temperature for growth		Ecology†	Pigmentation	Polysaccharide formed ± intracellular lipids	Urease activity	Fermentation	Nitrate	Nitrite	Salt tolerance (max. %)	Sugar tolerance (a_w)	Capsules and/or chlamydospores	Additional structures‡	G+C ratio
	Min. °C	Max. °C												
Basidiomycetous type organisms (cont.)														
Tr. cutaneum+	-10	41	An, S, F, FF	?	+	+	-	-	?	-	0·86	+	+	61
Tr. pullulans+	-8·5	34	An, S, F, FF	?	+	+	-	+	+	-	0·86	+	+	?
St. elviae+	?	37	An	-	-	+	-	-	?	?	?	-	-	52
St. halophilus+	?	30	M	-	-	-	-	-	?	>12	?	-	-	?
St. indicus+	?	30	M	-	-	-	-	-	?	>12	?	-	-	?
Ascomycetous type organisms														
Aureobasidium pullulans+	?	37	U	?	+	+	-	+	+	?	?	+	+	54
Torulopsis austromarina+	?	18	M	-	-	-	-	-	-	?	?	-	-	?
T. psychrophila+	?	20	M	-	-	-	-	-	+	0·7	?	-	-	36
Deb. hansenii	-12·5	37	U	-	-	-	+	-	-	12-24	0·75	-	-	37
H. anomala	2	37	B, F, FF	-	-	-	+	+	+	-	0·88	-	-	32-45
H. subpelliculosa	5	37	B, F, FF	-	-	-	+	+	-	-	0·85	-	-	34
L. anomalus	5	35	S	-	+	-	-	-	?	?	?	+	-	?
L. lipofekes	5	35	S	-	+	-	-	-	?	?	?	+	-	46
L. starkeyii	5	37	S	-	+	-	-	-	-	?	?	+	-	48
N. commutata	?	<30	S	-	-	-	-	-	-	?	?	-	-	?
N. elongata	?	<30	P	-	-	-	+	-	-	?	?	-	-	?
N. fulvescens	?	<30	P	-	-	-	+	-	-	?	?	-	-	?

	Temp. min (°C)	Temp. max (°C)	Habitat[†]	Add. structures[‡]						Ratio		G + C
Sch. alluvius	?	37	S	—	+	—	—	—	?	?	—	?
Schw. castellii	?	37	S	—	+	—	—	—	?	?	—	?
Schw. occidentalis	?	30	S	—	+	—	—	—	?	?	—	35
Schw. persoonii	?	30	S	—	+	—	—	—	?	?	—	?
Yeast-like organisms of unknown affinity												
Sympodiomyces parvus[+]	?	25	M	—	—	—	—	—	?	?	+	?
B. Optimum growth — temperature > 25°C												
Ascomycetous type organisms												
K. fragilis	8	47	An, F, FF	Pu	+	—	+	—	—	0·87	—	36–40
K. lactis	5	40	F, FF	Pu	+	—	+	—	—	0·96	—	—
Kl. apiculata[+] (H'spora valbyensis)	8	35	P, B, FF	—	—	—	—	—	—	—	—	32
Sacch. cerevisiae	5	38	B, F, P, FF	—	+	—	+	—	?	0·88	+	39
C. utilis	−2	43	F, FF	—	+	+	+	—	8	0·94	—	44
C. tropicalis	5	44	F, FF	—	+	—	+	—	13	0·92	+	35

*Data from references cited in Table 2.

[†]Most common habitats where species are found.

[‡]Additional structures = all forms of morphological characteristics (excluding dominant unicellular reproductive form, ascospores, ballistospores and primitive pseudomycelium) vegetative and sexual states.

?, characteristic not yet determined; [+], imperfect states; −, negative response; +, positive response; ±, variable response; C, carotenoid; Pu, Pulcherrimin; A, air; ANT, Antarctic sources (sea, soil, plant and animal); An, animal sources; B, beverages; F, foods; FF, frozen foods; M, marine sources; P, plant sources; S, soil; U, ubiquitous; G + C, guanine + cytosine ratio.

3. References

ALFORD, J. & PIERCE, D. A. 1961 Lipolytic activity of micro-organisms at low and intermediate temperature. III. Activity of microbial lipases at temperatures below 0°C. *Journal of Food Science* **26**, 518–524.

AYRES, J. C. 1960 Temperature relations and some other characteristics of the microbial flora developing on refrigerated beef. *Journal of Food Science* **25**, 1–18.

BARNETT, J. A. & PANKHURST, R. J. 1974 *A New Key to the Yeasts.* Amsterdam & London: Elsevier.

BARNETT, J. A., PAYNE, R. W. & YARROW, D. 1979 *A Guide to Identifying and Classifying Yeasts.* Cambridge: Cambridge University Press.

DAVENPORT, R. R. 1967 The microflora of cider apple fruit buds. *Report, Long Ashton Research Station for 1966* pp. 122–123.

DAVENPORT, R. R. 1968 *The origin of cider yeasts.* Thesis, Institute of Biology, London.

DAVENPORT, R. R. 1970 *Epiphytic yeasts associated with the developing grape vine.* M.Sc. Thesis. University of Bristol.

DAVENPORT, R. R. 1975 *The distribution of yeasts and yeast-like organisms in an English vineyard.* Ph.D. Thesis. University of Bristol.

DI MENNA, M. E. 1960 Yeasts from Antarctica. *Journal of General Microbiology* **23**, 295–300.

DI MENNA, M. E. 1966*a* Three new yeasts from Antarctic soils, *Candida nivalis, Candida gelida* and *Candida frigida* spp. nov. *Antonie van Leeuwenhoek* **32**, 15–28.

DI MENNA, M. E. 1966*b* Yeasts in Antarctic soils. *Antonie van Leeuwenhoek* **32**, 29–38.

EDDY, B. P. 1960 The use and meaning of the term 'psychrophilic'. *Journal of Applied Bacteriology* **23**, 189–190.

EKLUND, M. W., SPINELLI, J., MIYAUCHI, D. & GRONINGER, H. 1965 Characteristics of yeasts isolated from Pacific crab meat. *Applied Microbiology* **13**, 985–990.

EKLUND, M. W., SPINELLI, J., MIYAUCHI, D. & DASSOW, J. 1966 Development of yeast on irradiated Pacific crab meat. *Journal of Food Science* **31**, 424–431.

FELL, J. W. & PHAFF, H. J. 1970 Genus *Leucosporidium.* In *The Yeasts — A Taxonomic Study* 2nd edn, ed. Lodder, J. pp. 776–802. Amsterdam: North-Holland.

FELL, J. W., PHAFF, H. J. & NEWELL, S. Y. 1970 Genus *Rhodosporidium.* In *The Yeasts — A Taxonomic Study* 2nd edn, ed. Lodder, J. pp. 803–814. Amsterdam: North-Holland.

FELL, J. W. & STATZELL, A. C. 1971 *Sympodiomyces* gen. n. A yeast-like organism from southern marine waters. *Antonie van Leeuwenhoek* **37**, 359–367.

FELL, J. W., HUNTER, L. I. & TALLMAN, A. S. 1973 Marine basidiomycetous yeasts (*Rhodosporidium* spp. nov.) with tetrapolar and multiple allelic bipolar mating systems. *Canadian Journal of Microbiology* **19**, 643–657.

FELL, J. W. & HUNTER, L. I. 1974 *Torulopsis austromarina* sp. nov. A yeast isolated from the Antarctic Ocean. *Antonie van Leeuwenhoek* **40**, 307–310.

INNISS, W. E. 1975 Interaction of temperature and psychrophilic micro-organisms. *Annual Review of Microbiology* **29**, 445–465.

KOMAGATA, K. & NAKASE, T. 1965 New species of the genus *Candida* isolated from frozen foods. *Journal of General and Applied Microbiology* **11**, 255–267.

LODDER, J. ed. 1970 *The Yeasts — A Taxonomic Study* 2nd edn. Amsterdam: North-Holland.

MCMURROUGH, I. & ROSE, A. H. 1973 Effects of temperature variation on the fatty acid composition of a psychrophilic *Candida* species. *Journal of Bacteriology* **114**, 451–452.

MORITA, R. Y. 1975 Psychrophilic bacteria. *Bacteriological Reviews* **39**, 144–167.

MOSSEL, D. A. A. & ZWART, H. 1960 The rapid tentative recognition of psychrotrophic types among Enterobacteriaceae isolated from foods. *Journal of Applied Bacteriology* **23**, 185–188.

MOSSER, J. L., HERDRICH, G. M. & BROCK, T. C. 1976 Temperature optima for bacteria and yeasts from cold-mountain habitats. *Canadian Journal of Microbiology* **22**, 324–325.

MULCOCK, A. P. 1955 Spoilage in frozen peas. *New Zealand Journal of Science and Technology* Section B **37**, 15–19.

NAKASE, T. & KOMAGATA, K. 1969 DNA base composition of the genus *Hansenula. Journal of General and Applied Microbiology* **15**, 85–95.

NAKASE, T. & KOMAGATA, K. 1970*a* Significance of DNA base composition in the classification of yeast genera *Hanseniaspora* and *Kloeckera*. *Journal of General and Applied Microbiology* **16**, 241–250.

NAKASE, T. & KOMAGATA, K. 1970*b* Significance of DNA base composition in the classification of yeast genus *Pichia*. *Journal of General and Applied Microbiology* **16**, 511–521.

NAKASE, T. & KOMAGATA, K. 1971*a* Significance of DNA base composition in the classification of yeast genus *Debaryomyces*. *Journal of General and Applied Microbiology* **17**, 43–50.

NAKASE, T. & KOMAGATA, K. 1971*b* Further investigation on the DNA base composition of the genus *Hansenula*. *Journal of General and Applied Microbiology* **17**, 77–84.

NAKASE, T. & KOMAGATA, K. 1971*c* Significance of DNA base composition in the classification of yeast genera *Cryptococcus* and *Rhodotorula*. *Journal of General and Applied Microbiology* **17**, 121–130.

NAKASE, T. & KOMAGATA, K. 1971*d* Significance of DNA base composition in the classification of yeast genus *Torulopsis*. *Journal of General and Applied Microbiology* **17**, 161–166.

NAKASE, T. & KOMAGATA, K. 1971*e* Significance of DNA base composition in the classification of yeast genus *Saccharomyces*. *Journal of General and Applied Microbiology* **17**, 227–238.

NAKASE, T. & KOMAGATA, K. 1971*f* Significance of DNA base composition in the classification of yeast genus *Candida*. *Journal of General and Applied Microbiology* **17**, 259–279.

NAKASE, T. & KOMAGATA, K. 1971*g* DNA base composition of some species of yeasts and yeast-like fungi. *Journal of General and Applied Microbiology* **17**, 363–369.

NEWELL, S. Y. & HUNTER, L. I. 1970 *Rhodosporidium diobovatum* sp. nov. The perfect form of an asporogenous yeast (*Rhodotorula* sp.). *Journal of Bacteriology* **104**, 503–508.

PHAFF, H. J. 1970*a* Genus *Bullera*. In *The Yeasts — A Taxonomic Study* 2nd edn, ed. Lodder, J. pp. 815–821. Amsterdam: North-Holland.

PHAFF, H. J. 1970*b* Genus *Sporidiobolus*. In *The Yeasts — A Taxonomic Study* 2nd edn, ed. Lodder, J. pp. 822–830. Amsterdam: North-Holland.

PHAFF, H. J. 1970*c* Genus *Sporobolomyces*. In *The Yeasts — A Taxonomic Study* 2nd edn, ed. Lodder, J. pp. 831–862. Amsterdam: North-Holland.

RODRIGUES DE MIRANDA, L. 1975 Two species of the genus *Sterigmatomyces*. *Antonie van Leeuwenhoek* **41**, 193–199.

ROSE, A. H. 1962 Biochemistry of the psychrophilic habit: studies on the low maximum temperature. In *Recent Progress in Microbiology* Vol. 8, International Congress for Microbiology, Montreal. ed. Gibbons, N. E. Toronto: University of Toronto Press.

ROSE, A. H. 1968 Physiology of micro-organisms at low temperatures. *Journal of Applied Bacteriology* **31**, 1–11.

SCHMIDT-LORENZ, W. 1967 Behaviour of micro-organisms at low temperatures. *Bulletin of the International Institute of Refrigeration* Nos. 2 & 4, 1–59.

SCHMIDT-LORENZ, W. 1970*a* Vermenrung von Hefen bei Gefriertemperaturen. In *2nd Symposium Technische Mikrobiologie* ed. Dellweg, H. pp. 291–298. Berlin.

SCHMIDT-LORENZ, W. 1970*b* Psychrophile Mikroorganism und tiefgefrorene Lebensmittel. *Alimenta* **9**, 32–45.

SCHMIDT-LORENZ, W. & GUTSCHMIDT, J. 1968 Mikrobielle und senorische veränderungen gefrorener Lebensmittel bei Lagerung im Temperataturbereich von −2·5°C bis −10°C. *Lebensmittel-Wissenschaft Technologie* **1**, 26–43.

SCHMIDT-LORENZ, W. & GUTSCHMIDT, J. 1969 Mikrobielle und senorische veränderungen gefrorener Brathähnchen und Poularden bei Lagerung im Temperaturbereich von −2·5° bis −10°C. *Fleischwirtschaft* **49**, 1033–1038, 1041.

SEELIGER, H. P. R. 1956 Use of a urease test for screening and identification of cryptococci. *Journal of Bacteriology* **72**, 127–131.

SONCK, C. E. & YARROW, D. 1969 Two new yeast species isolated in Finland. *Antonie van Leeuwenhoek* **35**, 172–177.

STOKES, J. L. 1963 General biology and nomenclature of psychrophilic micro-organisms. In *Recent Progress in Microbiology* ed. Gibbons, N. E. pp. 187–192. Symposium held at the VIIIth International Congress for Microbiology, Montreal. Toronto: University of Toronto Press.

STOKES, J. L. & REMOND, M. L. 1966 Quantitative ecology of psychrophilic micro-organisms. *Applied Microbiology* **14**, 74–78.

VAN DER WALT, J. P. & HOPS-HAVU, V. K. 1976 A colour reaction for the differentiation of ascomycetous and hemibasidiomycetous yeasts. *Antonie van Leewenhoek* **42**, 157–163.

WATSON, K., ARTHUR, H. & SHIPTON, W. A. 1976 *Leucosporidium* yeasts: Obligate psychrophiles which alter membrane-lipid and cytochrome composition with temperature. *Journal of General Microbiology* **97**, 11–18.

The Pathogenic *Candida* Species and Diseases
Caused by Candidas in Man

ROSALINDE HURLEY

*Queen Charlotte's Maternity Hospital and Institute of Obstetrics and Gynaecology,
University of London, London, UK*

Contents

1. Introduction and Taxonomy

WINNER AND HURLEY (1964) reviewed much of the literature on candidosis, with especial reference to disease caused by *Candida albicans* in man, and they edited a series of papers on candida infections (1966). The pathogenic candida species were reviewed by Hurley (1967). A comprehensive review of English language contributions to knowledge of the entire field relating to candida and candidosis has been compiled by Odds (1979). Following the latter author and, as he states, for the sake of linguistic ease, the name 'candidosis' will be used in this chapter as a generalization for infections due to *Candida*, to yeasts formerly assigned to the genus *Torulopsis* and to *Rhodotorula* species. The names 'candida' and 'pathogenic yeasts' will be used similarly to refer to members of these genera. The term 'pathogenic yeasts' would ordinarily refer also to *Cryptococcus neoformans* and *Pityrosporum* spp. but here it will be used in the more narrow sense previously defined.

The principal pathogen of the genus *Candida* is the thrush fungus, *C. albicans*, discovered in 1839 by Langenbeck in the aphthae of a patient suffering from typhoid (typhus) fever, and believed by him to be causally related to the latter. The taxonomical position of the fungus was greatly confused until, in 1923, Berkhout suggested the generic name 'Candida', thus distinguishing the so-called 'medical monilias' from the fruit and leaf rotting moulds more properly placed in the genus *Monilia*. According to the rules of botanical nomenclature, the generic name *Monilia* was retained for the organism it first described, and *Candida* has been adopted internationally as the botanical *nomen conservandum* for the genus containing the thrush fungus. Many synonyms for *C. albicans* other than the botanically incorrect *Monilia albicans* have been used, van Uden & Buckley (1970) listing over 100, with fewer but still plentiful synonyms for the less notorious pathogens of the genus. The taxonomic confusion surrounding the yeast was reflected in the name used to describe disease caused by it. The term 'moniliasis' is gradually being expunged in favour of the more appropriate 'candidosis', but it is still used occasionally, especially in medical parlance.

The genera *Candida* and *Torulopsis* as formerly defined accommodate heterogeneous collections of asporogenous yeasts which do not qualify for inclusion in the more homogeneous genera of the Fungi imperfecti. The cross connections between the two genera are so numerous that *Candida* and *Torulopsis* (van Uden & Vidal-Leiria 1970) are discussed jointly in taxonomical treatises. Van Uden & Buckley (1970) pointed out the arbitrary nature and artificiality of the separation of the two genera on the criterion of the ability to form pseudohyphae, concluding that it would be logical to unite them. Yarrow & Meyer (1978) have proposed an amendment of the diagnosis of the genus *Candida* Berkhout to allow for nonhyphal species; in accordance with the amended description, the species currently classified in *Torulopsis* are transferred to the genus *Candida*. Thus, the yeast formerly known as *T. glabrata*, which is frequently isolated from man and is associated with disease, becomes *C. glabrata*. The proposal of Yarrow and Meyer will be adopted in this chapter, and no further reference to *Torulopsis* species will be made. Species of *Rhodotorula* Harrison (Phaff & Ahearn 1970) will be discussed with the *Candida* species, although they are far less frequently encountered in clinical practice. They are distinguished from *Candida* species principally by their synthesis of red and/or yellow carotenoid pigments in young malt agar cultures. The criteria of identification of all the species mentioned are those stated by Lodder (1970).

2. Ecological Considerations

The genus *Candida* comprises 81 species (van Uden & Buckley 1970) to which must be added a further 69 species, many formerly classified as *Torulopsis* (van Uden & Vidal-Leiria 1970) but now named as candidas (Yarrow & Meyer 1978) together with other species, such as *C. milleri* (Yarrow 1978) which have been described since 1970. There are nine species of *Rhodotorula* (Phaff & Ahearn 1970). Many of these yeasts have been isolated from man. Odds (1979) draws attention to the isolation from human sources, principally sputum, of 74 yeast species from 14 genera recorded by Lodder (1970), commenting that the majority of isolates were undoubtedly transient contaminants, but some might be regarded as normal human flora, and therefore as potential opportunistic pathogens. Hurley & de Louvois (1979) regarded the prevalence of the association of 25–33% of the known species of *Candida* with man, be it commonly or rarely, as noteworthy. They believed that potential for pathogenicity might be inherent in the genus, and that under favourable environmental conditions, it could find phenotypic expression as virulence, either at cellular or host level, in species not suspected of being actually or potentially pathogenic for man or mammals. Their experiments tended to confirm this view, in that they were able to induce virulence for laboratory mice in candidas from disparate ecological sources.

TABLE 1

Ecological sources* of 81 Candida *species*[†]

Man	27
Warm-blooded animals	20
Products of fermentation or	
decayed vegetation	21
Marine	9
Insects	10 + 1 variety
Plants or inanimate sources	66

*Groups not mutually exclusive.
[†]At least 13 are members or mating types of perfect species.

Table 1 shows the ecological sources of the 81 *Candida* species listed by Lodder (1970). Excluding species that are closely related to or, more properly, classified with the perfect yeasts, the genus splits readily into three fairly distinct ecological groups: those species that are associated predominantly with man or mammals; those that are isolated occasionally from man or warm-blooded animals, but not predominantly so; those that are isolated from soil, sea water, products of fermentation, vegetation, cold-blooded animals and various inanimate sources, but not from man or animals (Tables 2–4). The latter group is physiologically, also, fairly distinct from the other two. Fewer of the *Candida* species formerly classified as *Torulopsis* show a predominant association with man and warm-blooded animals, but *C. famata, glabrata* and *inconspicua* have been isolated principally from man, while *C. pintolopesii* is associated chiefly with rodents. Do Carmo-Sousa (1969) regarded *C. albicans, C. stellatoidea* and *C. glabrata* as obligatory animal saprophytes, classifying *C. guilliermondii, C. krusei, C. parapsilosis* and *C. tropicalis* as facultative saprophytes since

TABLE 2
Candida *species isolated principally from man or warm-blooded animals: Group 1*

Man		Other
aaseri	melibiosica	blankii
albicans*	membranaefaciens	ciferrii
brumptii	norvegensis	lusitaniae
catenulata	parapsilosis*	silvae
conglobata	pseudotropicalis*	slooffii
curvata	stellatoidea*	zeylanoides
guilliermondii*	tropicalis*	
krusei*	viswanathii*	
langeronii		

*Undoubted pathogens, also occur in other species of animals.

TABLE 3
Candida *species* isolated occasionally from man or warm-blooded animals: Group 2*

Man	Other
intermedia	lambica
lipolytica	
obtusa	
ravautii	
rugosa	
silvae	
tenuis	
utilis	

*Excluding those classified as *Hansenula, Leucosporidium, Metschnikowia, Kluyveromyces*.

TABLE 4
Candida *species* not reported isolated from man or warm-blooded animals: Group 3*

aquatica	diddensii	kefyr	salmonicola
beechii	diffluens	marina	santamariae
berthetii	diversa	maritima	scottii
bogoriensis	foliarum	melinii	shehatae
boidinii	freyschussii	mesenterica	silvicola
buffonii	friedrichii	mogii	solani
cacaoi	frigida	muscorum	valida
capsuligena	gelida	oregonesis	vartiovaarai
claussenii	glaebosa	rhagii	veronae
curiosa	ingens	sake	vini
	javanica	salmanticensis	

*Excluding those classified as *Hansenula, Leucosporidium, Metschnikowia, Kluyveromyces*.

they are frequently recovered from sources other than animals. None of the nine species of *Rhodotorula* (Phaff & Ahearn 1970) is associated predominantly with man or animals, although *Rh. aurantiaca, Rh. glutinis, Rh. minuta* and *Rh. rubra* have been isolated occasionally.

Yeasts of the *Candida* genus are abundantly distributed in nature on land and sea, in association with animals, plants and inanimate objects (Skinner & Fletcher 1960). The habitat of the pathogenic yeasts is more circumscribed. Scientific literature abounds in reports of their distribution in man and animals, many of which have been reviewed by Winner & Hurley (1964); Austwick *et al.* (1966); Hurley (1967); Gentles & La Touche (1969) and Odds (1979). The gastrointestinal tract is the most frequent source in the majority of animals. Most surveys attest the pre-eminent position of *C. albicans*, which has been recovered from a far wider range of animals than any other species. With few exceptions, it is also the yeast with the highest relative isolation rate. The principal data accumulated since 1960 have been extensively tabulated by Odds (1979) and Table 5, based on his figures, shows the percentage range of isolation of yeasts from various sites.

TABLE 5

Recovery of yeasts and/or C. albicans *from various sites in man*

Site	Status of subjects	Yeasts recovered % range	C. albicans recovered % range
Oral cavity	Healthy	36·9 – 2·0 (14·3)*	23·1 – 1·9 (10·3)
	Sick	76·2 – 12·7 (45·2)	69·6 – 6·0 (42·9)
Anus and rectum	Healthy	39·1 – 19·4 (23·7)	23·2 – 12·2 (14·6)
	Sick	62·2 – 5·7 (25·5)	48·0 – 8·5 (22·0)
Vagina	Healthy	13·3 – 8·1 (9·5)	11·3 – 0·0 (7·8)
	Sick or pregnant	67·9 – 7·8 (20·7)	52·2 – 4·5 (14·9)
	Local disease	76·5 – 25·0 (29·1)	84·4 – 14·2 (25·7)
Skin	Healthy	17·4 – 0·0	17·4 – 0·0
	Sick	16·5 – 4·7	12·4 – 0·4
Urine	Sick	17·6 – >3·0	4·4 – 1·6
Outer surface eye	Healthy or sick	8·7 – 0·0	0·7 – 0·0

*The figures in brackets are the calculated mean frequencies, weighted according to the number of subjects.
Data from Odds (1979).

There are few generalizations that can be made from these figures, and from the publications whence they arise, other than to remark that pathogenic yeasts are more frequently isolated from populations of patients than from healthy individuals; that there is a tendency for yeast colonization to increase with age, particularly in the case of the oral cavity; and that the isolation rate from the vagina increases both during pregnancy, and in the

presence of local morbidity. Yeasts are not denizens of healthy skin, although they may be isolated from the skin of those who attend patients with skin disorders, where, presumably, they are transient. If urine samples are not collected with meticulous care, yeasts from the vagina, perineum or gastro-intestinal tract may contaminate the specimen.

The relative incidence of yeasts isolated from man merits consideration. Hurley *et al.* (1973) compared their relative incidence in vaginal swabs studied prospectively, and in those sent for diagnosis of vulvovaginitis, finding that *C. albicans* predominated in both series, but that the frequency of its isolation, relative to *C. glabrata* and other less frequently encountered yeasts, was lower in the unselected population studied prospectively, while more recondite species were represented in greater variety and proportion. Their findings are summarized in Table 6 and show a greater prevalence of *C. albicans* among yeasts isolated from women with suspected vaginitis.

TABLE 6
Recovery of yeasts from the vagina

	Diagnostic specimens		Unselected specimens	
All yeasts	131		182	
C. albicans	118	(90%)	139	(76%)
C. albicans and				
C. glabrata	2		1	
C. glabrata	8	(6·0%)	27	(15%)
C. stellatoidea	2		3	
C. tropicalis	0		2	
C. parapsilosis	0		2	
C. holmii	0	(2·5%)	1	(8·5%)
C. inconspicua	0		1	
Sacch. cerevisiae	1		6	

Odds (1979, Tables 25 & 26) collated a number of studies on vaginal yeast flora, reaching the same conclusion. All the pathogenic candidas, other than *C. viswanathii* have been isolated from the vagina, as have *Rhodotorula* species. The data of Odds (1979, Tables 23, 24, 27) show that all, save *C. stellatoidea* and *C. viswanathii*, have been isolated from the oral cavity, faeces or anorectal swabs, and all save *C. stellatoidea*, *C. viswanathii* and *C. pseudotropicalis* have been isolated from the urine. *Rhodotorula* species have been recovered from urine, faeces or anorectal swabs and the oral cavity, accounting for less than 0·5% of yeasts in the mouth, but more than 12% of isolates from faeces or anorectal swabs. *Candida albicans* is the predominant yeast isolated, irrespective of site sampled.

In clinical practice, if a yeast is isolated from a superficial lesion of mouth

or vagina, there is a greater than 90% chance that it will prove to be *C. albicans*. In cases of vaginitis, there is a one in 20 chance that it will prove to be *C. glabrata. Candida parapsilosis* is isolated fairly frequently from skin lesions or paronychia, but other candidas are unlikely to be encountered from superficially diseased sites.

These observations, culled from experience and from analysis of the multitudinous data on yeast flora associated with man, have important consequences in routine clinical diagnostic practice. Most laboratory personnel have neither the skill nor the resources to engage on the labour-intensive exercise of identifying to specific status every yeast isolated from superficial sites. In the majority of cases, it suffices to state whether the candida isolated is, or is not, *C. albicans*, which can be established readily and rapidly from the colonial and morphological appearances, and the presence or absence of germ tubes and chlamydospores (MacKenzie 1966). Should further identification be warranted by the clinical circumstances, the isolate may be referred to an expert, and yeasts isolated from the blood-stream, other body fluids or the deep tissues should be fully identified. While this is usually beyond the range of non-specialist laboratories, kits are available commercially and their performance has been compared (de Louvois *et al.* 1979). The API 20C (Auxanogramme) kit should prove suitable for use in most laboratories.

3. Pathogenicity in the Genus

The thrush fungus was successfully transmitted in man in 1846 by Berg, who was able to reproduce the lesions, and Hausman (1875) elucidated the aetiology of vaginal candidosis by introducing scrapings from the mouth of an infant with thrush into the vagina of a pregnant woman, inducing vaginitis. Animal experimentation followed, and by 1898 de Stoechlin wrote that all experimentalists agreed that the thrush fungus was pathogenic for laboratory animals. The experiments of Redaelli (1924) and Benham (1931) established *C. albicans* as the only member of the genus of proven pathogenicity. They found that *C. albicans* injected intravenously into rabbits gave rise to a fatal disease, whereas other species were not pathogenic. The criterion of pathogenicity for the rabbit became accepted as part of the species identification of *C. albicans*, but the work of these experimentalists led to the dangerous and erroneous concept that other species of *Candida*, even when isolated from the bloodstream in the demonstrable presence of disease, were of no consequence. Discounting earlier reports where documentation was unsatisfactory, by the 1960s dis-seminated infection with *C. albicans, C. tropicalis* and *C. pseudotropicalis* had been reported (Hurley 1964; Winner & Hurley 1964). Endocarditis, ascribable to *C. albicans*, *C. parapsilosis*, *C. guilliermondii*, *C. krusei* and *C. tropicalis* had also been described (Winner & Hurley 1964).

Because of the prevalence of the false doctrine that species of *Candida* other than *C. albicans* were likely to be contaminants if isolated from the bloodstream, experimentalists took steps to establish that the putative pathogenic species obeyed Koch's postulates. In the main, they were successful. *Candida albicans*, *C. tropicalis* (Hasenclever 1959; Hasenclever & Mitchell 1961*a*; Hurley & Winner 1962), *C. stellatoidea* (Hasenclever & Mitchell 1961*b*, *c*; Hurley 1965), *C. pseudotropicalis* (Hurley & Winner 1964) and *C. viswanathii* (Sandhu *et al.* 1965) can all be shown to produce the typical lesions of systemic candidosis on intravenous injection in mice, or other laboratory animals. *Candida parapsilosis* produces visceral lesions in cortisone-treated mice (Goldstein *et al.* 1965) and typical lesions have been produced by *C. glabrata* in pregnant or cortisone-treated mice (Hasenclever & Mitchell 1963; Knudtson *et al.* 1973). No lesions ascribable to *C. guilliermondii* in intact hosts have been described. Partridge *et al.* (1971) showed that *C. albicans*, *C. tropicalis*, *C. pseudotropicalis* and *C. parapsilosis* produced lesions on chick chorioallantoic membrane whereas *C. glabrata* did not. Stanley & Hurley (1967) showed that the cells of five-day-old murine renal epithelial cells were destroyed by *C. albicans*, *C. tropicalis*, *C. stellatoidea* (Group 1), *C. krusei*, *C. parapsilosis*, *C. pseudotropicalis* (Group 2) and by *C. guilliermondii* (Group 3) in all instances more rapidly and more completely than by the control yeast, *Saccharomyces cerevisiae*. The yeasts were grouped according to the rapidity with which they destroyed the cultures, and the order given is the generally accepted descending order of virulence in these species (Table 7). It is not known where in this series *C. glabrata* or *Rhodotorula* spp. should rank, but Odds (1979) suggests that they are unlikely to rank higher in virulence than *C. guilliermondii*. The writer suspects that *C. glabrata* might

TABLE 7

The pathogenic Candida *species in descending order of virulence for man*

albicans
tropicalis
stellatoidea
glabrata
krusei
parapsilosis
pseudotropicalis
guilliermondii
viswanathii

well prove intermediate in virulence between the Group 1 and Group 2 yeasts of Stanley & Hurley (1967), for it is the only species other than *C. albicans* frequently associated with human vulvovaginitis and is, thus, numerically an important pathogen. The effects of strain variation in virulence are unlikely to be of consequence, for they are small indeed in these rather stable yeasts. The perfect states, known or postulated for some of these yeasts, do not seem to be associated with disease. Six 'non-pathogenic' yeasts, none isolatable from man, were inoculated into cultured mouse epithelial cells by Hurley & Stanley (1969). Five of them showed cytopathic effects of the same order as *C. guilliermondii*. Extension of experiments on four of these yeasts (*C. blankii*, *C. kefyr*, *C. diddensii*, *C. cacaoi*) after passage in mice, showed that the passaged strains became moderately virulent for intact hosts, suggesting that candidas from disparate ecological sources can adapt to parasitism (Hurley & de Louvois 1979).

4. The Spectrum of Candidosis

Candidosis exemplifies the whole range of states that may exist between the host and its indigenous parasites, from commensalism to fulminating systemic illness with an extremely poor prognosis if untreated. In most individuals harbouring pathogenic yeasts no harm results, but the fungi are able to invade and to damage tissues if host defences are impaired, either locally or generally. The disease states range from superficial, usually minor, infections of the buccal cavity (oral thrush), the aural canal or vagina to deep seated infections involving the viscera, which are often lethal. Oesophageal, laryngeal and bronchopulmonary candidosis may arise as extensions of overt or latent oral thrush. Hippocrates noticed that aphthae occurred in those subject to severe and debilitating disease and Parrot wrote of oral thrush in 1877: "le muguet est toujours la conséquence d'un état morbide antérieure et ne constitue pas une maladie" (Winner & Hurley 1964). Except under the most artificial circumstances (Krause *et al.* 1969) neither superficial nor deep seated candidosis arises in those subjects within the physiological norm. Table 8 outlines the principal factors that predispose to infection with pathogenic yeasts.

A. *Superficial candidosis*

By this term is usually meant infections involving the mouth, vagina, ear, skin or nails; in other words, those sites easily accessible to the examining physician (Table 9). Those of the mouth and vagina are amongst the most frequent infections encountered in medical practice. Ajello (1975) commented

TABLE 8
Principal factors predisposing to candidosis in man

Pathological or physiological	Severe microbial infections
	Endocrine disorders
	Defects in cell-mediated immunity
	Infancy, pregnancy
Mechanical	Trauma
	Local occlusion or maceration of tissues
Iatrogenic	Treatment with drugs altering or suppressing endogenous flora, or local immunity
	Intravenous infusions, transplant or prosthetic surgery

TABLE 9
Superficial candidosis

Oral thrush	Vulvovaginitis	Allergy
Denture stomatitis	Balanitis	Paronychia
Perlèche	Otitis externa	Onychia
'Leucoplakia'	Intertrigo	Congenital cutaneous candidosis
Glossitis	Napkin rash	Chronic mucocutaneous candidosis

that the true incidence and prevalence of mycotic disease remains unknown, and, in consequence, its socio-economic impact cannot be quantified. The superficial candida mycoses constitute a real public health hazard, and prove costly in terms of medical expertise and of money spent on treatment that is often ineffective.

Oral thrush is prevalent amongst the newborn and in infants, occurring also in sickly or debilitated adults. It is not infrequent after dental surgery or dental manipulations, and in those who have been receiving antibiotic therapy particularly for acute bacterial tonsillitis or asthma complicated by episodes of acute or chronic bronchitis. The clinical picture is characteristic. White patches, aphthae, appear as discrete lesions on the surfaces of the buccal mucosa or the tongue. They may extend to confluent pseudo-membranes, resembling curds. The lesions are painful, if ulceration occurs, but the local glands are not enlarged. Yeasts and pseudohyphae will be seen in smears from the lesions — *Candida albicans* is the aetiological agent.

Other lesions arising in or around the oral cavity are denture stomatitis (Cawson 1963), often ascribable to ill-fitting dentures, with or without perlèche (angular cheilitis). In the latter condition, painful fissures appear at the corners of the mouth. Candida leucoplakia, glossitis and glossodynia also occur, the latter two sometimes called 'antibiotic sore tongue'.

In the nineteenth century, oral thrush was commonplace in institutionalized infants, and assumes epidemic proportions today in maternity units where oral hygiene of the newborn is poor, and attention to aseptic assembly of feeds is less than scrupulous. The incidence of the disease is 1 : 200 at Queen Charlotte's Maternity Hospital, but rates of 40–60% are not unknown still in some European centres. The disease is frequent in elderly or debilitated patients.

Candida vulvovaginitis is extremely common, particularly in pregnant women. Characterized by pruritus and discharge, with white patches on the vulvar, vaginal or cervical epithelium, it afflicts a maximum of 16% of women during pregnancy at Queen Charlotte's Hospital. Pregnancy thrush may be the initiating event in longstanding vaginal thrush which often proves refractory to treatment (Hurley 1975). The fungus does not develop natural resistance to the polyene antibiotics which are ordinarily used; it is the disease itself that becomes refractory, for reasons yet to be explained. *Candida albicans*, or *C. glabrata*, is the usual aetiological agent, but the latter is not associated with the intractable form in the writer's experience.

Candida balanitis is, apparently, less common than candida vulvovaginitis, although Odds (1979, Fig. 14), analysing data from British venereal disease clinics, suggests a case incidence of at least $12/\text{annum}/10^5$ of population in 1971.

In normal subjects, pathogenic yeasts other than *C. albicans* are recovered only rarely from the vagina. The conflict of opinion on commensalism of *C. albicans* therein has been described elsewhere (Hurley *et al.* 1974), together with other aspects of candida vulvovaginitis.

Candida albicans is associated with otitis externa (Gregson & La Touche 1961; Gentles & La Touche 1969). It is usually treated with ointments or pastes containing nystatin.

Shelmire (1925) reviewed candida infections of the skin, and it is they that best exemplify the historic concept of the great physician and teacher, Trousseau (1869), who was the first explicitly to expound the view that disease caused by candida is always secondary to some host-determined defect. Summarizing his writings and lectures spanning 25 years, he described candidosis as "the local expression of a very bad state of the whole system". Other than oral thrush, skin thrush occurs characteristically in intertriginous areas, where locally moist conditions are caused by maceration and occlusion; lesions may also appear between the fingers or the toes, and some, such as barmaids or miners, may be at risk because of their occupations. Maibach & Kligman (1962) showed that *C. albicans* produced skin lesions in volunteers only when the site of inoculation was covered by an occlusive dressing. English and her colleagues (1971) were able to isolate *C. parapsilosis* from the moist skin surrounding leg ulcers, though not from the ulcers themselves, demonstrating the propensity of candidas to colonize unhealthy skin. The clinical features and treatment of skin candidosis arc discusscd by Winner & Hurley (1964) and Odds (1979). Napkin rash is an important and common manifestation. In recent years, several cases of congenital cutaneous candidosis (Sommerschein *et al.* 1960; Gellis *et al.* 1976) have been described. The postulated route of infection is the intra-uterine, and the disease is benign and self-limiting, clearing spontaneously after a few weeks. The rash is widely distributed.

Allergic 'id' reactions to candida have been reported, and the antigenic role of *C. albicans* in evoking symptomatology remote from sites of colonization, such as colitis, depression and psychiatric illness, and neurological syndromes has recently been proposed (Truss 1978). It is too early to evaluate this striking hypothesis.

Candida paronychia and onychia both occur as occupational diseases in housewives, fishmongers, fruit canners and others whose hands are macerated by prolonged immersion in water. Unlike candida lesions elsewhere, species other than *C. albicans* predominate, especially *C. parapsilosis* and *C. guilliermondii.*

Most arresting of all the skin or mucous membrane manifestations of candidosis is the syndrome known as chronic mucocutaneous candidosis (CMC) (Higgs 1973). The disease arises in children, usually in those under two years of age, occasionally presenting in the thirties or even in the fifties. It is extremely rare, often associated with genetic defect and endocrinological disorders, and almost certainly based on defects of cellular immunity, many of which have been demonstrated. This chronic superficial infection has been classified in various ways, but that of Wells is widely used clinically (Wells *et al.* 1972; Wells 1973). The four groups comprise: those with familial CMC with autosomal recessive inheritance, who have chronic oral thrush with minor lesions elsewhere; those with no obvious familial disposition but a general susceptibility to infection, including candidosis, in whom lesions are severe and extensive, and in whom candida granulomas occur; those who have single or multiple endocrinopathy involving the thyroid, parathyroid, adrenal, or pancreas, and in whom there may be autosomal recessive inheritance of the endocrinopathy; CMC of late onset. *Candida albicans* is the causative fungus in all cases, and treatment with amphotericin B or immune reconstitution methods has achieved success (Valdimarsson *et al.* 1972; Kirkpatrick & Smith 1974). Dominant inheritance has been described (Sams *et al.* 1979).

B. Deep-seated candidosis

Far less common but more severe than the superficial manifestations of candida mycosis are infections of single viscera, or widespread dissemination of candidas. Twenty to 40 deaths attributable to systemic candidosis are reported to the Registrar General each year for England and Wales. Even in countries where other pathogenic fungi are endemic and cause serious disease, the death rate from candidosis is appreciable, the projected rate of cases requiring admission to hospital being $0.4/10^6$ of population in the USA in 1970 (Hammerman *et al.* 1974). Candida accounted for 27 of 44 cases of fatal opportunistic fungal infection in women of childbearing age revealed from scrutiny of the files of the Armed Forces Institute of Pathology (Purtilo 1975).

The spread of candida lesions from the buccal cavity to involve other parts of the gastro-intestinal tract, notably the fauces, oesophagus, stomach and intestines has been recognized since the late eighteenth century. Oesophageal candidosis is a well recognized complication of leukaemia and other blood dyscrasias, partly in consequence of the immunosuppressive agents used in therapy. Diarrhoea associated with candida is also well recognized, usually occurring in seriously ill patients who have been treated extensively with antibacterial antibiotics. In a case seen by the writer, virtually pure cultures of *C. albicans* were isolated from the cholera-like stools. Holti (1964) believed that allergy to yeasts underlay many cases of the 'irritable colon syndrome'.

Not only can candida spread from the mouth distally into the gastro-intestinal tract proper, but the fungi may also invade the larynx, and, more commonly, the main bronchi. Less well documented, though often reported and all too frequently diagnosed on less than satisfactory evidence is involvement of pulmonary parenchyma in the mycotic process, in the absence of disseminated infection. Pulmonary allergy to candida was described by Castellani (1912) as "teataster's cough", and enhancement by candida of allergy to threshing dust has been shown. So potent are the antigenic properties of *C. albicans* that it is widely used experimentally as an allergen.

Although lesions of bone or joints, and isolated visceral lesions, such as endophthalmitis, meningitis and peritonitis can occur, as well as relatively minor infections of the urinary tract (cystitis, urethritis), the most serious infections caused by candidas are endocarditis, renal candidosis and acute disseminated (septicaemic) candidosis. These diseases share the common aetiology of being bloodborne. Few cases of candida endocarditis were recorded before the review of Seelig *et al.* (1974). It has to be admitted that the majority of cases are iatrogenic, following open heart surgery. Before the advent of this type of surgery, the disease occurred in mainline drug users who literally injected themselves with the pathogens contaminating the vehicles of addiction; some 20% of cases still arise in this way. The disease is an infective endocarditis arising primarily in patients with underlying disease of the valves, or in whom prostheses have been fitted or homografts undertaken, affecting the aortic valve more than the mitral, and rarely involving the tricuspid. The vegetations are large and friable, and, consequently, embolic phenomena occur, often involving the large arteries. The yeasts involved are shown in Table 10. The high incidence of *C. parapsilosis* is noteworthy, and may evidence either a selective affinity of this yeast for endocardial tissue, as Odds (1979) suggests, or may, as the writer believes, reflect the propensity of *C. parapsilosis* to colonize damaged skin, and thus to have opportunity to gain ingress along intra-vascular lines, which are usually taped at the site of entry.

TABLE 10
Species of yeasts associated with endocarditis

Species	%
C. albicans	54·8
C. guilliermondii	5·4
C. krusei	5·4
C. parapsilosis	25·8
C. stellatoidea	2·2
C. tropicalis	4·3
C. glabrata	1·1
Rh. pilimanae	1·1

Data from Odds (1979).

From study of isolated case reports and experimental work, Hurley & Winner (1963) adduced evidence of a variant form of the septicaemic disease, in which the renal parenchyma was the target organ, and the outcome a chronic fatal disease leading to hydronephrosis, sometimes with fungus ball in the pelvis of the ureter. The disease is uncommon. Experimentally, it can be induced by inocula smaller than that required to establish acute disseminated candidosis. The predilection of *C. albicans* for the kidney has long been known.

Twenty years ago, candida septicaemia and the syndrome of acute disseminated candidosis were unknown to all but a handful of experts, who admitted that on occasion candidas had been isolated from the bloodstream of seriously ill patients. Now the disease is well known to hospital practitioners, and the regrettable (and lethal) error of ignoring blood cultures that yield yeasts is made far less frequently. Hurley (1964) located only 48 cases that she regarded as adequately documented. On the subject of yeast-yielding blood cultures, Odds (1979) does not mince words; he advises practitioners to pay no heed to laboratories that describe them as contaminated. It seems likely that it is awareness of the disease, rather than genuine increase in its frequency that is responsible for the numbers of cases that are now recognized. The brain, the heart and the kidney are the primary target organs, but any viscus, particularly if already damaged by underlying disease, or chemical or physical poison, may provide lodgement for the fungi. The untreated prognosis is poor, but less so in children under one year of age than in adults. In both, it arises exclusively in debilitated hosts. The causal yeasts are shown in Table 11.

TABLE 11

Species of yeasts associated with acute disseminated candidosis

Species	%
C. albicans	66·4
C. guilliermondii	2·5
C. krusei	0·9
C. parapsilosis	3·1
C. pseudotropicalis	0·3
C. stellatoidea	1·6
C. tropicalis	10·4
C. glabrata	0·9
Rhodotorula species	1·3
Other (unidentified)	12·6

Data from Odds (1979).

5. Conclusion

In the 1930s, disease in man ascribable to candida species, even so common a disease as vulvovaginitis (Plass *et al.* 1931), was regarded as a great rarity, and the deep-seated candida mycoses were virtually unknown. Fifty years later, all authorities agree that, numerically speaking, superficial candidosis ranks amongst the most common of all infectious diseases. Both superficial and deep candidosis are important public health hazards which are encountered throughout the world, not only in man, but in wild and domestic animals (Odds 1979). Their economic importance is great, and progress in understanding their prevalence has led to the introduction of a plethora of antifungal drugs. Discussion of the latter is outside the scope of this review.

6. References

AJELLO, L. 1975 In *The Epidemiology of Human Mycotic Disease,* ed. Al-Doory, Y. p. 290. Springfield, Illinois: Charles Thomas.

AUSTWICK, P. K. C., PEPIN, G. A., THOMSON, J. C. & YARROW, D. 1966 *Candida albicans* and other yeasts associated with animal disease. In *Symposium on Candida Infections,* ed. Winner, H. I. & Hurley, R. pp. 89–100. London: Livingstone.

BENHAM, R. W. 1931 Certain monilias parasitic on man: their identification by morphology and by agglutination. *Journal of Infectious Diseases* **49**, 183–215.

BERG, T. F. 1846 *Om Torsk Hos Barn.* Stockholm: Hjerta.

BERKHOUT, C. M. 1923 *De schimmelgeschlachten* Monilia, Oidium, Oospora *en* Torula. Dissertation, University of Utrecht.

CASTELLANI, A. 1912 Observations on the fungi found in tropical bronchomycosis. *Lancet* **i**, 13–15.

CAWSON, R. A. 1963 Denture sore mouth and angular cheilitis. Oral candidiasis in adults. *British Dental Journal* **115**, 441–449.

DE LOUVOIS, J., MULHALL, A. & HURLEY, R. 1979 The biochemical identification of clinically important yeasts. *Journal of Clinical Pathology* **32**, 715–718.

DE STOECHLIN, H. 1898 Recherches cliniques et expérimentales sur le rôle des levures trouvées dans les angines suspectées de diphtérie. *Archiv für experimentelle Medizin* **10**, 1–41.

DIDDENS, H. A. & LODDER, J. 1942 *Die Anascosporogenen Hefen* 2 Halfte. Amsterdam: North-Holland.

DO CARMO-SOUSA, L. 1969 Distribution of yeasts in nature. In *The Yeasts* Vol. 1, ed. Rose, A. H. & Harrison, J. S. pp. 79–105. London: Academic Press.

ENGLISH, M. P., SMITH, R. J. & HARMAN, R. R. M. 1971 The fungal flora of ulcerated legs. *British Journal of Dermatology* **84**, 567–581.

GELLIS, S. S., FEINGOLD, M., KOZINN, P. J., TARIQ, A. A., REALE, M. R. & RUDOLPH, N. 1976 Picture of the month; congenital cutaneous candidiasis. *American Journal of Diseases of Childhood* **130**, 291–292.

GENTLES, J. C. & LA TOUCHE, C. J. 1969 Yeasts as human and animal pathogens. In *The Yeasts* Vol. 1, ed. Rose, A. H. & Harrison, J. S. pp. 108–192. London: Academic Press.

GOLDSTEIN, E., GRIECO, M. H., FINKEL, G. & LOURIA, D. B. 1965 Studies on the pathogenesis of experimental *Candida parapsilosis* and *Candida guilliermondii* infections in mice. *Journal of Infectious Diseases* **115**, 293–302.

GREGSON, A. E. & LA TOUCHE, C. J. 1961 Otomycosis: a neglected disease. *Journal of Laryngology* **75**, 45–69.

HAMMERMAN, K. J., POWELL, K. E. & TOSH, F. E. 1974 The incidence of hospitalised cases of systemic mycotic infections. *Sabouraudia* **12**, 33–45.

HASENCLEVER, H. F. 1959 Comparative pathogenicity of *Candida albicans* for mice and rabbits. *Journal of Bacteriology* **78**, 105–109.

HASENCLEVER, H. F. & MITCHELL, W. O. 1961*a* Pathogenicity of *C. albicans* and *C. tropicalis. Sabouraudia* **1**, 16–21.

HASENCLEVER, H. F. & MITCHELL, W. O. 1961*b* Antigenic studies of *Candida III.* Comparative pathogenicity of *Candida albicans* Group A, Group B and *Candida stellatoidea. Journal of Bacteriology* **82**, 574–577.

HASENCLEVER, H. F. & MITCHELL, W. O. 1961*c* Antigenic studies with *Candida III.* Comparative pathogenicity of *C. albicans* Group A, Group B, and *C. stellatoidea. Mycopathologia et mycologia applicata* **14**, 230–236.

HASENCLEVER, H. F. & MITCHELL, W. O. 1963 Pathogenesis of *Torulopsis glabrata* in physiologically altered mice. *Sabouraudia* **2**, 87–95.

HAUSMAN, D. 1875 *Parasites des Organes Sexuels Femelles de l'Homme et de Quelques Animaux, avec une Notice sur Développement de l'Oidium albicans Robin,* translated by Walker, P. E. Paris: Baillière.

HIGGS, J. M. 1973 Muco-cutaneous candidiasis: historical aspects. *Transactions of St. John's Hospital Dermatological Society* **59**, 175–194.

HOLTI, G. 1964 Candida allergy. In *Symposium on Candida Infections* ed. Winner, H. I. & Hurley, R. pp. 73–81. London: Livingstone.

HURLEY, R. 1964 Acute disseminated (septicaemic) monilias in adults and children. *Postgraduate Medical Journal* **40**, 644–653.

HURLEY, R. 1965 The pathogenicity of *Candida stellatoidea. Journal of Pathology and Bacteriology* **90**, 351–354.

HURLEY, R. 1967 The pathogenic candida species: a review. *Review of Medical and Veterinary Mycology* **6**, 159–176.

HURLEY, R. 1975 Inveterate vaginal thrush. *Practitioner* **215**, 753–756.

HURLEY, R. & WINNER, H. I. 1962 The pathogenicity of *Candida tropicalis*. *Journal of Pathology and Bacteriology* **85**, 33–38.

HURLEY, R. & WINNER, H. I. 1963 Experimental renal moniliasis in the mouse. *Journal of Pathology and Bacteriology* **86**, 75–82.

HURLEY, R. & WINNER, H. I. 1964 Pathogenicity in the genus *Candida*. *Mycopathologia et mycologia applicata* **24**, 337–346.

HURLEY, R. & STANLEY, V. C. 1969 Cytopathic effects of pathogenic and non-pathogenic species of *Candida* on cultured mouse epithelial cells: relation to the growth rate and morphology of the fungi. *Journal of Medical Microbiology* **2**, 63–74.

HURLEY, R., LEASK, B. G. S., FAKTOR, J. A. & DE FONSEKA, C. I. 1973 Incidence and distribution of yeast species of *Trichomonas vaginalis* in the vagina of pregnant women. *Journal of Obstetrics and Gynaecology of the British Commonwealth* **80**, 252.

HURLEY, R., STANLEY, V. C., LEASK, B. G. S. & DE LOUVOIS, J. 1974 Microflora of the vagina during pregnancy. In *The Normal Microbial Flora of Man* ed. Skinner, F. A. & Carr, J. G. pp. 155–182. London: Academic Press.

HURLEY, R. & DE LOUVOIS, J. 1979 Ecological aspects of yeast-like fungi of medical importance: pathogenic potential in the genus *Candida*. In *Medical Mycology,* ed. Prenseek, H. J. *Zentralblatt für Bakteriologie, Parasitenkunde, Infektionskrankheiten und Hygiene.* Abt. I, Suppl. 8.

KIRKPATRICK, C. H. & SMITH, T. K. 1974 Chronic mucocutaneous candidiasis: immunological and antibody therapy. *Annals of Internal Medicine* **80**, 310–320.

KNUDTSON, W. U., WOHLGEMUTH, K., KIRKBRIDE, C., ROBL, M. & KIEFFER, M. 1973 Pathogenicity of *Torulopsis glabrata* for pregnant mice. *Sabouraudia* **11**, 175–178.

KRAUSE, W., MATHEIS, H. & WULF, K. 1969 Fungaemia and funguria after oral administration of *Candida albicans*. *Lancet* **i**, 598–599.

LANGENBECK, B. 1839. Auffingung von Pilzen aus der Schleimhaut der Speiseröhre einter Typhus-leiche Neue Notizen auf dem Gebiete der Natur-und-Heilkunde (Froriep) **12**, 145–147.

LODDER, J. ed. 1970 *The Yeasts — A Taxonomic Study* 2nd edn. Amsterdam: North-Holland.

MACKENZIE, D. W. R. 1966 Laboratory investigations of candida infections. In *Symposium on Candida Infections* ed. Winner, H. I. & Hurley, R. pp. 26–44. London: Livingstone.

MAIBACH, H. I. & KLIGMAN, A. M. 1962 The biology of experimental human cutaneous moniliasis (*Candida albicans*). *Archives of Dermatology* **85**, 233–255.

ODDS, F. C. 1979 *Candida and Candidosis*. Leicester: Leicester University Press.

PARTRIDGE, B. M., ATHAR, M. A. & WINNER, H. I. 1971 Chick embryo inoculation as a pathogenicity test for candida species. *Journal of Clinical Pathology* **24**, 645–648.

PHAFF, H. J. & AHEARN, D. G. 1970 *Rhodotorula* Harrison. In *The Yeasts — A Taxonomic Study* 2nd edn, ed. Lodder, J. pp. 1187–1223. Amsterdam: North-Holland.

PLASS, E. D., HESSSELTINE, H. C. & BORTS, I. H. 1931 Monilia vulvovaginitis. *American Journal of Obstetrics and Gynaecology* **21**, 320.

PURTILO, D. T. 1975 Opportunistic mycotic infections in pregnant women. *Gynaecology* **122**, 607–610.

REDAELLI, P. 1924 Experimental moniliasis. *Journal of Tropical Medicine and Hygiene* **27**, 211–213.

SAMS, W. M., JORIZZO, J. L., SNYDERMAN, R., JEGOSOTHY, B. V., WARD, F. E., WEINER, M., WILSON, J. G., YOUNT, W. J. & DILLARD, S. B. 1979 Chronic mucocutaneous candidiasis. *American Journal of Medicine* **67**, 948–959.

SAUDHU, R. S., RANDHAWA, H. J. & GUPTA, I. M. 1965 Pathogenicity of *Candida viswanathii* for laboratory animals. A preliminary study. *Sabouraudia* **4**, 37–40.

SEELIG, M. S., SPETH, C. P., KOZINN, P., TASCHDIJIAN, C. L., TONI, E. F. & GOLDBERG, P. 1974 Patterns of Candida endocarditis following cardiac surgery: importance of early diagnosis and therapy (an analysis of 91 cases). *Progress in Cardiovascular Disease* **17**, 125–160.

SHELMIRE, B. 1925 Thrush infections of the skin. *Archives of Dermatology and Symphilology, Chicago* **12**, 789–813.

SKINNER, C. E. & FLETCHER, D. W. 1960 A review of the genus *Candida, Bacteriological Reviews* **24**, 397–416.

SOMMERSCHEIN, H., CLARK, H. L. & TASCHDJIAN, C. L. 1960 Congenital cutaneous candidosis in a premature infant. *American Journal of Diseases of Childhood* **107**, 20–266.

STANLEY, V. C. & HURLEY, R. 1967 Growth of *Candida* species in cultures of mouse epithelial cells. *Journal of Pathology and Bacteriology* **94**, 301–315.

TROUSSEAU, A. 1869 Lectures on Clinical Medicine, delivered at the Hotel-Dieu, Paris. Translated from the edition of CORMACK, J. R. (1868) Vol. 2. London: New Sydenham Society.

TRUSS, C. O. 1978 Tissue injury induced by *Candida albicans.* Mental and neurological manifestations. *Journal of Orthomolecular Psychiatry* **7**, 17–37.

VAN UDEN, N. & BUCKLEY, H. R. 1970 *Candida* Berkhout In *The Yeasts — A Taxonomic Study* 2nd edn, ed. Lodder, J. Amsterdam: North-Holland.

VAN UDEN, N. & VIDAL-LEIRIA, M. 1970 *Torulopsis* Berlese. In *The Yeasts — A Taxonomic Study* 2nd edn, ed. Lodder, J. p. 1235. Amsterdam: North-Holland.

VALDIMARSSON, H., MOSS, P. D., HOLT, P. J. L. & HOBBS, J. R. 1972 Treatment of chronic mucocutaneous candidiasis with leucocytes from HLA compatible sibling. *Lancet* **i**, 469–472.

WELLS, R. S. 1973 Chronic mucocutaneous candidiasis: a clinical classification. *Proceedings of the Royal Society of Medicine* **66**, 801–802.

WELLS, R. S., HIGGS, J. M., MACDONALD, A., VALDIMARSSON, H. & HOLT, P. J. L. 1972 Familial chronic muco-cutaneous candidiasis. *Journal of Medical Genetics* **9**, 302–310.

WINNER, H. I. & HURLEY, R. 1964 Candida albicans. London: Churchill.

WINNER, H. I. & HURLEY, R. eds 1966 *Symposium on Candida Infections.* London: Livingstone.

YARROW, D. 1978 *Candida milleri* sp. nov. *International Journal of Systematic Bacteriology* **28**, 608–610.

YARROW, D. & MEYER, S. A. 1978 Proposal for amendment of the diagnosis of the genus *Candida* Berkhout nom. cons. *International Journal of Systematic Bacteriology* **28**, 611–615.

Cryptococcus neoformans — Pathogen and Saprophyte

C. K. CAMPBELL AND D. W. R. MACKENZIE

Mycological Reference Laboratory
London School of Hygiene and Tropical Medicine, London, UK

Contents

1. Introduction

PROBABLY NO YEAST has produced as many surprises during the past 15 years as has *Cryptococcus neoformans*. From the time of its first recorded involvement in human disease in 1894 until the mid 1960s it was known as a simple budding yeast, the cause of a rare and progressive infection of the brain from which there was little chance of recovery. Today, largely through work in the USA, we know it as consisting of two species of a totally new division of the Basidiomycetes and that primarily it causes infections which are benign and common rather than rare and lethal. Other surprises have emerged in its saprophytic life history. The present paper is an attempt to summarize these changes in knowledge of *Cr. neoformans*, and to illustrate advances in the study of the disease it produces in man and animals.

2. *Cryptococcus neoformans* as a Pathogen

The symptoms and pathological features of cryptococcosis have been reviewed many times in medical literature. The patient is typically one already debilitated by some chronic disease such as lymphoma, systemic lupus, sarcoid etc. or by long-term steroid therapy. The most familiar type of infection involves primarily invasion of the brain and spinal chord. It becomes established via the blood stream from a pre-existent focus elsewhere in the body. Once the yeast cells have penetrated the meninges, they become disseminated over the surface of the brain by currents in the cerebro-spinal fluid (CSF) (Littman 1959). In most individuals the disease develops insidiously. The patient experiences periods of apathy and abnormal behaviour accompanied over weeks or months by increasing severity and frequency of headache. In advanced stages, neck stiffness and other meningeal symptoms lead to coma and, eventually, death (Emmons *et al.* 1977). The prognosis in untreated patients is very grave and cryptococcosis is rightly regarded as one of the most feared of all the deep-seated fungal infections.

Other parts of the body which may be infected are bones and skin. Bone lesions, accompanied by painful swelling, occur in between 5 and 10% of all cases (Nottebart *et al.* 1974; Emmons *et al.* 1977). Cutaneous lesions may exceptionally result by direct implantation of the yeast from a saprophytic source, but dissemination from some deeper focus of infection is probably much more frequent. There is now considerable circumstantial evidence which suggests that cryptococcal infections of all types originate by haematogenous spread from a primary focus in the lungs.

That inhalation of airborne *Cr. neoformans* may initiate the disease has been a theme repeated by many writers since the discovery of its saprophytic source in dry accumulations of pigeon droppings 24 years ago (Emmons 1955). Haugen & Baker (1954) drew attention to a number of cases in which the yeast was seen in single subpleural nodules. These soft nodules were very small, less than 1·5 cm in diameter, and were only found by diligent searching of lungs *post mortem*, in patients who died of unrelated causes as well as in those with overt cryptococcosis elsewhere. These findings have been substantiated by others (Greer 1978). In addition, later workers have come to recognize the 'cryptococcoma', apparently a development from such small nodules, in which the host is less able to contain the yeast. Such lesions may be several centimetres in diameter, and contain masses of actively growing yeast cells. This condition appears to be seen rarely in Britain.

Most investigators believe that pulmonary infection may not be uncommon and is characteristically subclinical and transient (Emmons *et al.* 1977; Salfelder 1978). Parallels are evident between this concept and the much better documented studies on the epidemiology of other systemic fungal pathogens, *Coccidioides* and *Histoplasma*, where skin tests have indicated a high incidence of exposure and subclinical infection in areas where the diseases by these fungi are endemic.

Figures for cryptococcosis in many countries have shown a tendency to increase during the past 15 years. For instance, only some 300 cases were on record worldwide by 1956, whereas for the period 1965–1977, 1264 proven cases were described in the USA alone (Kaufman & Blumer 1978). The figures for Britain, although smaller than those for the USA, also show a similar increase. Data gathered at the Mycological Reference Laboratory in London show that from 1963 to 1967 there was a maximum of 5 cases/year (average 1·6). From 1973 to 1977 diagnosis has been more frequent, 3–9 cases/year (average 6·2).

Such figures depend very much on the frequency with which the clinician includes cryptococcosis as a diagnostic possibility, and by the efficiency of the tests employed in laboratory examination. The advent of efficient and sensitive tests for this disease has undoubtedly placed the clinician in a better position to interpret clinical signs than was possible 15 years ago. A valuable account of diagnostic and prognostic parameters was given by Diamond & Bennett (1974) in a study of 111 patients with proven cryptococcosis.

3. Laboratory Tests in Diagnosis

Advances in laboratory testing have been made in both culture and serology. The clinical laboratory can now identify an unknown yeast as *Cr. neoformans* by using a small battery of cultural tests. An accurate preliminary identification can be made less than 2 d after the specimen has been received from the patient. Classical assimilation and fermentation procedures are used to confirm the identification.

One test that has become invaluable is the production by *Cr. neoformans* of melanin-like pigment when grown in the presence of melanin precursors (*o*-diphenols). The method originates in the intriguing observations by Staib (1962, 1963) that *Cr. neoformans* produces brown colonies when grown on extracts of bird manure only if the birds had been fed with seeds of the Niger plant (*Guizotia abyssinica*). Incorporation into agar of extracts of the seeds themselves gave similar browning and resulted in a valuable selective differential medium (Shields & Ajello 1966; Botard & Kelley 1968; Jennings *et al.* 1968). Later workers have described similar pigmentation when *Cr. neoformans* is grown in the presence of extracts of potatoes or carrots (Shaw & Kapica 1972), caffeic acid [(3:4 dihydroxycinnamic acid) (Korth & Pulverer 1971)] or DOPA (Chaskes & Tyndall 1975). The postulated mechanism of action of all these is that the yeast produces phenyloxidase which catalyses synthesis of melanin from the diphenols whether as pure compounds or in the natural substrates. Using caffeic acid impregnated discs the test has been refined to give results within 2–6 h.

The urease test, so important as a first screen in identifying *Cr. neoformans*, normally takes 3 or 4 days. Recently a micro-test was described by

Roberts *et al.* (1978) which reduces this time to 2–4 h. This is probably not yet widely used but promises to speed up provisional reporting in the clinical laboratory.

Serological aids have now become of prime importance in diagnosis and management of cryptococcal infections. Until the mid 1960s most patients with cryptococcosis were thought to be serologically inert, since conventional methods of antibody detection failed in more than 50% of proven cases.

A major advance in the serodiagnosis of the disease followed the realization that in many patients with active cryptococcosis, antigen derived from the conspicuous capsules which are usually formed *in vivo* is more commonly found than antibody. Bloomfield *et al.* (1963) described an inert particle agglutination test which is both sensitive and specific for detection of this capsular polysaccharide. Latex particles coated with specific antibody will agglutinate in the presence of the antigen. The amount of antigen present (titre) can be assessed by determining the extent to which CSF or serum can be diluted and still elicit a positive latex agglutination (LA) test.

By combining the results of the LA test with antibody detection methods such as whole cell agglutination test and indirect fluorescent antibody test, the correlation between positive serological tests and proven infection can reach 95% (Kaufman & Blumer 1968).

The use of the LA test has greatly enhanced the contribution the laboratory can make towards diagnosis and management. A positive result is virtually diagnostic of cryptococcosis and the titres bear a direct relationship to the severity of infection. Diamond & Bennett (1974) showed that persistently high antigen titres at the end of therapy were correlated with treatment failure and death. In contrast, successfully treated patients gradually lost antigen from the CSF and serum, and this was followed by appearance and rise in antibody titre.

4. Virulence

The part played by the capsule of the yeast in enabling it to resist phagocytosis and destruction has been a source of interest and experimentation for many years. Littman & Tsubura (1959) obtained thin and thick capsular variants on different media and found no difference in their virulence towards experimental animals. It was observed, however, that yeasts with thin capsules developed large capsules following their introduction into the test animals. Gadesbusch (1960) removed the capsules by digesting with bacterial enzymes and found the yeasts were more readily phagocytosed by mouse polymorphonuclear leukocytes *in vitro*. Despite several early reports to the contrary (Hasenclever & Mitchell 1960; Kaze & Metzgar 1962), most recent work has supported the early attempts to relate capsule to virulence.

Bulmer and co-workers (Bulmer *et al.* 1967; Bulmer & Sans 1967, 1968) induced non-encapsulated mutants by u.v. irradiation and found them to be avirulent in mice. In contrast the parent wild strains caused death in eight days. Such mutants gradually redeveloped capsules on repeated subculture and concomitantly regained virulence. The same authors showed that virulence could be correlated with *in vitro* phagocytosis of the yeasts by leukocytes. The addition of soluble capsular polysaccharide to the mutants under these conditions gave a three-fold inhibition of phagocytosis. This did not apply to the phagocytosis of other yeast species or bacteria, suggesting that the free polysaccharide was being actively absorbed into the cells of *Cr. neoformans*. A similar inhibition of phagocytosis was noted by Kozel & Mastoianni (1976). Diamond *et al.* (1972) showed that the size of the capsule, rather than the amount of free polysaccharide in solution, was related to the inhibition of phagocytosis and killing.

The development of capsular material by poorly encapsulated cells during animal passage, together with the observation that capsules tend to be lost if the yeast is kept in soil, led Ishaq *et al.* (1968) and Fereshteh *et al.* (1970) to postulate that the saprophytic phase may be non-encapsulated. Such cells would be easily destroyed by phagocytosis after inhalation but in subjects with deficient phagocytosis mechanisms, the resultant delay in yeast killing might enable capsular synthesis to occur, thereby promoting establishment of infection.

5. Mycology of *Cryptococcus neoformans*

Early reports suggested that *Cr. neoformans* might possess an ascomycete perfect state. Todd & Herrmann (1936) reported sexual fusion by short conjugation tubes and Benham (1955) described the formation of asci and ascospores. These reports have not been confirmed by later workers, although the findings of Gordon & Devine (1970) appear partially supportive and have not been satisfactorily explained. During their experimental studies with poorly encapsulated strains they noticed apparent fusion of cells by conjugation tubes and development of intracellular spore-like structures. However, these failed to stain as ascospores. The strains only produced these structures following mouse passage. Possibly such conjugation tubes were yet another expression of the aberrant morphology described in the strains before mouse passage as rudimentary hyphal formation.

Strains forming hyphae had been noted earlier by Shadomy & Utz (1966). In 1970 Shadomy described basidiomycete-like clamp connections in two of these strains and suggested on this basis that *Cr. neoformans* should be transferred to *Leucosporidium* as a heterobasidiomycete yeast. Some such relationship had seemed possible since Slodki *et al.* (1966) had described similarities in extracellular polysaccharides of *Cr. laurentii* and *Tremella*, and in view of DNA base studies of *Cryptococcus* species (Storck *et al.* 1969).

The discovery of true basidia and basidiospores by Kwon-Chung (1975) in four experimental crosses of *Cr. neoformans* strains indicated that the parent strains should not be referred to *Leucosporidium*. These basidia were non-septate and produced terminal basidiospores. In these respects they resembled those of *Filobasidium*, described by Olive (1968) and later equated with *Cr. albidus* (Moore & Kreger-van Rij 1972). Kwon-Chung accordingly placed this sexual state of *Cr. neoformans* in a new genus, *Filobasidiella*. In a more detailed account, she described conjugation of two cells by tubes 1–2 μm long, preceeding the development of a clamped dikaryotic mycelium. Chains of basidiospores were budded basipetally from four points on the basidium, the nuclei remaining within the basidium and giving rise to daughter nuclei in the spore-chains. No capsule was formed by these structures, though basidiospore germination gave rise to typical encapsulated yeasts. Ultrastructural studies (Kwon-Chung & Popkin 1976) showed the occurrence of dolipore septa resembling those of *Filobasidium* species (Moore & Kreger-van Rij 1972) in lacking a parenthosome membrane system. These two genera remain the only members of the Ustilaginales to possess a dolipore.

The existence of a few strains in which the *Filobasidiella* state occurred without experimental crossing (Erke 1976) was shown by Kwon-Chung (1978) to be due to diploidization rather than to heterokaryon formation. These strains were unstable and reverted to a single mating type (α) on prolonged culture, possibly by action of a lethal gene affecting the opposite (a) mating component. Heterothallism in the majority of strains was confirmed by observing segregation of a 5-fluorocytosine-resistant genetic marker to be independent of mating type.

Three serotypes (A, B, C) of *Cr. neoformans* were recognized by Evans (1962) on the basis of agglutination and precipitin reactions and a fourth (D) was later described by Wilson *et al.* (1968). Kwon-Chung (1976*b*) reported a correlation between these serotypes and sexuality. *Filobasidium neoformans* could only be obtained by crossing suitable mating types within serotypes A & D. These failed to produce sexual states when crossed with serotypes B & C, though some hyphal development occurred, indicating the presence of mating types here too. Crosses between strains allocated to serotypes B & C were eventually achieved by the use of a different medium, and resulted in a distinct type of *Filobasidiella* in which the basidiospores were rod-shaped and smooth, unlike the globose, finely roughened spores of *Filobasidiella neoformans*. This second species was named *Filobasidiella bacillispora* (Kwon-Chung 1976*b*).

Other *Cryptococcus* species of which the basidiomycete states were known have so far all been referable to *Filobasidium* (Kwon-Chung 1977) so *Filobasidiella* at present has two species sharing a single asexual state.

6. Ecology

The occurrence of *Cr. neoformans* in soil, reported as early as 1951 (Emmons 1951) is apparently sporadic. By contrast, studies of its occurrence in old dried accumulations of bird droppings have repeatedly pointed to these substrates as sources of human infection (Littman & Schneierson 1959; Emmons 1960; Partridge & Winner 1965). Emmons (1962) demonstrated the presence of heavy fungal growth by examination of sectioned pigeon guano.

Several investigators have studied the conditions favouring growth of the yeast in bird faeces. Staib (1963) showed that the nitrogen-rich purines, uric acid, guanine and creatinine all supported good growth of *Cr. neoformans in vitro*. He also showed that whilst cells of the yeast could survive more than 12 months in dry, hard bird faeces, it was inhibited after only a few days if seeded into fresh moist faeces. This was attributed to rapid alkalinization by bacteria. Hubalek (1975) and Hubalek & Prikazsky (1975) have reported that droppings from sunlit exposed sites harboured fewer yeast cells than those from sheltered, shaded sites. Experimental u.v. irradiation of the droppings gave an increase in the level of inhibitory peroxides.

In examining the relationship of *Cr. neoformans* to the pigeon itself Littman & Borok (1968) concluded that the high body temperature of the birds was probably a factor in protecting them from infection.

In a recent reappraisal of the distribution of *Cr. neoformans* in pigeon excreta and in other saprophytic substrates, Bennett *et al.* (1977) found no B or C serotypes among 117 such isolates. Only two saprophytic isolations of serotype C (Denton & Di Salvo 1968; Walter & Coffee 1968) and none of serotype B are known, suggesting that pigeon excreta, while it may be a frequent source of *Filobasidiella neoformans*, is not the normal habitat for *Filobasidiella bacillispora*. B and C serotypes are known to be uncommon in infections at least in the majority of USA states and in a few other countries where this relationship has been studied. There have been a few notable and intriguing exceptions. For instance, B/C serotypes made up 25 of 49 isolates from infections in southern California. Similarly (although numbers were small) isolates from cryptococcosis in Thailand yielded 3 B or C serotypes of six cases (Mondana & Thasnakorn 1971).

Distribution of the fungus according to the mating type has also been shown to be disproportionate. Kwon-Chung & Bennett (1978) studied 105 saprophytic isolates and 233 clinical isolates of *Cr. neoformans* and found ratios of the mating types (*a*:*a*) as 40:1 and 30:1, respectively. Suggested reasons for the supposed demise of mating type (*a*) were the existence of lethal mutation or vulnerability to some factor of saprophytic existence.

7. References

BENHAM, R. W. 1955 *Cryptococcus neoformans*: "An ascomycete". *Proceedings of the Society for Experimental Biology and Medicine* **89**, 243–245.

BENNETT, J. E., KWON-CHUNG, K. J. & HOWARD, D. H. 1977 Epidemiologic differences among serotypes of *Cryptococcus neoformans*. *American Journal of Epidemiology* **105**, 582–586.

BLOOMFIELD, N., GORDON, M. A. & ELMENDORF, D. F. 1963 Detection of *Cryptococcus neoformans* antigen in body fluids by latex particle agglutination. *Proceedings of the Society for Experimental Biology and Medicine* **114**, 64–67.

BOTARD, R. W. & KELLEY, D. C. 1968 Modified Littman oxgall agar to isolate *Cryptococcus neoformans*. *Applied Microbiology* **16**, 689–690.

BULMER, G. S. & SANS, M. D. 1967 *Cryptococcus neoformans*. II. Phagocytosis by human leukocytes. *Journal of Bacteriology* **94**, 1480–1483.

BULMER, G. S. & SANS, M. D. 1968 *Cryptococcus neoformans*. III. Inhibition of phagocytosis. *Journal of Bacteriology* **95**, 5–8.

BULMER, G. S., SANS, M. D. & GUNN, C. M. 1967 *Cryptococcus neoformans*. I. Non-encapsulated mutants. *Journal of Bacteriology* **94**, 1475–1479.

CHASKES, S. & TYNDALL, R. L. 1975 Pigment production by *Cryptococcus neoformans* by *para* and *ortho*-diphenols: effect of the nitrogen source. *Journal of Clinical Microbiology* **1**, 509–514.

DENTON, J. F. & DI SALVO, A. F. 1968 The prevalence of *Cryptococcus neoformans* in various natural habitats. *Sabouraudia* **6**, 213–217.

DIAMOND, R. D. & BENNETT, J. E. 1974 Prognostic factors in cryptococcal meningitis. *Annals of Internal Medicine* **80**, 176–180.

DIAMOND, R. D., ROOT, R. K. & BENNETT, J. E. 1972 Factors influencing killing of *Cryptococcus neoformans* by human leukocytes *in vitro*. *Journal of Infectious Diseases* **125**, 367.

EMMONS, C. W. 1951 Isolation of *Cryptococcus neoformans* from soil. *Journal of Bacteriology* **62**, 685–690.

EMMONS, C. W. 1955 Saprophytic sources of *Cryptococcus neoformans* associated with the pigeon (*Columbia livia*). *Americal Journal of Hygiene* **62**, 227–232.

EMMONS, C. W. 1960 Prevalence of *Cryptococcus neoformans* in pigeon habitats. *Public Health Reports* **75**, 362–365.

EMMONS, C. W. 1962 Natural occurrence of opportunistic fungi. *Laboratory Investigations* **11**, 1026–1032.

EMMONS, C. W., BINFORD, C. H., UTZ, J. P. & KWON-CHUNG, K. J. 1977 *Medical Mycology* 3rd edn. Philadelphia: Lea & Febiger.

ERKE, K. H. 1976 Light microscopy of basidia, basidiospores and nuclei in spores and hyphae of *Filobasidiella neoformans* (*Cryptococcus neoformans*). *Journal of Bacteriology* **128**, 445–455.

EVANS, E. E. 1962 Reactivity of the cryptococcal capsule. In *Fungi and Fungous Diseases* ed. Daldorf, G. Springfield, Illinois: Charles Thomas.

FERESHTEH, F., BULMER, G. S. & TACKER, J. R. 1970 Cryptococcus neoformans. IV. The not-so-encapsulated yeast. Infection and Immunity 1, 526–531.

GADESBUSCH, H. H. 1960 Specific degradation of Cryptococcus neoformans 3723 capsular polysaccharide by a microbial enzyme. Journal of Infectious Diseases 107, 402–405.

GORDON, M. A. & DEVINE, J. 1970 Filamentation and endogenous sporulation in Cryptococcus neoformans. Sabouraudia 8, 227–234.

GREER, D. L. 1978 Cryptococcosis in Colombia: epidemiological and clinical aspects. In The Black and White Yeasts. Proceedings of the IVth International Conference on the Mycoses, Pan American Health Organisation.

HASENCLEVER, H. F. & MITCHELL, W. O. 1960 Virulence and growth rates of Cryptococcus neoformans in mice. Annals of the New York Academy of Science 89, 156–162.

HAUGEN, R. K. & BAKER, R. D. 1954 The pulmonary lesions in cryptococcosis with special reference to subpleural nodules. American Journal of Clinical Pathology 24, 1381–1390.

HUBALEK, Z. 1975 Distribution of Cryptococcus neoformans in a pigeon habitat. Folia Parasitologica, Praha 22, 73–79.

HUBALEK, Z. & PRIKAZSKY, Z. 1975 Growth of Cryptococcus neoformans in UV-irradiated excreta of pigeons. Folia Microbiologica 20, 231–235.

ISHAQ, C. M., BULMER, G. S. & FELTON, F. G. 1968 An evaluation of various environmental factors affecting the propagation of Cryptococcus neoformans. Mycopathologia et mycologia applicata 35, 81–90.

JENNINGS, A., BENNETT, J. E. & YOUNG, V. 1968 Identification of Cryptococcus neoformans in a clinical laboratory. Mycopathologia et mycologia applicata 35, 256–264.

KAUFMAN, L. & BLUMER, S. 1968 Value and interpretation of serological tests for the diagnosis of cryptococcosis. Applied Microbiology 16, 1907–1912.

KAUFMAN, L. & BLUMER, S. 1978 Cryptococcosis, the awakening giant. In The Black and White Yeasts. Proceedings of the IVth International Conference on the Mycoses, Pan American Health Organisation.

KAZE, A. & METZGAR, J. F. 1962 Isolation of a dry variant from a mucoid strain of Cryptococcus neoformans and preliminary comparative studies. Journal of Bacteriology 83, 926–928.

KORTH, H. & PULVERER, G. 1971 Pigment formation for differentiating Cryptococcus neoformans from Candida albicans. Applied Microbiology 21, 541–542.

KOZEL, T. R. & MASTOIANNI, R. P. 1976 Inhibition of phagocytosis by cryptococcal polysaccharide: dissociation of the attachment and ingestion phases of phagocytosis. Infection and Immunity 14, 62.

KWON-CHUNG, K. J. 1975 A new genus. Filobasidiella, the perfect state of Cryptococcus neoformans. Mycologia 67, 1197–1200.

KWON-CHUNG, K. J. 1976a Morphogenesis of Filobasidiella neoformans, the sexual state of Cryptococcus neoformans. Mycologia 68, 821–833.

KWON-CHUNG, K. J. 1976b A new species of Filobasidiella, the sexual state of Cryptococcus neoformans B and C serotypes. Mycologia 68, 942–946.

KWON-CHUNG, K. J. 1977 Perfect state of Cryptococcus uniguttulatus. International Journal of Systematic Bacteriology 27, 293–299.

KWON-CHUNG, K. J. 1978 Heterothallism vs. self fertile isolates of Filobasidiella neoformans (Cryptococcus neoformans). In The Black and White Yeasts. Proceedings of the IVth International Conference on Mycoses, Pan American Health Organisation.

KWON-CHUNG, K. J. & BENNETT, J. E. 1978 Distribution of α and a mating types of Cryptococcus neoformans among natural and clinical isolates. American Journal of Epidemiology 108, 337–340.

KWON-CHUNG, K. J. & POPKIN, T. J. 1976 Ultrastructure of septal complex in Filobasidiella neoformans (Cryptococcus neoformans). Journal of Bacteriology 126, 524–528.

LITTMAN, M. L. 1959 Cryptococcosis (Torulosis). Current concepts and therapy. American Journal of Medicine 27, 976–998.

LITTMAN, M. L. & BOROK, R. 1968 Relation of the pigeon to cryptococcosis. Natural carrier state, heat resistance and survival of Cryptococcus neoformans. Mycopathologia et mycologia applicata 35, 329–345.

LITTMAN, M. L. & SCHNEIERSON, S. 1959 *Cryptococcus neoformans* in pigeon excreta in New York City. *American Journal of Hygiene* **69**, 49–59.

LITTMAN, M. L. & TSUBURA, E. 1959 Effect of degree of encapsulation upon virulence of *Cryptococcus neoformans*. *Proceedings of the Society for Experimental Biology and Medicine* **101**, 773–777.

MONDANA, G. & THASNAKORN, P. 1971 In vitro sensitivity to amphotericin B of *Cryptococcus neoformans* isolated in Thailand. *Journal of the Medical Association of Thailand* **54**, 405–410.

MOORE, R. T. & KREGER-VAN RIJ, N. J. W. 1972 Ultrastructure of *Filobasidium* Olive. *Canadian Journal of Microbiology* **18**, 1949–1951.

NOTTEBART, H. C., MCGEHEE, R. F. & UTZ, J. P. 1974 *Cryptococcus neoformans* osteomyelitis. Case report of two patients. *Sabouraudia* **12**, 127–132.

OLIVE, L. S. 1968 An unusual new heterobasidiomycete with Tilletia-like basidia. *Journal of the Elisha Mitchell Scientific Society* **84**, 261–266.

PARTRIDGE, B. M. & WINNER, H. I. 1965 *Cryptococcus neoformans* in bird droppings in London. *Lancet*, May 15, 1060–1061.

ROBERTS, G. D., HORSTMEIER, C. D., LAND, G. A. & FOXWORTH, J. H. 1978 Rapid urea broth test for yeasts. *Journal of Clinical Microbiology* **7**, 584–588.

SALFELDER, K. 1978 Pathology of Cryptococcosis, Candidiasis and Torulopsosis. In *The Black and White Yeasts*. Proceedings of the IVth International Conference on the Mycoses, Pan American Health Organisation.

SHADOMY, H. J. 1970 Clamp connections in two strains of *Cryptococcus neoformans*. Spectrum Monograph Series in the Arts and Sciences, Georgia State University, Atlanta. Vol. 1, pp. 67–72.

SHADOMY, H. J. & UTZ, J. P. 1966 Preliminary studies on a hyphae forming mutant of *Cryptococcus neoformans*. *Mycologia* **58**, 383–390.

SHAW, C. E. & KAPICA, L. 1972 Production of diagnostic pigment by phenoloxidase activity of *Cryptococcus neoformans*. *Applied Microbiology* **24**, 824–830.

SHIELDS, A. B. & AJELLO, L. 1966 Medium for selective isolation of *Cryptococcus neoformans*. *Science* **151**, 208.

SLODKI, M. E., WICKERHAM, L. J. & BANDONI, R. J. 1966 Extracellular heteropolysaccharides from *Cryptococcus* and *Tremella*: a possible taxonomic relationship. *Canadian Journal of Microbiology* **12**, 489–494.

STAIB, F. 1962 *Cryptococcus neoformans* and *Guizotia abyssinica* (syn. *G. oleifera* D.C.). *Zeitschrift für Hygiene* **148**, 466–475.

STAIB, F. 1963 New concepts in the occurrence and identification of *Cryptococcus neoformans*. *Mycopathologia et mycologia applicata* **19**, 143–144.

STORCK, R., ALEXOPOULOS, C. J. & PHAFF, H. J. 1969 Nucleotide composition of deoxyribonucleic acid of some species of *Cryptococcus, Rhodotorula* and *Sporobolomyces*. *Journal of Bacteriology* **98**, 1069–1072.

TODD, R. L. & HERRMANN, W. W. 1936 The life cycle of the organism causing yeast meningitis. *Journal of Bacteriology* **32**, 89–103.

WALTER, J. E. & COFFEE, E. G. 1968 Distribution and epidemiologic significance of the serotypes of *Cryptococcus neoformans*. *American Journal of Epidemiology* **87**, 167–172.

WILSON, D. E., BENNETT, J. E. & BAILEY, J. W. 1968 Serologic grouping of *Cryptococcus neoformans*. *Proceedings of the Society for Experimental Biology and Medicine* **127**, 820–823.

Media and Methods for Growing Yeasts: Proceedings of a Discussion Meeting

Contents

Introduction

F. W. BEECH

*University of Bristol, Long Ashton Research Station,
Long Ashton, Bristol, UK*

THE PROCEEDINGS began with several short talks, the majority of which are printed here, together with a summary of the salient points arising from the lively discussion that followed.

Speakers and delegates alike emphasized the need to use the microscope at all stages of any study on yeasts. Firstly, early examination of substrates was an essential step to determine the approximate size and composition of the microbial population and its location. This would act as a check on results obtained later using culture media.

Secondly, the morphological and sexual characteristics of the yeast as it exists in the substrate, may not be achieved on artificial media. Hence observations based on the latter would not allow true identification of the specimen. Most contributors agreed that it was doubtful whether yeasts could be identified solely by the results of biochemical tests.

There was a considerable discussion on media, emphasizing that it should be considered as a micro-environment in which the growth and habit of the yeast could be affected by such factors as the diluent used, the chemical and physical nature of the substrate, the presence of selective inhibitors and the temperature and atmosphere of incubation.

The constituents of a medium are not static, but can change during the course of incubation. Many speakers gave their differing experiences using rose bengal as a selective inhibitor. Not only is it photosensitive, but there are interactions between rose bengal and constituents of some media to which it is added. Again a_w values are of paramount importance, particularly when isolating or counting osmophilic (xerotolerant) yeasts which can otherwise suffer stress or damage. The need for studies on resuscitation procedures was emphasized, particularly when isolating yeasts from a hostile environment, such as those containing preservatives or much alcohol.

Several workers reported inconsistencies between yeast counts made on membranes and on agar media. All agreed that membrane thickness and pore size, diffusion of medium through the pores, the composition of the substrate and the microflora were some of the factors affecting the use of membranes.

An Outline Guide to Media and Methods for Studying Yeasts and Yeast-like Organisms

R. R. DAVENPORT

*University of Bristol, Long Ashton Research Station,
Long Ashton, Bristol, UK*

This guide lists media (Appendix Tables 1–8) and a selection of methods suitable for yeast studies. In addition some practical details are described. These are of particular relevance where investigations are directed toward the 'role' of media as well as their identity.

Comments on the use of media

The use of media for yeasts should be carefully considered before attempting to enumerate and isolate these organisms. One must remember that a medium is only part of an environmental system which includes chemical composition, cultural conditions, yeast characteristics (Ch. 1), interaction with other micro-organisms and carry-over of substrate materials. An important consideration is the choice of isolation medium (with or without additives) because the dominant yeast cultured may not be the main organism responsible for the specific activity being investigated, or the presence of inhibitors may affect certain yeasts. Thus if selective and differential media are used it is highly recommended that a corresponding set of general purpose media are used in parallel. This should be continued until it is shown that no incomplete results are obtained for yeast isolation studies. The key to any investigation is whether the objective is the presence of one yeast, all yeasts or all micro-organisms present. Finally, yeast strain behaviour may vary, often giving difficulties of interpretation both with industrial problems and identification of isolates. Many of these problems may be solved by the application of some of the techniques described in the following section.

Commercial media for single medium and double layer gradient plates

Many commercial bacteriological media can be used for differentiating yeasts (Davenport 1968, 1975) and Table 1 gives some examples of this application. Another extension of this work is the combination of two dissimilar media where one medium acts as a partial inhibitor and the other as a control. The two media can be combined either by mixing different proportions or as a gradient plate. This technique is to pour the inhibitor medium into a sloped 10 cm diam. sterile plastic Petri dish. When this layer

TABLE 1

The use of commercial media for the differentiation of yeast species and strains within selected genera, used as a single medium and/or bottom wedge of gradient plates*

Medium	Conventional usage	Brettanomyces	Candida	Hansenula	Pichia	Rhodotorula	Saccharomyces	Trichosporon
		Growth for selected strains within the yeast genera						
Bismuth Sulphite	*Salmonella* and	+					+	+
MacConkey	*Shigella*	+		+			+	
Violet Red Bile		+	+	+	+	+	+	+
Eosin Methylene Blue	*Escherichia coli* and enterobacter	+	+	+	+	+	+	+
Kanamycin Aesculin Azide	Group D streptococci	+						
Streptococcus Selective	Streptococci	+						
Mannitol Salt Phenol Red	Staphylococci	+						
WL differential	Acid tolerant bacteria	+	+	+	+	+	+	+
Littmans Ox Gall Crystal Violet	Dermatophytic fungi	+	+			+	+	+
Rose Bengal Chloramphenicol	Yeasts and moulds	+	+	+	+	+	+	+

*Data based on Davenport (1968, 1975 & unpublished).

is set, a corresponding amount of a different medium is poured on top of the slope, with the dish lying flat. The gradient plate is best inoculated 48 h after pouring, to allow time for diffusion to take place between the agar layers and for the surface to dry. Inoculation can be made either by streaking or by a multipoint-pin technique. This gives a better result since the end-point is more readily seen (i.e. last colony to form rather than the streak inoculum end — see Figs 1 & 2). The gradient technique can be used for a number of applications including isolation, species and strain recognition. (Fig. 1) and product challenge testing.

Integration of isolation and identification methods

Modifications of standard methods can be of great advantage in assessing the role and/or potential significance of a yeast from an environment, and in identifying it. Moreover, case history and initial observation of both gross and microscopic habitats and materials can determine many yeast features. These include: pellicle formation; fermentation; odours and taints of estery compounds, mousiness and acetic acid formation; resistance to high osmotic pressure; growth at high and low temperatures and sometimes lipolysis as well as cell shapes and modes of reproduction. All such characteristics are of great value, firstly as an isolation aid and to ensure that the required objective is reached, and secondly, so that a short list of tentative yeast identities can be obtained. This list can subsequently be assessed by a selection of tests rather than an extensive examination to determine all known yeast properties. For example, there is a restricted number of yeast species which commonly spoil commodities of high sugar and high salt content. Such spoilage organisms are best detected on osmophilic media where the concentrations of sugar or salt are adjusted to similar values to those of the product. A suitable general guide is to use medium containing concentrations of either 10–15% (w/w) sodium chloride (for halophilic yeasts) or glucose [40% (w/w) for osmotolerant and 60% (w/w) for osmophilic yeasts], so that identification of these isolates can be simply achieved (Table 2). There are other instances where the same principle may be applied, for example, cycloheximide-resistant yeasts can be detected on WL Differential (Difco) or Actidione Agar (Oxoid) when routine tests are being made for acid-tolerant bacteria. A useful isolation-challenge technique can be used with liquid samples, particularly if inhibitors are present. Sterile liquefied 4% washed agar is added to each sample, previously warmed to 48–50°C and pour plates produced. These can either be used as a gradient system (described earlier) or as individual plates containing different proportions of the samples. This allows the detection of those organisms that can multiply in the sample as well as possible inhibitor-dependent strains (i.e. using a product as medium to detect any special adaptive strains and other strains which are likely to cause spoilage of the product being examined). The yeasts detected can then be isolated, counted and subjected to a selection of identification tests as outlined in the next section (non-conventional techniques) or examined by classical methods (Lodder 1970). Media for all these yeast studies are summarized in Appendix Tables 1–8.

Other features may also be observed more readily before starting classification procedures (e.g. ascospores and ballistospores) again making this task less time-consuming and more economic. In addition it is wise to check all cultures for the presence of ascospores before discarding rather than preparing a wide selection of sporulation media.

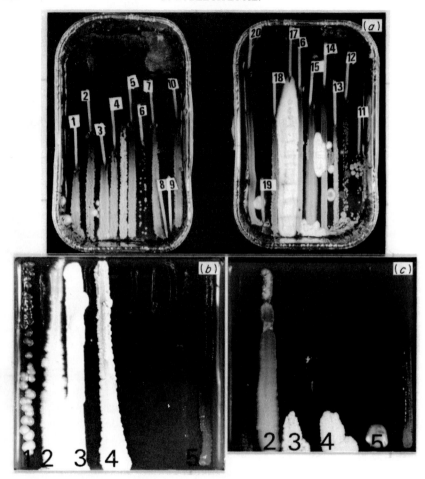

Fig. 1. Yeasts on agar gradient plates.

(a) Top wedge = yeast extract, 1% (w/w); glucose, 1%. Bottom wedge = yeast extract, 1% (w/w), glucose, 65% (w/w). Both media at pH 6·8 with 3% of agar. 1. *Sacch. cerevisiae* (wine strain); 2. *Br. anomalus** (NCYC 449); 3. *Sacch. fermentati* (CBS 813); 4. *S'codes ludwigii* (CBS 821); 5. *Sacch. cerevisiae* (Champagne strain); 6. *C. valida*/P. membranaefaciens* (NCYC 101); 7. *C. krusei** (CBS 573); 8. *Tr. capitatum** (soil); 9. *Deb. hansenii/T. famata** (CBS 116); 10. *C. utilis** (NCYC 737); 11. *Schiz. pombe* (CBS 351); 12. *Schiz. octosporus* (NCYC 131); 13. *Sacch. bailii* (acid food strain); 14. *Sacch. bailii* (wine strain); 15. *Sacch. bailii* (wine strain); 16. *Sacch. bailii* (NCYC 464); 17. *Hyphopichia burtonii* (NCYC 439); 18. *Sacch. cerevisiae* (port wine strain); 19. *H. anomala* (NCYC 18); 20. *Zygosacch. rouxii* (NCYC 568).

(b) Top wedge = Eosin Methylene Blue Agar (Oxoid). Bottom wedge = YM Agar (Difco). Both media at pH 6·8. 1. *Zygosacch. bailii* (acid food strain); 2. *C. utilis* (NCYC 737); 3. *Tr. capitatum* (soil); 4. *C. krusei* (CBS 573); 5. *Sacch. cerevisiae* (wine strain).

(c) Top wedge = Bismuth Sulphite Agar, pH 7·6 (Oxoid). Bottom wedge = YM Agar pH 6·8 (Difco). 1. *Zygosacch. bailii* (acid food strain); 2. *Schiz. pombe* (CBS 351); 3. *C. utilis* (NCYC 737); 4. *Tr. capitatum* (soil); 5. *C. krusei* (CBS 573); 6. *Sacch. cerevisiae* (wine strain).

*Imperfect states.

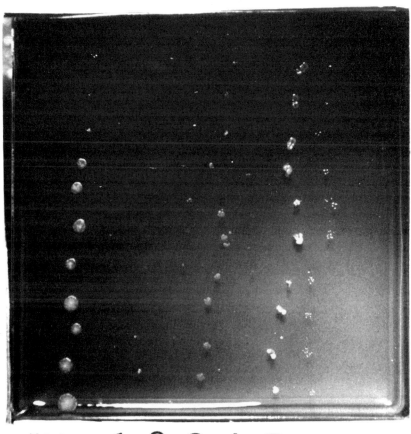

KAA
MYGP
B4/3
1 2 3 4 5 6 7 8

Fig. 2. Growth response of *Br. bruxellensis* strains, on a gradient plate. Top wedge = YM Agar (Difco). Bottom wedge = Kanamycin Aesculin Azide Agar (Oxoid). E, end points of growth on gradient when inoculated with multipoint-pin technique.

TABLE 2

Simplified identification of the common osmotolerant and osmophilic yeasts

| | Observations | | | | | Characteristics | | | | | | |
| | Salt content (w/w) | Sugar content (w/w) | | Acid formed | Preservative resistance | Cell division mode | Ascospore frequency | Assimilation | | | Dominant habitat(s) |
	15%	40%	60%					NA	NI	RA	
Zygosacch. bisporus	-	+	+	-	-	B	R	-	-	-	HSP
Zygosacch. bailii	-	+	±	+	+	B	C	-	-	±	HSP,W,FB,FP
Zygosacch. rouxii	+	+	+	-	-	B	R	-	-	-	HSP
H. anomala	±	+	±	-	-	B	C	+	-	+	HSP,FP,FB
Deb. hansenii	+	±	±	-	-	B	C	-	+	±	BP
T. lactis-condensii	-	+	±	-	?	B	NF	+	-	+	CM
Schiz. pombe	-	+	+	+	+	F	C	-	-	+	HSP

*Preservative resistance (at pH 3·5), SO_2 50 parts/10^6 (free); benzoic acid > 350 parts/10^6 and sorbic acid > 350 parts/10^6.
B, budding; F, fission; R, rare; C, common; NF, not formed; NA, nitrate; NI, nitrite; RA, raffinose; HSP, high sugar products; W, wines; FB, fruit beverages; FP, food products; CM, condensed milk.

Appendix

APPENDIX TABLE 1
Alphabetical list of selected media with suppliers and/or references

Medium	BBL	Difco	Merck	Oxoid	Reference*
Apple juice yeast extract agar					8, 9
Arbutin agar					14, 56, 51
Biggy Agar	11026	0635			
Bismuth Sulphite Agar	11031	0073	5418	CM201	
Blood Agar Base	11036/11048	0044/0696	10886	CM55/271	
Brain Heart Agar	11056/11065	0499		CM375	
Buffered Yeast Agar				CM153	
Candida Medium			10456		
Cassia fistula extract agar (Youssef)					57
Charcoal Gelatin Discs				BRIO	14
Chlamydospore Agar	11110	0513			
Corn Meal Agar	11132	0386		CM103	
Corn Meal-Tween agar					
Czapek Dox Agar	11140		5460	CM97	
Dextrose Agar	11165	0067			
Dextrose Peptone Agar		0068		CM13	
Eugon Agar	11229	0589			
Fermentation medium (Wickerham)					51
Fermentation/assimilation medium (Davenport)					17, 18
Fowell's acetate agar					51
Fungus Agar			5414		
Germ-tube production liquid medium					33
Glucose-calcium carbonate agar					51
Glucose-yeast extract-peptone agar					51
Glucose-yeast extract-peptone water					51
Glucose *or* fructose (40% & 60% w/w)					7, 51
yeast extract agars					8, 9
Grape juice agars — various					8, 9, 15, 18
Guizotia abyssinica extract media					10, 46
Kleyn's acetate agars					14, 15, 51
Levine Eosin-Methylene Blue Agar	11221	0005	1342	CM69	55
Littman's Ox Gall Agar	11349	0249	5415		16, 18
Littman's ox gall agar, modified					10, 45

APPENDIX TABLE 1 (*cont'd*)

Medium	BBL	Difco	Merck	Oxoid	Reference*
		Suppliers Code No.			
Liu-Newton Agar	11351				
Lysine Agar				CM191	
Malt Agars & Broths — various	11403	0186/0112	5398	CM59	
Medium for lyophilization (Carr)					8, 9
Molybdate Agar			10513		25
MRS Agar		10660	10660	CM361	16, 18
Mycobiotic Agar			0689		
Mycophil Agar	11456/11449				
Mycosel Agar	11462				
Orange Serum Agar	11486	0521	10673		
Plate Count Agar		0479		CM183	
Potato agars — various					42
Potato Dextrose Agar	11549	0013	10130	CM139	
Potato Malt Agar		0625			
Rice Extract Agars	11567	0899			51
Rose Bengal Chloramphenicol Agar	LAB 36†				26
Sabhi Agar		0797			
Sabouraud Agars — various	11583	0110	7315	R23	
	11588	0109	5438	CM41	
	11580	0747	5439	CM41a	12
Sodium chloride–glucose medium					51
Soil extract agars					19
Trybutyrin Agar				PM4	14, 18
Tryptone Soya Agars — various		0369		CM131	
Universal beer agars					8, 54
V8 agar					51
Vitamin Free Yeast Base		0394			51
Whey Agar		0034		CM309	
WL Agar	11817	0424			
WLD Agar or Actidione Agar	11815	0425			
Wort Agar	11825	0111	5448	CM247	
Yeast Carbon Base	11839	0391		PM118	51
Yeast Extract Agar (Adjust to pH 4·8)				CM19	
Yeast Extract–Malt Extract– Glucose-Peptone Broth/Agar		0711 0712			
Yeast Morphology Agar		0393			51
Yeast Nitrogen Base	12101				51
Zein Agar	11840				

*References follow Appendix.

†LAB M product (London Analytical & Bacteriological Media Ltd, Ford Lane, Salford, UK).

APPENDIX TABLE 2

General media for isolation and enumeration of yeasts from different sources

| Medium | Application | | | |
	General	Natural Crop Habitats	Beverages	Foods
Apple juice yeast extract agar	+	+	+	+
Buffered yeast agar	+			
Cassia fistula extract agars	+			
Dextrose agar			+	+
Grape juice agars — various		+	+	
Littman ox gall agar		+	+ (B)*	
Lysine agar			+ (B)	
Malt extract agars and broths	+	+	+	+
MRS agar		+	+	
Orange serum agar			+	+
Potato agars — various	+	+	+	+
Potato dextrose agar	+	+	+	+
Plate count agar		+		
Rose bengal chloramphenicol agar				+
Sabouraud agars — various	+			
Soil extract agars		+		
Tryptone soya agars — various		+		
Universal beer agars			+ (B)	
Whey agar		+	+	+
WL	+	+	+	+
WLD	+	+	+	+
Yeast extract agar (pH <4·8)	+			

*(B), suitable for specific applications in the brewing industry.

APPENDIX TABLE 3

Selected media for the isolation and enumeration of pathogenic yeasts*
and yeast-like fungi

Medium	*C. albicans*	*Cr. neoformans*	*T. glabrata*	*B. dermatitidis*	*Co. immitis*	*H. capsulatum*
Biggy agar	+					
Blood agar base	+	+		+ +	+	
Brain heart agars				+	+	+
Candida medium	+					
Chlamydospore agar	+					
Corn meal agar ± modification	+					
Corn meal-Tween agar	+					
Czapek Dox agar	+					
Dextrose peptone agar	+	+	+			
Eugon agar				+		+
Fungus agar	+	+	+			
Germ-tube production/liquid medium	+					
Guizotia abyssinica extract media		+				
Levine eosin-methylene blue agar	+					
Littman ox gall agar	+					
modified	+	+				
Liu-Newton agar	+					
Malt agars - various	+	+	+			
Molybdate agar	+	+	+	+	+	+
Mycobiotic agar					+	
Mycophil agar					+	
Mycosel agar	+				+	
Potato dextrose agar	+	+				
Potato malt agar			+			
Rice extract agar	+					
Sabouraud agars (various)	+	+	+	+	+	+
Sabhi agar				+		+
Zein agar	+					

*Many common yeasts may become 'pathogenic' under special conditions — these yeasts can be grown on the media in Appendix Tables 2 & 4.

APPENDIX TABLE 4
Selective and differential media

Selective	Differential
Malt or Wort agars, pH 4·5 & 3·5	*Acid producers* Malt or wort agars + chalk
Potato dextrose agars, pH 3·5	*Osmophiles* Malt or wort agar + high sugar content or yeast extract + high sugar
Sabouraud agars	*Pulcherrimin yeasts* yeast extract–malt extract agar + biotin + ferric ammonium citrate
WL nutrient agar	*Brewery yeasts* Lysine agar
Buffered yeast agar	*'Wild yeasts'* / *Other industrial spoilage yeasts* } WLD-actidione agar

The selective media will support most yeasts, some other microfungi and acid-tolerant bacteria, while non-acid-tolerant bacteria are suppressed. Further selection may be obtained by altering cultural conditions (Beech & Davenport 1969, 1971; Davenport 1972). Differential media are more restrictive in the types of yeasts supported but some growth of acid-tolerant bacteria and moulds can be expected. Various additives (Appendix Table 5) may be incorporated in some circumstances to make these media more efficient. Commercial media, not necessarily designed for yeasts, can often be useful, e.g. Littman Ox Gall — Crystal Violet Agar (Davenport 1972, 1975). See Beech & Davenport (1969, 1971) for general accounts of selective and differential media suitable for yeasts.

APPENDIX TABLE 5

Additives for media

Additives	Reference*
Anti-bacterial agents	
Actinomycin (2)[†] + aureomycin (50)	8, 9
Chloramphenicol (25)	22
Chloramphenicol (20) + streptomycin (20) + aureomycin (100)	8, 9, 41
Chloramphenicol (30) + streptomycin (50)	6, 12, 37
Chloramphenicol (10) + rose bengal (50)	26
Gentamicin (50)	37
Penicillin (200 i.u./ml)	27
Penicillin (60) + streptomycin (100)	8, 9, 50
Penicillin (20) + streptomycin (40)	8, 9
Rose bengal (30)	1, 2, 3, 8, 9, 32, 34
Rose bengal (50) + chloramphenicol (10) (Lab 36, LAB M-London Analytical & Bacteriological Media Ltd)	26
Rose bengal (0·7%) + kanamycin (100)	8, 9, 32
Streptomycin (250)	27
Anti-fungal + anti-bacterial + semi-selective for certain yeasts	
Aureomycin (50) + chloramphenicol (50) + actidione (100)	52
Anti-yeast agents	
Actidione (range of concentrations used 1 – > 100)	8, 9, 14, 18, 54
(can be used to select/identify actidione-resistant yeasts of potential industrial significance)	
Actidione (10) + 8-hydroxyquinoline (250)	8, 9
(to suppress yeasts and determine levels of acid-tolerant bacteria)	
Anti-fungal agents	
Diphenyl (100)	8, 9, 24
Calcium propionate (250)	8, 9, 11, 14, 18
Sodium propionate (250)	8, 9, 14

*References follow Appendix.
[†]Figures in brackets give concentrations in media expressed as parts/10^6, unless otherwise specified.

APPENDIX TABLE 6
Media and methods for maintenance and storage of cultures

	Reference
Media	
Yeast extract–malt extract agar	8, 9, 53
Yeast extract–malt extract broth	8, 9, 14, 15, 16, 18, 51
Malt extract agar	8, 9, 53
Medium for lyophilization (Carr)	8, 9, 14, 15, 16, 18, 51
Wort agar	8, 9, 53
Sabouraud medium	8, 9, 53
Corn meal agar	8, 9, 14, 15, 16, 18, 51
Other preservation media and methods	
Liquid paraffin on broth and agar slopes	8, 9
Silica gel	56
Silica gel/paper replica techniques	5
Distilled water	39
Special methods for yeasts having particular characteristics	
Acid-producing yeasts — agar media + calcium carbonate	8, 9, 51
Osmophilic/osmotolerant strains	8, 9, 18, 51
Pulcherrimin pigment-producing yeasts	8, 9, 14, 15, 51
Cyniclomyces (formerly *Saccharomycopsis*)	19, 53
Pityrosporum	47

APPENDIX TABLE 7

Selected media for recognition, identification and classification

Name	Recognition tests*	Identification and/or classification characteristics
Arbutin agar†		Ascospore formation
Bismuth sulphite agar	+ (H$_2$S production)	
Charcoal gelatin discs	+ (strain differences)	
Corn meal agar	+ (strain differences)	Mycelium/pseudomycelium
Levine eosin methylene blue agar		
Fermentation medium (Wickerham)		Fermentation
Fermentation/assimilation medium (Davenport)		Fermentation and assimilation
Fowell's acetate agar		Ascospores
Glucose–calcium carbonate agar		Acid production
Glucose–yeast extract–peptone agar		Morphology
Glucose–yeast extract–peptone water		Morphology + *pellicle + pigment + gelatin disc
Glucose or fructose, yeast extract agar 40% (w/w) & 60% (w/w)		Osmotolerant (40%w/w) strains, osmophilic (60%w/w)
Kleyn's acetate agars		Ascospores
Littman's ox gall agar	+ (selection of spp. and strains)	
Malt or wort agar		Morphology
Malt or wort broth		Morphology
Morphology agar		Colonial & cellular features
Potato dextrose agar		Mycelium pseudomycelium/ascospores
Rice extract agar		Chlamydospores
Sodium chloride–glucose medium		Moderate osmotic pressure test
Trybutrin agar		Fat splitting
WL		Non-actidione resistant strains
WLD		Actidione resistant strains
Yeast extract–glucose–peptone–malt extract broth or agar		Morphology and a base medium for, e.g. salt, dyes, tolerance/resistance/killing, alcohol etc., effects
V8		Ascospores
Vitamin free broth		Vitamin requirements
Vitamin solutions		
Yeast carbon base		Single nitrogen compounds — assimilation tests
Yeast nitrogen base		Single carbon compounds — assimilation and fermentation tests

*Unpublished data.

†As a diagnostic test (i.e. hydrolysis) this is of little value but spores, from *Saccharomyces* and some other genera, are readily formed.

APPENDIX TABLE 8

Some media for the study of miscellaneous yeast activities

Yeasts/activities	References
Wine yeasts — basal medium ± alcohol or sulphur dioxide	43
Brewery yeasts — efficiency of various media and comparison or detection of wild yeasts	29
Yeast using crystal violet and lysine agar	44
Vitamin deficiencies in yeasts	13
Vitamin requirements and nutritional imbalances in psychrophobic yeasts	48
Sulphur metabolism of brewing yeasts and spoilage bacteria	4
Environmental studies, i.e. substrate and habitat considered for media and cultural conditions, e.g. herbicides and pecticides as nutrients for selected yeasts	Ch. 1, this volume

Appendix References

1. ADAMS, A. M. 1960 Yeasts in horticultural soils. *Report of the Horticultural Experiment Station and Products Laboratory, Vineland Station, Ontario, Canada for 1959-60* pp. 79-82.
2. ADAMS, A. M. 1963 The collection and screening of airborne yeasts from horticultural sites. *Report of the Horticultural Experiment Station and Products Laboratory, Vineland Station, Ontario, Canada, for 1963* pp. 68-71.
3. ADAMS, A. M. 1964 Airborne yeasts from horticultural sites. *Canadian Journal of Microbiology* **10**, 641-646.
4. ANDERSON, R. J., HOWARD, G. A. & HOUGH, J. S. 1971 The sulphur metabolism of brewing yeasts and spoilage bacteria. *Proceedings of the 13th European Brewery Convention* pp. 253-257.
5. BASSEL, J., CONTOPOULOU, R., MORTIMER, R. & FOGEL, S. 1977 Comparison of the silica gel and paper replica techniques with respect to the long term storage of strains of *Saccharomyces cerevisiae*. *United Kingdom Federation of Culture Collections Newsletter* No. 2 p. 7.
6. *BBL Manual of Products and Laboratory Procedures, 1968* 5th edn. Cockeysville, Maryland 21030, USA: Becton, Dickinson and Company.
7. BEECH, F. W., BURROUGHS, L. F., CARR, J. G., DAVENPORT, R. R., JORGENSEN, K. H., SCHILDMANN, J. A. & WERNER, J. 1974 Provisional methods for microbiological examination of fruit juice products of pH value less than 3·7. *Confructa* **19**, 281-285.
8. BEECH, F. W. & DAVENPORT, R. R. 1969 The isolation of non-pathogenic yeasts. In *Isolation Methods for Microbiologists* ed. Shapton, D. A. & Gould, G. W. pp. 71-88. Society for Applied Bacteriology Technical Series No. 3. London: Academic Press.
9. BEECH, F. W. & DAVENPORT, R. R. 1971 Isolation, purification and maintenance of yeasts. In *Methods in Microbiology* Vol. 4, ed. Booth, C. pp. 153-182. London: Academic Press.
10. BOTARD, R. W. & KELLEY, D. C. 1968 Modified Littman oxgall agar to isolates *Cryptococcus neoformans*. *Applied Microbiology* **16**, 689-690.
11. BOWEN, J. F. 1962 *The ecology of cider yeasts*. Ph.D. Thesis, University of Bristol.
12. BUCKLEY, H. R., CAMPBELL, C. K. & THOMPSON, J. C. 1969 Techniques for the isolation of pathogenic fungi. In *Isolation Methods for Microbiologists* ed. Shapton, D. A. & Gould, G. W. Society for Applied Bacteriology Technical Series No. 3. London: Academic Press.
13. BURKHOLDER, P. R. 1943 Vitamin deficiencies in yeasts. *American Journal of Botany* **30**, 206-211.
14. DAVENPORT, R. R. 1968 *The origin of cider yeasts*. Thesis, Institute of Biology, London.
15. DAVENPORT, R. R. 1970 *Epiphytic yeasts associated with the developing grape vine*. M.Sc. Thesis, University of Bristol.
16. DAVENPORT, R. R. 1972 Vineyard yeasts (an environmental study). In *Sampling — Microbiological Monitoring of Environments* ed. Board, R. G. & Lovelock, D. pp. 143-174. Society for Applied Bacteriology Technical Series No. 7. London: Academic Press.
17. DAVENPORT, R. R. 1974 Mycology and taxonomy of fungi in fruit juices. Lectures and Laboratory Course Manual, pp. 1-250. Washington D.C.: Organisation of the American States and Sao Paulo, Brazil: Institute de Botanica.
18. DAVENPORT, R. R. 1975 *The distribution of yeasts and yeast-like organisms in an English vineyard*. Ph.D. Thesis, University of Bristol.
19. DI MENNA, M. E. 1957 The isolation of yeasts from soil. *Journal of General Microbiology* **17**, 678-688.
20. DI MENNA, M. E. 1966 Yeasts in Antarctic soils. *Antonie van Leeuwenhoek* **32**, 29-38.
21. *Difco Manual, 1966* 9th edn. Detroit, Michigan 48232, USA: Difco Laboratories.
22. FEO, M. 1973 Diagnostico rapido de *Candida albicans*. *Revista Latinoamericana de Microbiologia, Mexico* **15**, 217-218.

23. *Handbook of Microbiology, 1975*. E. Merck, Darmstadt, FDR.
24. HERTZ, M. R. & LAVINE, M. 1942 Fungistatic medium for enumerating yeasts. *Food Research* 1, 430–441.
25. HOLLAND, M. L. & KUNG, L. J. 1961 Evaluation of molybdate agar as a selective differential medium for yeasts. *Journal of Bacteriology* 81, 869–874.
26. JARVIS, B. 1973 Comparison of an improved rose bengal-chlortetracycline agar with other media for the selective isolation and enumeration of moulds and yeasts in foods. *Journal of Applied Bacteriology* 36, 723–727.
27. KAHANPÄÄ, A. 1971 *Bronchopulmonary occurrence of fungi in adults especially according to cultivation material*. Academic Dissertation, University of Helsinki.
28. KOZULIS, J. A. & PAGE, H. E. 1968 A new universal beer agar medium for the enumeration of wort and beer microorganisms. *American Society of Brewing Chemists Proceedings* 52–58.
29. LIN, Y. 1975 Detection of wild yeasts in the brewery — efficiency of differential media. *Journal of the Institute of Brewing* 81, 410–418.
30. LODDER, J. ed. 1970 *The Yeasts — A Taxonomic Study* 2nd edn. Amsterdam: North-Holland.
31. LUND, A. 1954 *Studies in the ecology of yeasts*. Ph.D. Thesis, University of Copenhagen.
32. MARTIN, J. P. 1950 Use of acid, rose bengal and streptomycin in the plate method for estimating soil fungi. *Soil Science* 69, 215.
33. MARDEN, D. N., HURST, S. K. & BALISH, E. 1971 Germ-tube production by *Candida albicans* in minimal liquid culture media. *Canadian Journal of Microbiology* 17, 851–856.
34 MILLER, J. J. & WEBB, N. S. 1954 Isolation of yeasts from soil with the aid of acid, rose bengal and ox gall. *Soil Science* 77, 197–204.
35. MOSSEL, D. A. A., VISSER, M. & MENGERINK, W. H. J. 1962 A comparison of media for the enumeration of moulds and yeasts in foods and beverages. *Laboratory Practice* 2, 109–112.
36. MOSSEL, D. A. A., KLEYMAN-SEMMELING, A. M. C., VINCENTIE, H. M., BEERENS, H. & CATSARAS, M. 1970 Oxytetracycline-glucose yeast extract agar for selective enumeration of moulds and yeasts in foods and clinical material. *Journal of Applied Bacteriology* 33, 454–457.
37. MOSSEL, D. A. A., VEGA, C. L. & PUT, H. M. C. 1975 Further studies on the suitability of various media containing antibacterial antibiotics for the enumeration of moulds in food and food environments. *Journal of Applied Bacteriology* 35, 15–22.
38. ODDS, F. C. 1966 Preservation of fungal stocks under distilled water. *United Kingdom Federation of Culture Collections Newsletter No. 2* p. 6.
39. *Oxoid Manual, 1971* 3rd edn. Basingstoke, UK: Oxoid Ltd.
40. PHAFF, H. J. *Saccharomycopsis* Schiönning. In *The Yeasts* ed. Lodder, J. Amsterdam: North-Holland.
41. ROSS, S. S. & MORRIS, E. O. 1965 An investigation of the yeast flora of marine fish from Scottish coastal waters and a fishing ground off Iceland. *Journal of Applied Bacteriology* 28, 224–234.
42. SASAKI, Y. & YOSHIDA, T. 1959 Distribution and classification of the wild yeast or budding fungi on the fresh fruits in Hokkaido. *Journal of the Faculty of Agriculture, Hokkaido University, Sapporo* 51, 194–220.
43. SCHEFFER, W. R. & MRAK, E. M. 1951 Characteristics of yeasts causing clouding of dry white wines. *Mycopathologia et mycologia applicata* 5, 236–249.
44. SEIDEL, H. 1972 Differenzierung zwischen Brauerei-kulturhefen und "Wilden Hefen" Teil I: Erfahrungen beim nachweis von "Wilden Hefen" auf Kristallviolettagar und Lysinagar. *Brauwissenschaft* 25, 384–389.
45. SHAPIRO, E. M., MULLINS, J. R. & PINKERTON, E. M. 1956 Direct identification of *Candida albicans* on Littman's ox gall agar. *Journal of Investigative Dermatology* 26, 77–80.
46. SHIELDS, A. B. & AJELLO, L. 1966 Medium for selective isolation for *Cryptococcus neoformans*. *Science, New York* 151, 208–209.

47. SLOOF, W. C. 1970 *Pityrosporum sabouraud*. In *The Yeasts — A Taxonomic Study* 2nd edn, ed. Lodder, J. Amsterdam: North-Holland.
48. TRAVASSOS, L. R. & MENDONCA, L. 1972 Vitamin requirements and induced nutritional imbalances as criteria in speciating psychrophobic yeasts. *Antonie van Leeuwenhoek* **38**, 379–389.
49. WALKER, L. & HUPPERT, M. 1960 Corn meal-Tween agar: an improve medium for the identification of *Candida albicans*. *American Journal of Clinical Pathology* **33**, 190–194.
50. VAN UDEN, N. & DO CARMO SOUSA, L. 1962 Quantitative aspects of the intestinal yeast flora of swine. *Journal of General Microbiology* **27**, 35–41.
51. VAN DER WALT, J. P. 1970 Criteria and methods used in classification. In *The Yeasts — A Taxonomic Study* 2nd edn, ed. Lodder, J. Amsterdam: North-Holland.
52. VAN DER WALT, J. P. & VAN KERKEN, A. E. 1961 The wine yeasts of the Cape. V, Studies on the occurrence of *Brettanomyces intermedius* and *Brettanomyces schanderii*. *Antonie van Leeuwenhoek* **27**, 81–90.
53. VAN DER WALT, J. P. & SCOTT, D. B. The genus *Saccharomycopsis* Schiönning. *Mycopathologia et mycologia applicata* **43**, 279–288.
54. VAN OEVELEN, D. 1979 Microbiology and biochemistry of the natural wort fermentation in the production of lambic and gueze. *Agricultura (Heverlee)* **26**, 353–505.
55. WELD, T. J. 1952 *Candida albicans* rapid identification in pure cultures with carbon dioxide on modified eosin-methylene blue medium. *Archives of Dermatology and Syphilology* **66**, 691–694.
56. WOODS, R. 1976 Preservation of yeasts on silica gel. *United Kingdom Federation of Culture Collections Newsletter No. 2* p. 5.
57. YOUSSEF, K. A. 1975 Selective and enrichment medium for the identification and propagation of yeasts and fungi. US Patent No. 3,870,600.

Experience with Methods for the Enumeration and Identification of Yeasts Occurring in Foods

D. A. A. MOSSEL, K. E. DIJKMANN AND MIRÈSE KOOPMANS

Department of the Science of Food of Animal Origin,
Faculty of Veterinary Medicine,
The University of Utrecht,
Utrecht, The Netherlands

Introduction

During the last few decades there has been much useless discussion on the functioning of methods in food microbiology. Often the opposing views seem to be both right. However, the factor frequently overlooked is that the functioning of a medium cannot possibly be divorced from the food substrate to which it is applied.

The reason for this is two-fold. First and foremost, the microbial associations of various foods are completely different (Mossel 1971). Therefore both the organisms looked for and those to be inhibited, often indicated as the 'background flora', vary tremendously in character and numbers. In addition intrinsic attributes of a food, such as a_w, pH and gross composition, as well as the extrinsic conditions of storage i.e. temperature, pO_2, relative humidity etc., vary greatly. These differences result in a considerably different degree of sublethal stressing of the microbial populations to be recovered and therefore in the functioning of a given medium (Mossel & Corry 1977).

In this presentation general statements will be avoided and only specific experience summarized.

Effects peculiar to organisms

The yeasts occurring in most foods are to some extent damaged due to properties of the foods themselves, their mode of processing or the conditions of storage. This calls for extreme caution in the use of media, of the type introduced by Waksman (1922), which owe their selectivity to reduced pH (Nelson 1972). A second shortcoming of acid media is their inadequate inhibition of aciduric bacteria (Holwerda 1952; Koburger 1971; Hup & Stadhouders 1972). Both defects can be corrected by the use of antibacterial antibiotics that will not interfere with the recovery of yeasts (Beech & Carr 1955, 1960; Flannigan 1974). Although some reassuring data are available for moulds (Beuchat & Jones 1978) the possibility cannot be ruled out that antibacterial antibiotics may affect the recovery of sublethally damaged yeasts even on media of neutral pH. To overcome this it may be necessary to use a so-called resuscitation step (Allen *et al.* 1952). Data in the literature indicate (Stevenson & Richards 1976; Graumlich & Stevenson 1978, 1979) that this may require several hours in a non-inhibitory medium. The situation prevailing should be investigated for the particular set of circumstances being studied. One method which can be used relies on replica plating at various times from master plates of resuscitation agar onto the selective medium under study (Mossel *et al.* 1965), another on overlayering such repair plates in due course with the selective medium (Ray 1979).

With these reservations in mind, oxytetracycline seems to be the antibiotic of choice for general purposes (Cooke 1954; Mossel *et al.* 1962; Buttiaux & Catsaras 1965; Sainclivier & Roblot 1966; Mossel *et al.* 1970; O'Toole & O'Neill 1974; Joyce & Ould 1978). As stated in the Introduction a different choice of antibacterial antibiotics may be preferred for the examination of special foods.

When thermotrophic or thermophilic yeasts are to be enumerated it may be better not to rely on the use of the thermolabile antibiotics chloro- or even oxytetracycline (Put 1964, 1974), but instead use chloramphenicol for example (Koburger & Farhat 1975; Blaser 1978; Beuchat 1979*a*). However, when examining particular foods chloramphenicol alone may not be adequately bacteriostatic (Smith & Worrel 1950; Pepper & Kiesling 1963; Holt 1967; Ingram & Hassan 1975; Gaffney *et al.* 1978; Grehn & von Graevenitz 1978; Kelch & Lee 1978). In this case the addition of the equally thermostable antibiotic gentamicin (Mossel *et al.* 1970, 1975; Koburger & Rodgers 1978) may be effective as illustrated in Table 1 (Dijkmann *et al.* 1979).

TABLE 1

Recovery of pure cultures of yeasts on various mycological media

Strain	Mean $\log_{10}$ c.f.u. ml after 3 days at 22°C on		
	DY*	OGDY	CGDY
C. curvata[2]	4·45	4·42	4·51
C. lambica[1]	5·08	5·09	5·20
C. sake[2]	5·18	5·28	5·21
Cr. albidus[2]	5·30	5·34	5·32
Deb. hansenii[1]	4·90	4·67	4·79
Debaromyces MP FRSL	5·16	5·16	5·10
Rh. graminis[2]	4·84[†]	4·53[†]	4·59[†]
Sacch. lipolytica[1]	4·58	4·63	4·57
T. norvegica	5·19	5·22	5·14
Tr. beigelii[2]	5·09	5·14	5·14

*DY, dextrose yeast extract agar; OGDY, DY + oxytetracycline (100 μm/ml) and gentamicin (50 μg/ml); CGDY, DY + chloramphenicol (50 μg/ml) + gentamicin (50 μg/ml).
[†]After six days incubation.
[1], Strains from Delft culture collection; 2, isolates from minced meats.

When slow growing yeasts are to be isolated quantitatively, thermostable antibiotics will again have to be used. In addition, it may be necessary to use a compound such as rose bengal to inhibit the rapid and copious colony development by some types of mould (Smith & Dawson 1944; Martin 1950; Ottow & Glathe 1968; Overcast & Weakley 1969) although it has been shown repeatedly to inhibit yeasts in a somewhat erratic way (Freeman & Giese 1952; King et al. 1979; Dijkmann et al. 1979). In addition it is photosensitive (Kramer & Pady 1961) and as illustrated in Table 2 seems to reduce the bacteriostatic effect of antibiotics (Dijkmann et al. 1979). Therefore it may be advisable to use a rather low concentration of rose bengal, e.g. 25 μg/ml, and combine it with 2,6 dichloro-n-nitroaniline (Beuchat 1979b; King et al. 1979). Some control of mould may also be achieved by adding calcium propionate to the medium (Beech and Davenport 1971) or by placing two drops of a saturated alcoholic solution of diphenyl on the lid of Petri dishes (Ingram 1955).

TABLE 2

Recovery of yeasts and bacteria on a medium with and without rose bengal

Type of minced meat	CDP*		CRDP[†]	
	yeasts $\log_{10}$ c.f.u./g	bacteria c.f.u./plate	yeasts $\log_{10}$ c.f.u./g	bacteria c.f.u./plate
Beef, fresh	4·38	*ca.* 5	4·30	*ca.* 50
Beef & pork, fresh	3·67	*ca.* 5	3·66	*ca.* 50
Beef & pork, frozen	3·55	*ca.* 5	3·68	*ca.* 50

*CDP, chloramphenicol–dextrose–peptone agar.
[†]CRDP, CDP + 50 μg/ml rose bengal.

For the selective recovery of psychrotrophic yeasts from chilled proteinaceous staple foods, the combination of chloramphenicol with rose bengal seems to have many disadvantages as shown in Table 3 (Dijkmann et al. 1979). Oxytetracycline and gentamicin seems to be the bacteriostatic combination of choice for this purpose as also illustrated by Table 3 (Dijkmann et al. 1979).

The enumeration of xerotolerant ('osmophilic') yeasts requires a medium of reduced a_w to inhibit xerosensitive organisms. Almost all bacteria of common occurrence in foods are inhibited by even the highest a_w currently chosen for this purpose (Mossel 1975) so that there is no need for the use of antibacterial antibiotics in this case. It has been demonstrated that only a_w values below 0·85 will fully inhibit non-xerotolerant yeasts (Mossel & Bax 1967; Pitt 1975). Fructose at a concentration of 60% (w/w) is the best agent for this purpose: it is readily soluble at this concentration (Mossel & Bax

TABLE 3

The functioning of various mycological media for the enumeration of yeasts in chilled minced meat

Minced meat sample	$\log_{10}$ c.f.u./g obtained on media		
	OGDY	CGDY	CRDP
065	4·66	4·26 [+]	5·27 [+ + +]
070-I	4·73	4·65 [+]	5·19 [+ + + +]
070-II	5·39	5·23 [+ +]	5·17 [+ + + +] .
075	5·05	5·00 [+]	5·11 [+ + + +]
080-I	5·35	5·17	*ca.* 5·5 [+ + + +]
080-II	5·89	5·88 [+]	5·89 [+ + + +]
085-I	7·0	7·0	4·78 [+ + + +]
085-II	3·26 [+]	3·04 [+]	3·61 [+ + + +]
090-I	5·79	5·75 [+]	5·85 [+ + +]
090-II	5·46	5·26 [+ +]	5·60 [+ + + +]
100-I	4·86	4·81 [+]	4·89 [+ + +]
100-II	6·00	5·96	5·00 [+ + +]
100-III	3·91	3·69	4·81 [+ + + +]

[+] 3–10 bacterial colonies per plate; [+ +] 10–100; [+ + +] 100–300; [+ + + +] >300; all levels making yeast colony counts often dubious.

For composition of media, see legends to Tables 1 & 2.

1967; Windisch *et al.* 1978) and causes least stress (Brown 1978). However, at this low a_w xerotolerant yeasts require 10 days or more to develop discernible colonies and plates have to be protected from drying out by special measures. Consequently, for some purposes a less selective medium that allows results to be obtained much faster may be preferred (Ingram 1959; Ch. 9, this volume).

Parameters of the food to be plated

The character of the food to be examined will in many instances determine the choice of the diluent and medium to be used. While this is often taken into account in so far as the diluent is concerned, it is frequently overlooked with regard to the medium used.

For instance, if the portion plated out could carry over substantial concentrations of protein or heavy metals, oxytetracycline may not be an effective bacteriostatic agent as its active HO-group will be complexed by some of these food components, thereby inactivating the agent (Put 1974). Chloramphenicol, when necessary in combination with gentamicin (Dijkmann *et al.* 1979), is the selective agent of choice for this purpose, as illustrated by the data in Tables 1 and 3.

Because the type of food determines its fungal association (Mossel 1971) it is clear that it also dictates the incubation temperature to be used in any particular instance. Generally speaking the range 17–20°C will be used for psychrotrophic organisms (Koburger 1973; Mossel & Ratto 1973), but when thermotrophic yeasts are to be enumerated a temperature well over 30°C is required (Watson *et al.* 1978).

Precautions to be observed with media

In general terms the mode of preparation, sterilization and inoculation of selective enumeration media may often exert a more profound influence on the results than its gross composition. This, in the case of media for the selective enumeration of yeasts, applies particularly to the way in which antibacterial antibiotics are added. They may be autoclaved in the medium, added aseptically (Jarvis 1973; Mossel *et al.* 1975) or added to the basal ingredients and heated to boiling without further sterilization being required (Dijkmann *et al.* 1979). Furthermore, when rose bengal is used, measures to avoid its photodecomposition are required (Kramer and Pady 1961). It cannot be repeated too many times that only slight inaccuracies in this respect can influence the productivity as well as the selectivity of the culture media dramatically.

It makes a considerable difference whether media are used for surface plating or as poured plates (Christensen 1946). The a_w, pO_2 and pCO_2 of the growth environment differ substantially in these instances, as may the pH. It has also been observed repeatedly that the thermal shock resulting from the exposure of the food dilution to the tempered medium greatly influences recoveries. If only slight errors occur, e.g. if the medium temperature has not been carefully checked to ensure that it does not exceed 48°C *throughout* the bulk, misleading results can be obtained (Nash & Sinclair 1968; Nash & Grant 1969; Flannigan 1973). It is no less true that careless handling of plates to be surface-inoculated can lead to fully unreliable results. This applies particularly to the use of plates that have not been prepared and monitored in such a way that the presence of 'micro-colonies' of contaminants (Mossel & van de Moosdijk 1964) is absolutely precluded. This aspect is particularly important for plates used in mycological work where airborne contamination with spores can so easily occur.

All these potential interferences with adequate functioning of selective culture media for yeasts call for regular monitoring of media before use. Quantitative procedures, wherein a selection of the organisms to be recovered and inhibited are used, are mandatory. These should in all instances include robust as well as fastidious strains of both groups of test organisms. For the first selection of media the modification of the Miles-Misra technique (1938) suggested by Slack & Wheldon (1978) is the method of choice. For rapid routine monitoring of accepted media we found the ecometric principle, (Fig. 1; Mossel *et al.* 1978, 1979) also quite appropriate.

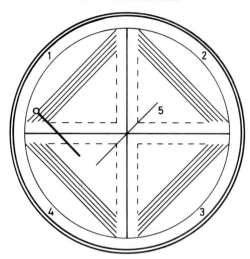

Fig. 1. Scheme of the ecometric technique for the monitoring of selective culture media for yeasts. Media are poured into standard Petri dishes and thoroughly dried after solidification. The inoculum is a fully grown culture in yeast extract glucose broth containing approximately 10^6 c.f.u./ml.

Identification of yeasts isolated from foods

However incredible it may sound to mycologists, the need for a microscopic examination of isolates cannot be sufficiently stressed. In a recent survey we found that many applied microbiologists rely merely on the colony appearance of isolates in order to decide which commercially available set of tests to use for their further examination (Mossel *et al.* 1980). Another factor, often overlooked, is that the morphology of an isolate can be profoundly influenced by the growth medium. Hence, prior to microscopic examination and biochemical study, isolates should be subcultured on a medium generally relied on in mycology, such as an appropriate preparation of malt extract agar (Davenport 1975).

Subsequently the isolates should be examined for the key criteria used in their primary taxonomic grouping. These include study of the formation of pseudomycelium and ascospores, mode of attack on glucose and assimilation of nitrate (Davenport 1975).

Tentative identification steps to be undertaken in food microbiology (Barnett *et al.* 1979) have been greatly simplified recently. First of all there is the well validated API 20 'Candida' system (Buesching *et al.* 1979; Land *et al.* 1979) though its name is somewhat misleading. It allows both assimilation and fermentation tests to be carried out but, as indicated before, it can only be used in conjunction with the assessment of morphological characters.

Equally convenient is the combined assimilation/fermentation technique described by Davenport (1975). It relies on the use of sugar-yeast extract broth in 5 ml Bijou bottles with Durham tubes. Sufficient access of oxygen to allow copious development of the more aerobic types of yeasts can be achieved by the use of injection vials with oxygen-permeable septa. Oxidative attack on glucose can be detected by the addition of phenol red to the medium, adjusted to pH 7 in a concentration of 80 mg/l.

Final identification of isolates should always be entrusted to experts — as is the current procedure with regard to moulds, *Salmonella*, enteropathogenic *Escherichia coli* and many other organisms of significance in food microbiology.

References

ALLEN, L. A., PASLEY, S. M. & PIERCE, M. S. F. 1952 Conditions affecting the growth of *Bacterium coli* on bile salts media. Enumeration of this organism in polluted waters. *Journal of General Microbiology* 7, 257-267.

BARNETT, J. A., PAYNE, R. W. & YARROW, D. 1979 *A Guide to Identifying and Classifying Yeasts.* Cambridge: Cambridge University Press.

BEECH, F. W. & CARR, J. G. 1955 A survey of inhibitory compounds for the separation of yeasts and bacteria in apple juices and ciders. *Journal of General Microbiology* 12, 85-94.

BEECH, F. W. & CARR, J. G. 1960 Selective media for yeasts and bacteria in apple juice and cider. *Journal of the Science of Food and Agriculture* 11, 38-40.

BEECH, F. W. & DAVENPORT, R. R. 1971 Isolation, purification and maintenance of yeasts. In *Methods in Microbiology* Vol. 4, ed. Booth, C. pp. 153-182. London: Academic Press.

BEUCHAT, L. R. 1979a Survival of conidia of *Aspergillus flavus* in dried foods. *Journal of Stored Products Research* 15, 25-31.

BEUCHAT, L. R. 1979b Comparison of acidified and antibiotic-supplemented potato dextrose agar from three manufacturers for its capacity to recover fungi from foods. *Journal of Food Protection* 42, 427-428.

BEUCHAT, L. R. & JONES, W. K. 1978 Effects of food preservatives and antioxidants on colony formation by heated conidia of *Aspergillus flavus*. *Acta Alimentaria* 7, 373-384.

BLASER, P. 1978 Vergleichende Untersuchungen zur quantitativen Erfassung des Schimmelpilzbefalls bei Lebensmitteln. III. Mitteilung: Vergleich verschiedener Nährmedien zur kulturellen Koloniezahlbestimmung. *Zentralblatt für Bakteriologie, Parasitenkunde, Infektionskrankheiten und Hygiene Abt. I, Originale B*, 167, 146-164.

BROWN, A. D. 1978 Compatible solutes and extreme water stress in eukaryotic micro-organisms. *Advances in Microbial Physiology* 17, 181-242.

BUESCHING, W. J., KUREK, K. & ROBERTS, G. D. 1979 Evaluation of the modified API 20 C system for identificatiion of clinically important yeasts. *Journal of Clinical Microbiology* 9, 565-569.

BUTTIAUX, R. & CATSARAS, M. 1965 L'analyse bactériologique des bières. *Annales de l'Institut Pasteur, Lille* 16, 167-171.

CHRISTENSEN, C. M. 1946 The quantitative determination of molds in flour. *Cereal Chemistry* 23, 322-329.

COOKE, W. B. 1954 The use of antibiotics in media for the isolation of fungi from polluted water. *Antibiotics and Chemotherapeutics* 4, 657-662.

DAVENPORT, R. R. 1975 *The distribution of yeasts and yeast-like organisms in an English vineyard.* Ph.D. Thesis, University of Bristol.

DIJKMANN, K. E., KOOPMANS, M. & MOSSEL, D. A. A. 1979 The recovery and identification of psychrotrophic yeasts from chilled and frozen comminuted fresh meats. *Journal of Applied Bacteriology* 47, ix.

FLANNIGAN, B. 1973 An evaluation of dilution plate methods for enumerating fungi. *Laboratory Practice* **22**, 530–531.

FLANNIGAN, B. 1974 The use of acidified media for enumeration of yeasts and moulds. *Laboratory Practice* **23**, 633–634.

FREEMAN, P. F. & GIESE, A. C. 1952 Photodynamic effects on metabolism and reproduction in yeast. *Journal of Cellular and Comparative Physiology* **39**, 301–322.

GAFFNEY, D. F., FOSTER, T. J. & SHAW, W. V. 1978 Chloramphenicol acetyltransferases determined by R plasmids from Gram-negative bacteria. *Journal of General Microbiology* **109**, 351–358.

GRAUMLICH, T. R. & STEVENSON, K. E. 1978 Recovery of thermally injured *Saccharomyces cerevisiae*: effects of media and storage conditions. *Journal of Food Science* **43**, 1865–1870.

GRAUMLICH, T. R. & STEVENSON, K. E. 1979 Respiration and viability of thermally injured *Saccharomyces cerevisiae*. *Applied and Environmental Microbiology* **38**, 461–465.

GREHN, M. & VON GRAEVENITZ, A. 1978 Search for *Acinetobacter calcoaceticus* subsp. *anitratus*: enrichment of fecal samples. *Journal of Clinical Microbiology* **8**, 342–343.

HOLT, R. 1967 The bacterial degradation of chloramphenicol. *Lancet* **i**, 1259–1260.

HOLWERDA, K. 1952 The importance of the pH of culture media for the determination of the numbers of yeasts and bacteria in butter. *Netherlands Milk and Dairy Journal* **6**, 36–52.

HUP, G. & STADHOUDERS, J. 1972 Comparison of media for the enumeration of yeasts and moulds in dairy products. *Netherlands Milk and Dairy Journal* **26**, 131–140.

INGRAM, J. M. & HASSAN, H. M. 1975 The resistance of *Pseudomonas aeruginosa* to chloramphenicol. *Canadian Journal of Microbiology* **21**, 1185–1191.

INGRAM, M. 1955 *An Introduction to the Biology of Yeasts.* London: Pitman.

INGRAM, M. 1959 Comparisons of different media for counting sugar tolerant yeasts in concentrated orange juice. *Journal of Applied Bacteriology* **22**, 234–247.

JARVIS, B. 1973 Comparison of an improved rose bengal-chloratetracycline agar with other media for the selective isolation and enumeration of moulds and yeasts in foods. *Journal of Applied Bacteriology* **36**, 723–727.

JOYCE, D. A. & OULD, A. J. L. 1978 Observations on some techniques for the microbiological testing of butter. *Journal of the Society of Dairy Technology* **31**, 227–230.

KELCH, W. J. & LEE, J. S. 1978 Antibiotic resistance patterns of Gram-negative bacteria isolated from environmental sources. *Applied and Environmental Microbiology* **36**, 450–456.

KING, A. D., HOCKING, A. D. & PITT, J. I. 1979 Dichloran-rose bengal medium for enumeration and isolation of molds from foods. *Applied and Environmental Microbiology* **37**, 959–964.

KOBURGER, J. A. 1970 Fungi in foods. I. Effect of inhibitor and incubation temperature on enumeration. *Journal of Milk and Food Technology* **33**, 433–434.

KOBURGER, J. A. 1971 Fungi in foods. II. Some observations on acidulants used to adjust media pH for yeasts and mold counts. *Journal of Milk and Food Technology* **34**, 475–477.

KOBURGER, J. A. 1973 Fungi in foods. V. Response of natural populations to incubation temperatures between 12 and 32°C. *Journal of Milk and Food Technology* **36**, 434–435.

KOBURGER, J. A. & FARHAT, B. Y. 1975 Fungi in foods. VI. A comparison of media to enumerate yeasts and molds. *Journal of Milk and Food Technology* **38**, 466–468.

KOBURGER, J. A. & RODGERS, M. F. 1978 Single or multiple antibiotic-amended media to enumerate yeasts and molds. *Journal of Food Protection* **41**, 367–369.

KRAMER, C. L. & PADY, S. M. 1961 Inhibition of growth of fungi on rose bengal media by light. *Transactions of the Kansas Academy of Science* **64**, 110–116.

LAND, G. A., HARRISON, B. A., HULME, K. L., COOPER, B. H. & BYRD, J. C. 1979 Evaluation of the new API 20 C strip for yeast identification against a conventional method. *Journal of Clinical Microbiology* **10**, 357–364.

MARTIN, J. P. 1950 Use of acid, rose bengal and streptomycin in the plate method for estimating soil fungi. *Soil Science* **69**, 215–232.

MILES, A. A. & MISRA, S. S. 1938 The estimation of the bactericidal power of the blood. *Journal of Hygiene* **38**, 732–749.

MOSSEL, D. A. A. 1951 Investigation of a case of fermentation in fruit products rich in sugars. *Antonie van Leeuwenhoek* **17**, 146–152.

MOSSEL, D. A. A. 1971 Physiological and metabolic attributes of microbial groups associated with foods. *Journal of Applied Bacteriology* **34**, 95–118.

MOSSEL, D. A. A. 1975 Water and micro-organisms in foods — A synthesis. In *Water Relations of Foods*, ed. Duckworth, R. B. pp. 347–361. London: Academic Press.

MOSSEL, D. A. A. & BAX, A. W. 1967 Selektive Zählung von züchtbaren osmophilen Hefen in Lebensmitteln niedrigen a_w-Wertes. *Mitteilungen aus dem Gebiete der Lebensmittelunter-suchung und Hygiene* **58**, 154–155.

MOSSEL, D. A. A. & CORRY, J. E. L. 1977 Detection and enumeration of sublethally injured pathogenic and index bacteria in foods and water processed for safety. *Alimenta* **16**, Special Issue Microbiology, 19–34.

MOSSEL, D. A. A. & RATTO, M. A. 1973 Wholesomeness of some types of semi-preserved foods. *Journal of Food Technology* **8**, 97–103.

MOSSEL, D. A. A. & VAN DE MOOSDIJK, A. 1964 The practical significance in the microbiological examination of cold stored foods of the allegedly low heat resistance among psychro-trophic micro-organisms. *Journal of Applied Bacteriology* **27**, 221–223.

MOSSEL, D. A. A., VISSER, M. & MENGERINK, W. H. J. 1962 A comparison of media for the enumeration of moulds and yeasts in food and beverages. *Laboratory Practice* **11**, 109–112.

MOSSEL, D. A. A., JONGERIUS, E. & KOOPMAN, M. J. 1965 Sur la nécessité d'une revivification préalable pour le dénombrement des Enterobacteriaceae dans les aliments déshydratés, irradiés ou non. *Annales de l'Institut Pasteur, Lille* **16**, 119–125.

MOSSEL, D. A. A., KLEYNEN-SEMMELING, A. M. C., VINCENTIE, H. M., BEERENS, H. & CATSARAS, M. 1970 Oxytetracycline-glucose–yeast extract agar for selective enumeration of moulds and yeasts in foods and clinical material. *Journal of Applied Bacteriology* **33**, 454–457.

MOSSEL, D. A. A., VEGA, C. L. & PUT, H. M. C. 1975 Further studies on the suitability of various media containing antibacterial antibiotics for the enumeration of moulds in food and food environments. *Journal of Applied Bacteriology* **39**, 15–22.

MOSSEL, D. A. A., EELDERINK, I., KOOPMANS, M. & VAN ROSSEM, F. 1978 Optimalisation of a McConkey type medium for the enumeration of *Enterobacteriaceae*. *Laboratory Practice* **27**, 1049–1050.

MOSSEL, D. A. A., VAN ROSSEM, F. & RANTAMA, A. 1979 Ecometric monitoring of agar immersion plating and contact (AIPC)-slides used in assuring the microbiological quality of perishable foods. *Laboratory Practice* **28**, 1185–1187.

MOSSEL, D. A. A., RICHARD, N., GAYRAL, J. P. & BISSUEL, C. 1980 Etude comparative des tests préliminaires à l'identification des bactéries. *Annales de Biologie Clinique*, in press.

NASH, C. H. & GRANT, D. W. 1969 Thermal stability of ribosomes from a psychrophilic and a mesophilic yeast. *Canadian Journal of Microbiology* **15**, 1116–1118.

NASH, C. H. & SINCLAIR, N. A. 1968 Thermal injury and death in an obligately psychrophilic yeast, *Candida nivalis*. *Canadian Journal of Microbiology* **14**, 691–697.

NELSON, F. E. 1972 Plating medium pH as a factor in apparent survival of sublethally stressed yeasts. *Applied Microbiology* **24**, 236–239.

O'TOOLE, D. K. & O'NEILL, G. H. 1974 Comparison of an antibiotic medium with acidified malt extract medium for the enumeration of fungi in butter. *Australian Journal of Dairy Technology* **29**, 115–116.

OTTOW, J. C. G. & GLATHE, H. 1968 Rose bengal-malt extract-agar, a simple medium for the simultaneous isolation and enumeration of fungi and actinomycetes from soil. *Applied Microbiology* **16**, 170–171.

OVERCAST, W. W. & WEAKLEY, D. J. 1969 An aureomycin rose bengal agar for enumeration of yeast and mold in cottage cheese. *Journal of Milk and Food Technology* **32**, 442–445.

PEPPER, E. H. & KIESLING, R. L. 1963 Differential media for the isolation of bacteria and fungi from plated barley kernels. *Cereal Chemistry* **40**, 191–193.

PITT, J. I. 1975 Xerophilic fungi and the spoilage of foods of plant origin. In *Water Relations of Foods*, ed. Duckworth, R. B. pp. 273–307. London: Academic Press.

PUT, H. M. C. 1964 A selective method for cultivating heat resistant moulds, particularly those of the genus *Byssochlamys*, and their presence in Dutch soil. *Journal of Applied Bacteriology* **27**, 59–64.

Put, H. M. C. 1974 The limitations of oxytetracycline as a selective agent in media for the enumeration of fungi in soil, feeds and foods in comparison with the selectivity obtained by globenicol (chloramphenicol). *Archiv für Lebensmittel hygiene* **25**, 73-83.

Ray, B. 1979 Methods to detect stressed micro-organisms. *Journal of Food Protection* **42**, 346-355.

Sainclivier, M. & Roblot, A. M. 1966 Choix d'un milieu de culture pour le dénombrement des levures et moisissures dans le beurre. *Annales de l'Institut Pasteur, Lille* **17**, 181-187.

Silver, S. A. & Sinclair, N. A. 1979 Temperature induced atypical morphogenesis of the obligately psychrophilic yeast, *Leucosporidium stokesii. Mycopathology* **67**, 59-64.

Slack, M. P. E. & Wheldon, D. B. 1978 A simple and safe volumetric alternative to the method of Miles, Misra and Irwin for counting viable bacteria. *Journal of Medical Microbiology* **11**, 541-545.

Smith, G. N. & Worrel, C. S. 1950 The decomposition of chloromycetin (chloramphenicol) by microorganisms. *Archives of Biochemistry* **28**, 232-241.

Smith, N. R. & Dawson, V. T. 1944 The bacteriostatic action of rose bengal in media used for plate counts of soil fungi. *Soil Science* **58**, 467-471.

Stevenson, K. E. & Richards, L. J. 1976 Thermal injury and recovery of *Saccharomyces cerevisiae. Journal of Food Science* **41**, 136-137.

Waksman, S. A. 1922 A method for counting the number of fungi in the soil. *Journal of Bacteriology* **7**, 339-341.

Watson, K., Arthur, H. & Morton, H. 1978 Thermal adaptation in yeast: obligate psychrophiles are obligate aerobes, and obligate thermophiles are facultative anaerobes. *Journal of Bacteriology* **136**, 815-817.

White, A. H. & Hood, E. G. 1931 A study of methods for determining numbers of moulds and yeasts in butter. 2. The influence of temperatures and time of incubation. *Journal of Dairy Science* **14**, 494-507.

Windisch, S., Kowalski, S. & Zander, I. 1978 Nachweis von osmotoleranten Hefen in Mandeln. *Zucker-Süsswarenwirtschaft* **31**, 177-180.

Enumeration of Yeasts in the Presence of Moulds

J. DE JONG AND HENRIËTTE M. C. PUT

*Carl C. Conway Laboratories Thomassen & Drijver — Verblifa N. V.,
Deventer, The Netherlands*

Materials and methods

Decimal dilutions of the sample are inoculated, either in pour plates, or on streak plates of Mycophil agar (BBL) which consists of (g/l): phytone peptone, 10; glucose, 10; agar, 16; final pH $7·0 \pm 0·1$.

Depending on the type of sample and the number of viable micro-organisms likely to be present, decimal dilutions can be membrane filtered (Sartorius SM 13005 ACN) and inoculated on prepared wort broth pads (Sartorius SM 14058 ACN). The wort broth consists of (g/l): malt extract, 15; peptone, 0·78; maltose, 12·75; dextrin, 2·75; glycerol, 2·35; K_2HPO_4, 1; NH_4Cl, 1; final pH $4·8 \pm 0·2$.

If bacteria are present then 100 μg/ml of Globenicol (Gist-Brocades, Delft, The Netherlands) may be added to inhibit these micro-organisms (Jarvis 1973; Put 1974; Mossel *et al.* 1975).

Incubation procedure

Plates and membrane filters are pre-incubated at 25°C for 3 d under strictly anaerobic conditions in a BBL GasPak anaerobic jar (BBL 1968). Pre-incubation is followed by aerobic incubation during another 2 d at 25°C, after which yeast colony counts are made.

Results and discussions

During the anaerobic conditions only pin-point colonies of certain yeasts develop. In the subsequent aerobic incubation period these colonies enlarge within 1 d to colonies of normal size. Other yeasts, the obligately aerobic types, develop to pin-point colonies after 1 d aerobic incubation and to colonies of normal size after 2 d without the full development of common moulds.

Thus this combined anaerobic/aerobic method allows both enumeration and examination of yeasts in the presence of moulds which otherwise often completely cover the surface of plates or membranes. Neither the mould nor yeast counts are decreased by the anaerobic pre-incubation. Even the yeast colony morphology remains unaltered as compared to aerobic incubation only. It must be taken into account however that some facultative anaerobic

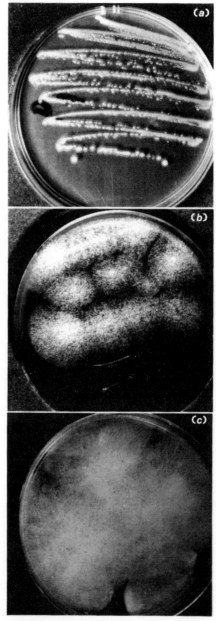

Fig. 1. *Mucor* sp. (*a*) after 3 d anaerobic incubation, (*b*) after 2 d aerobic incubation, (*c*) after 3 d incubation, all at 25°C.

moulds are dimorphic (*Mucor* spp.), which indicates that yeast-like colonies may be formed during anaerobic incubation (Smith & Berry 1974; Sheperd *et al.* 1978) (see Fig. 1*a*, *b*).

Anaerobic conditions alone may be used in conjunction with suitable media for the selective isolation and enumeration of *Saccharomyces* spp. (Longley *et al.* 1978).

It was found that some gas jar methods are suitable, e.g. evacuation of an anaerobic jar followed by washing the jar for 30 min with 0·75 l of oxygen-free N_2 gas/min. (residual O_2 = *ca.* 0·08%, Seip *et al.* 1976), while others were unsatisfactory. No mould inhibition was obtained with the BBL GasPak CO_2 system (residual O_2 = 10·15%); or a candle jar (residual O_2 = *ca.* 1%, Holdeman *et al.* 1977).

Some compounds added to limit mould growth, e.g. diphenyl (100 μg/g), propionate (250 μg/g) or rose bengal (10–50 μg/g), are insufficiently inhibitive and at the same time yeast growth can be decreased (Beech & Davenport 1969; King *et al.* 1979).

Molybdate agar (Holland & Kung 1961) which is highly inhibitory to bacteria does not even allow full development of all types of yeasts. Anti-fungal antibiotics like actidione, amphotericin B, griseofulvin or nystatin (10–100 μg/g) have the same disadvantages (Beech & Davenport 1970; Kurytowicz 1976).

References

BBL Manual of Products and Laboratory Procedures, 1968 5th edn. Becton, Dickinson and Company, Cockeysville, Maryland 21030, USA.

BEECH, F. W. & DAVENPORT, R. R. 1969 The isolation of non-pathogenic yeasts. In *Isolation Methods for Microbiologists* ed. Shapton, D. A. & Gould, G. W. pp. 71–88. Society for Applied Bacteriology Technical Series No. 3. London: Academic Press.

BEECH, F. W. & DAVENPORT, R. R. 1970 Isolation, purification and maintenance of yeasts. In *Methods in Microbiology* Vol. 4, ed. Booth, C. pp. 153–182. London: Academic Press.

HOLDEMAN, L. V., CATO, E. P. & MOORE, W. E. C. eds 1977 *Anaerobe Laboratory Manual* 4th edn. Blacksburg, Virginia: Southern Printing.

HOLLAND, M. L & KUNG, L. J. 1961 Evaluation of molybdate agar as a selective differential medium for yeasts. *Journal of Bacteriology* 81, 869–874.

JARVIS, B. 1973 Comparison of an improved rose bengal-chlortetracycline agar with other media for the selective isolation and enumeration of mould and yeasts in foods. *Journal of Applied Bacteriology* 36, 723–727.

KING, D. A. JR, HOCKING, A. D. & PITT, J. I. 1979 Dichloran-Rose Bengal medium for enumeration and isolation of moulds from foods. *Applied and Environmental Microbiology* 37, 959–964.

KURYTOWICZ, W. 1976 *Antibiotics: a Critical Review.* Washington D. C.: American Society for Microbiology.

LONGLEY, R. P., DENNIS, R. R., HEYER, M. S. & WREN, J. J. 1978 Selective *Saccharomyces* media containing ergosterol and Tween 80. *Journal of the Institute of Brewing* 84, 341–345.

MOSSEL, D. A. A., VEGA, C. L. & PUT, H. M. C. 1975 Further studies on the suitability of various media containing antibacterial antibiotics for the enumeration of moulds in food and food environments. *Journal of Applied Bacteriology* **35**, 15–22.

PUT, H. M. C. 1974 The limitations of oxytetracycline as a selective agent in media for the enumeration of fungi in soil, feeds and foods in comparison with the selectivity obtained by globenicol (chloramphenicol). *Archiv für Lebensmittelhygiene* **25**, 73–83.

SEIP, W. F., EVANS, G. L. & ABRAMSON, J. 1976 Electrometrically measured voltage potentials in culture media in the GasPak system. *Abstracts of the Annual Meeting of the American Society for Microbiology* **C89**, 40.

SHEPERD, P. S., BORGIA, P. T. & PAZNOKAS, J. L. 1978 Biochemistry of dimorphism in the genus *Mucor*. *Advances in Microbial Physiology* **18**, 67–104.

SMITH, J. E. & BERRY, D. R. 1974 *An Introduction to Biochemistry of Fungal Development* pp. 125–134. London: Academic Press.

Isolation of Xerotolerant Yeasts

R. H. TILBURY

Tate & Lyle Limited, Group Research & Development, Reading, Berkshire, UK

Habitat of xerotolerant yeasts

High sugar and high salt content foods.

Isolation medium

Scarr's osmophilic agar (Scarr 1959).

Fifty grams of Oxoid Wort Agar (CM 247) are dissolved in 1 litre of a 45° Brix syrup containing (g): sucrose, 35; glucose, 10. The medium is sterilized at 115°C for 15 min. pH is normally *ca.* 4·8. Reinforcement of gel strength by addition of plain agar (10 g/l) is desirable when plates are to be used for streaking or spreading. The medium should not be re-melted more than once prior to use.

Incubation conditions

$30 \pm 2°C$, aerobic. Count colonies after 3 and 5 days.

Colony characteristics

Colonies are usually well-defined and slightly or distinctly opaque. Xerotolerant *Saccharomyces* spp. form colonies 1–2 mm diam. after 3 d, e.g. *Sacch. rouxii*, *Sacch. bisporus* var. *mellis*. Slower-growing *Torulopsis* spp. require 5 d.

Contaminants that interefere

Xerophilic moulds will grow slowly on this medium but are readily distinguished by their mycelial morphology. Bacteria do not usually grow on this medium. Non-xerotolerant or slightly xerotolerant yeasts develop more slowly than the xerotolerant yeasts. Some *Sacch. cerevisiae* strains will grow after 5 d.

Reference

SCARR, M. P. 1959 Selective media used in the microbiological examination of sugar products. *Journal of the Science of Food and Agriculture* **10**, 678–681.

Abstracts of papers read at meetings of the Society are published without verification of their scientific content.

Selected Abstracts Presented at the Summer Conference

Cold Tolerant Yeasts*

R. R. DAVENPORT

*University of Bristol, Long Ashton Research Station,
Long Ashton, Bristol BS18 9AF, UK*

Many yeasts and yeast-like organisms have been isolated from a variety of cold natural, crop and food environments. Some species are restricted to cold habitats while other species have strains which have been isolated from cold environments but are able to grow at higher temperatures.

Low temperature incubation (*i.e.* other than 25°C, or the dominant standard yeast incubation temperature) is essential for these cold tolerant organisms.

Incubation temperatures

Two important considerations were discussed:
1. the presence of yeasts with other organisms in cold environments;
2. problems with taxonomic procedures where isolates fail to grow on sub-culturing, from isolation plates, either at the same or higher temperatures. In the latter instance this is very important as the taxonomy of yeasts and yeast-like organisms requires tests performed at much higher temperatures.

Genealogical Age Distribution of Unequally Dividing Cells
of *Saccharomyces cerevisiae*

P. G. LORD AND A. E. WHEALS

*Microbiology Group, School of Biological Sciences, The University,
Bath, BA2 7AY, UK*

Recent evidence points to division in budding yeast being unequal (Hartwell, L. H. & Unger, M. 1977 *Journal of Cell Biology* **75**, 427–435; Wheals, A. E. & Lord, P. G. — this meeting), with the population consisting of daughter cells with longer cycle times (D) and parents with shorter cycle times (P). This means that the genealogical age distribution for symmetrically dividing cells does not apply. We have formulated the genealogical age distribution for asymmetrically dividing yeast cells to

*See p. 215, this volume.

be $e^{-\alpha P}$ for the fraction of daughter cells and $(e^{-\alpha P})^{N-1}$ $(1-e^{e\alpha P})^2$ for the fraction of parent cells with n bud scars ($n = 1, 2, \ldots$), a bud scar being produced on a parent cell at each bud separation.

We have analysed the bud-scar distribution (rendered visible by u.v. fluorescence of calcofluor stain) on over a 1000 cells of strain S288C taken from exponentially growing batch cultures at 30°C over a range of growth rates, produced by alterations in the composition of the medium.

We have found: (i) the equations for the genealogical age distribution for asymmetrically dividing yeast cells to be a good fit to the data at all growth rates (16 different rates from $\tau = 76'$ to $\tau = 260'$); (ii) the fraction of first generation parents is always less than 25% and decreases to 17% at slow growth rates; (iii) the fraction of second generation parents is usually less than 12½% and decreases to 10% at slow growth rates;(iv) the fraction of cells of increasing age decreases exponentially, up to 17 scars being seen on one cell, and (v) the pattern of scars occurs in a definite spiral from one pole of the cell, except at slow growth rates where scars may occur at both poles.

Cycle Times of Unequally Dividing Cells of *Saccharomyces cerevisiae* obtained by Bud-Scar Analysis

A. E. WHEALS AND P. G. LORD

Microbiology Group, School of Biological Sciences, The University, Bath BA2 7AY, UK

L. H. Hartwell & M. Unger (1977 *Journal of Cell Biology* **75**, 422–435) recently proposed that cell division in *Saccharomyces cerevisiae* was asymmetric, producing small daughters which have longer cycle times (D) and larger parents which have shorter cycle times (P), the two being related to each other by $e^{-\alpha D} + e^{-\alpha P} = 1$ and $D > P > B > O$ where B is the duration of the budded phase, and $\alpha = \mu = 1n2/\tau$.

It is possible to estimate these values by bud-scar analysis, where the number of bud scars, shown by u.v. fluorescence of calcofluor-stained cells, is counted and cells are classified as budded or unbudded daughters (no scars) and budded or unbudded parents (at least one scar). The total fraction of budded cells is also noted. Peter Green (Mathematics Department, University of Durham) has devised equations which give maximum likelihood estimates of P, D and B from these measurements, subject to the constraints above. Our aim was to use an independent method to check the validity of the model and obtain estimates of the duration of these phases over a range of growth rates obtained by altering the composition of the medium.

We found that: (i) the data were fully consistent with the Hartwell & Unger model at all growth rates; (ii) all parameters measured (including the duration of the budded phase) increased in length with increasing doubling time in an approximately linearly correlated fashion; (iii) the duration of the period from start of the parent cycle to bud initiation increased from 7 min to longer values at slower growth rates, and (iv) the values of P and D converge to equality (*i.e.* symmetrical division) at $\tau = 65$ min.

Effect of Temperature on the Ultrastructure of a Haploid
Saccharomyces cerevisiae

KAILASH C. SRIVASTAVA

Fundacao Universidade Estadual de Londrina, Centro do Ciencias Rurais e de Tecnologia, Dep. de Tecnologia de Alimentos e Medicamentos, Campus Universitario, Cx. Postal 2111, Londrina, PR, Brazil

Little is known on the effect of temperature on the ultrastructure of yeast (Miyake, S., Iwamoto, Y., Nagao, M., Sugimora, T. & Osumi, M. 1972 *Journal of Bacteriology* **109**, 409–415).

With the increase in the incubation temperature, in the range of 15–45°C, the thickness of the cell wall of *Saccharomyces cerevisiae* 18 haploid grown in glucose (DP) and nutrient broth (NB) increased, but when grown in glycerol (GP), it decreased. In DP-grown cells at 15°C and 20°C, the bud did not separate from the mother and in GP-grown cells at 25°C, the bud showed fibre-like structures on either side of the neck. In all the three media, larger mitochondria were seen at temperatures lower than (LT) optimum. In NB-grown cells and in DP at 15°C and 20°C, the cristae were well-developed. At 15°C, in DP, giant mitochondria were common. Also at LTs, increase in the size of nucleus, number of nuclear pores, multilobate nucleus in dividing cells, larger arrays of endoplasmic reticulum and bigger number of vesicles were observed. A wavy nuclear membrane at both LTs and HTs, heterogeneous nucleoplasm and larger vacuoles at HTs were common. The vacuoles contained electron-dense material. Exceedingly high numbers of lipid storage granules were observed at 35°C. Results presented here on the number and morphology of mitochondria are in agreement with earlier work cited above.

The Effects of Metal Ions on Yeast Cells

MAGDALENA EGER AND J. HRSAK

Institute for Medical Research and Occupational Health, Zagreb, Yugoslavia

A haploid yeast *Saccharomyces cerevisiae* was grown under aerobic conditions in a complete nutritive medium to which ions of either lead, zinc or cadmium were added in the form of chlorides in a final concentration of $4{\cdot}0 \times 10^{-5}$ M or $8{\cdot}0 \times 10^{-5}$ M respectively. During a 24 h period of growth at 30°C cadmium ions slowed down the growth of the cell culture and, depending on the concentration of the ions, decreased the density of the population in the stationary phase of growth. Under the same conditions of growth the effect of lead in zinc ions was negligible. The effect of cadmium ions on the growth of the culture was reversible if the cells, after growing in the presence of cadmium, were inoculated in a fresh nutritive medium which contained no cadmium ions or in a medium to which lead or zinc ions had been added. Previous growth of cells in the medium which contained lead or zinc ions did not modify the effect of cadmium ions.

By quantitative biochemical analysis it was shown that in the cells which were grown for 24 h in the medium containing cadmium the amount of DNA doubled. This is an indication of a possible reproductive capacity of the treated cells although their further division in these conditions was inhibited.

Unequal Division in Chemostat Culture of *Saccharomyces cerevisiae*

P. W. THOMPSON AND A. E. WHEALS
*Microbiology Group, School of Biological Sciences,
University of Bath, Bath BA2 7AY, UK*

Batch-grown cultures of yeast divide unequally and have an asymmetrical genealogical age distribution (Lord & Wheals; Wheals & Lord — this meeting). We have examined the same strain of *Saccharomyces cerevisiae* in a glucose-limited chemostat with cells grown on yeast extract, peptone and glucose at 30°C at dilution rates of from $0{\cdot}05$/h to $0{\cdot}3$/h. We found that: (i) unequal division was seen at all growth rates; (ii) the relationship between daughter cycle times (D), parent cycle time (P) and the population doubling time (τ) was curvilinear, especially at faster growth rates so that D, P and τ approached each other asymptotically; (iii) the budded interval increased linearly in duration by $0{\cdot}16$ min for every minute increase in τ, and (iv) the distribution of parent cells of different ages (as measured by bud-scar analysis) was predicted by the genealogical age distribution except for a shortage of cells of older ages.

We conclude that there are differences in cell cycle behaviour between batch- and chemostat-grown cells but these are quantitative rather than qualitative in character and explain some of the differences seen in the published literature.

Effects of Osmotic Pressure on Ethanol and Glycerol Production by a Brewing Strain of *Saccharomyces uvarum (carlsbergensis)*

C. J. PANCHAL AND G. G. STEWART

Brewing R & D, Labatt Breweries of Canada Ltd, London, Ontario, Canada

The objective of the study reported here, was to obtain a better understanding of the processes involved in the production, excretion and tolerance to its own product, ethanol, by a brewing yeast strain of *Saccharomyces uvarum (carlsbergensis)*. Several physiological parameters have been investigated, including the effect of osmotic pressure, and the results of this study are reported. The non-metabolizable sugar, sorbitol, was used together with sucrose, the carbon and energy source, at different concentrations to vary the osmotic pressure of the culture medium. When the intracellular and extracellular levels of ethanol were compared, it was found that during the first 24 h of fermentation the proportion of intracellular ethanol increased with increasing osmotic pressure. Concomitantly, the viability of the cells decreased with increasing osmotic pressure, presumably due to the toxic effect of elevated intracellular concentrations of ethanol. The ethanol, however, diffused out of the cells during the course of the fermentation, the rate of diffusion being slower with higher osmotic pressure.

Analysis of glycerol levels indicated that the proportion of intracellular glycerol remained almost constant with changes in osmotic pressure. However, the total amount of glycerol produced increased significantly with increasing osmotic pressure of the culture medium.

Transformation of Maltotriose Uptake Ability into a Haploid Strain of *Saccharomyces* spp.

G. G. STEWART AND I. RUSSELL

Brewing R & D, Labatt Breweries of Canada Ltd, London, Ontario, Canada

Transformation is a non-sexual technique for achieving genetic recombination. It involves the implantation of DNA material obtained from a donor cell into another cell usually called the recipient.

A maltose positive/maltotriose negative haploid strain of *Saccharomyces* was spheroplasted and tested with native DNA isolated from a production ale strain of *Saccharomyces cerevisiae*; a plasmid vector was not used in these experiments. After cell-wall regeneration, a maltotriose positive variant was isolated and found to possess all the characteristics of the recipient except that it could ferment both maltose and maltotriose. Using hybridization techniques and the mutagen ethidium bromide, the transformed maltotriose uptake trait has been found to be located not in the nucleus but in an extra-nuclear or cytoplasmic position. In the wild-type state, however, the uptake of maltotriose in *Saccharomyces* is controlled at the nuclear level.

The Recovery and Identification of Psychrotrophic Yeasts from Chilled and Frozen, Comminuted Fresh Meats

K. E. DIJKMANN, MIRÈSE KOOPMANS AND D. A. A. MOSSEL
Faculty of Veterinary Medicine, The University of Utrecht, The Netherlands

Meats, particularly those stored at refrigeration temperatures tend to become colonized by psychrotrophic yeasts, in addition to psychrotrophic bacteria. At sufficiently high numbers of CFUs, yeasts may give rise to off-odours. Several currently used media of neutral pH, made selective by the addition of antibacterial antibiotics were examined for their suitability to recover yeasts from chilled and frozen minced meats. These included: oxytetracycline gentamicin dextrose yeast extract (OGDY) agar, chloramphenicol agar, with and without Rose Bengal (CRDP, resp. CDP), chloramphenicol streptomycin ('phytone yeast extract' — CSDYP) agar and a newly developed one: chloramphenicol (50 μg/ml) gentamicin (50 μg/ml) dextrose yeast extract (CGDY)-agar, with and without 50 μg/ml Rose Bengal.

Neither OGDY nor CGDY appeared to suppress populations of yeasts, stressed by refrigeration or storage at —20°C. Inhibition of psychrotrophic bacteria was rather complete on CGDY and still better on OGDY; the few bacterial colonies noted on CGDY could be distinguished from yeast colonies by an experienced analyst. Rose Bengal tended to reduce the recovery of some types of yeasts on these media, as was noted earlier. CSDYP grew many psychrotrophic Gram negative rod shaped bacteria. CRDP also showed insufficient selectivity; furthermore distinction between bacterial and yeast colonies was considerably more difficult than on media without Rose Bengal.

Isolation of Yeasts from Forensic Clinical Specimens, Especially those Containing Sodium Fluoride

JANET E. L. CORRY, JANE C. MARCH AND RUCHI H. GUNEWARDENE
*The Metropolitan Police Forensic Laboratory, 109 Lambeth Road,
London SE1 7LP, UK*

Sodium fluoride at levels which vary from 0·4–2% (m/v) is widely used to preserve samples of blood and urine to be analysed for ethanol. Delay frequently occurs before analysis and high numbers of yeasts can sometimes be detected. These have apparently multiplied in spite of the presence of fluoride.

A survey of a selection of yeasts isolated from preserved and unpreserved samples has shown that many are able to multiply in media containing 0·6% NaF and one strain was found that was able to multiply in 1·6% NaF.

Yeasts able to multiply in the presence of fluoride were also able to change the concentration of ethanol.

Sublethal damage caused by NaF and chilling has also been investigated.

Yeast Spoilage of Fortified Wines with Emphasis on Sherry

V. M. ARJUN

Heaton Avenue, Romford, Essex RM3 7HR, UK

Yeast spoilage has occurred in many types of fortified wine, sherry, port, madeira and vermouth, but the greatest incidence has been in sherry. Rare instances of secondary fermentation by 'Jerez' strains of *Saccharomyces ellipsoideus* and *S. oviformis* have been reported every year. In 1968, however, there were many cases of secondary fermentation by *Rhodotorula* which was later traced to inadequately cleaned casks and some items of bodega equipment.

Since 1972 there have been many instances of yeast turbidity, reaching a peak in 1976 and 1977, involving several yeast genera including *Candida, Rhodotorula* and *Brettanomyces* spp. with *Candida* spp. as the most often encountered. The alcohol tolerance of the strains has varied from 17–18·5% (v/v). The sherry most vulnerable to yeast spoilage is the pale dry type with a residual sugar content in the range of 8–20 g/l. Investigations have shown that there are other factors that enhance the propensity of the wine to yeast spoilage.

Subject Index

Acetic acid, inhibition of yeasts by, 80
Acetic acid preserves, yeast spoilage of, 128
Active-dried yeast, production of, 105
Adaptive maltase system, 136, 138
Aerobic incubation, of yeasts in the presence of moulds, 289, 291
Alcoholic beverages
  importance of flocculation in the production of, 110, 111
  strains of *Saccharomyces cerevisiae* used in the production of, 110, 111
Ambrosiozyma, 30, 38
American Bottler's Association standard, 164
Anaerobic incubation, of yeasts, 289, 291
Anionic groups, involvement of in flocculation, 111, 112
Antarctica, yeasts found in, 217, 222
Antibiotics
  antifungal, 291
  inhibition of yeasts by, 80
  thermolabile, 280
  thermostable, 280, 281
API 20 *Candida* system, 284
API 20C kit, 237
Apiotrichum, 58, 59
Arthroascus, 38
Arthrospores, formation of, 48
Asci, formation of, 38, 43, 56, 59, 64, 74, 253
Ascigenic phase, of *Candida ingens*, 74
Ascoidea, 59
Ascomycetes, 56, 59
Ascomycetous yeasts, 29, 31, 43, 46, 48, 49, 87–95, 183, 217
Ascospores, 181–208
  checking cultures for, 263
  D and *z* values of, 185–187, 207
  decimal dilutions of, 184, 190, 191
  density of, 184, 185
  formation of, 187, 189, 190
  heat survival curves of, 200, 206, 207
  influence of sporing medium on heat resistance of, 191
  of *Cryptococcus neoformans*, 253
  of *Kluyveromyces* spp., 189–195
  of *Saccharomyces* spp., 184–190
  shapes and numbers of, 3, 33, 37, 42, 43, 59, 91
  ultrastructure of, 30, 31
Ascosporogenous yeasts, genera of, 32–35, 37–39, 42–44, 46, 183–208, 232
Ascus formation, 37–39

Ashbya, 59
Aspergillus oryzae, 176
Assimilation/fermentation technique, 285
ATP, synthesis of, 109, 110
Aureobasidium pullulans, 5, 12, 129, 218
a_w, *see* Water activity

Baker's yeast
  development of new strains of, 136–138
  spoilage of pastry products by, 141
Balanitis, 241
Ballistospores, formation of, 45, 219
Ballistosporogenous yeasts, 96
Bark beetles, yeasts associated with, 89, 90, 92, 97
Bartnicki-Garcia classification, 56
Basidiomycetes, 55, 56, 249
Basidiomycetous yeasts, 3, 29–31, 44, 46, 48, 49, 95, 96, 216
Basidiosporogenous yeasts, genera of, 32, 36
Beer yeasts, 105, 107
Bees, yeasts distributed by, 91
Benzoate, as a preservative in foods, 163
Biodeterioration, of high solute foods, 175
Bird faeces, yeasts isolated from, 96, 250, 251, 255
Bones, cryptococcal infections of, 250
Bread
  post-baking contamination of, 139
  spoilage of, 138–140
Brettanomyces spp., 6, 7, 47, 80, 85, 88, 89
Brettanomyces
  anomalus, 8, 10
  bruxellensis, 15
Brewer's yeast, 106, 107, 136
Brined vegetables, spoilage of, 124, 128
Bud-scar analysis, 296, 297
Budding, 54–56, 60
Bulk fermentation process, 136, 138
Bullera, 46

Cabbage, yeast spoilage of, 126, 127
Cactus-specific yeasts, 81–84, 93, 99
Cadavers, human, *Hansenula* sp. isolated from, 90
Calcium-mediated bridge, 113
Candida spp., 46, 49, 56–59, 83, 97, 98, 221, 223
  allergic 'id' reactions to, 242
  diarrhoea associated with, 243
  ecological sources of, 233–237